GRAIN BOUNDARIES: THEIR CHARACTER, CHARACTERISATION AND INFLUENCE ON PROPERTIES

GRAIN BOUNDARIES: THEIR CHARACTER, CHARACTERISATION AND INFLUENCE ON PROPERTIES

Proceedings of a Workshop
Held at Birmingham University, UK
16–17 September 1999
To Mark the 70th Birthday of
Professor R. E. Smallman

Edited by
I. R. HARRIS
and
I. P. JONES

Book 743
First published in 2001 by
IOM Communications Ltd
1 Carlton House Terrace
London SW1Y 5DB

IOM Communications Ltd
is a wholly-owned subsidiary of
The Institute of Materials

ISBN 1–86125–121–1

Typeset in the UK by
Fakenham Photosetting, Norfolk

Printed and bound in the UK at
The University Press, Cambridge

Contents

Contents

INTERMETALLICS

NON-METALS

Introduction

The papers in this volume were presented at a two-day Workshop on Grain Boundaries: Their Character, Characterisation and Influence on Properties, held on the 14th/15th of September 1999 in the School of Metallurgy and Materials, The University of Birmingham. The meeting was held to mark two special occasions, the celebration of Professor Smallman's 70th Birthday (and 40 years on the Staff) and the opening of the new University Centre for Electron Microscopy. This consists of a new Life Sciences Electron Microscopy Laboratory, the SEM and TEM Laboratories in Metallurgy and Materials and the new Midlands FEGTEM (Philips Tecnai) which is a joint facility between the Universities of Birmingham, Warwick and Loughborough.

The Centre was opened officially by Sir Peter Hirsch FRS at an informal ceremony attended by the three vice-chancellors of the sponsoring universities, special guests and the delegates to the Workshop. Forty years earlier in 1959, Ray Smallman introduced electron microscopy in the form of an EM6G into what was then the Department of Physical Metallurgy and maintained the position of Birmingham as a leading centre of electron microscopy, through the various stages of microscopy development. This included the one million volt High Voltage Electron Microscope (HVEM) and the 400kV Intermediate Voltage Analytical Electron microscopy (IVAEM) projects. The introduction of the FEGTEM continues this development of electron microscopy at Birmingham, with an instrument capable of providing X-ray microanalysis from areas down to 0.5nm across, and with energy loss spectroscopy (PEELS) down to 0.2nm, via a very intense small probe of electrons. Many of the advanced materials of today depend critically on the properties and behaviour of grain boundaries and interfaces and the new FEGTEM increases very significantly our ability to study the structure and chemistry of such defect regions. This was one of the reasons why the celebratory meeting took the form of a two-day workshop on grain boundaries.

The Workshop demonstrated the wide-ranging importance of grain boundaries and interfaces in a variety of materials including structural metals and ceramics, magnetic and electronic materials. Sir Alan Cottrell contributed a paper on the electronic aspects of grain boundary strengths. A significant part of the first day was concerned with grain boundaries and interfaces in metals and alloys and included segregation in steels, nickel-based superalloys, oxidation embrittlement, stress corrosion cracking and creep, as well as a discussion on studying grain boundary composition by atom probe microanalysis.

The second day was devoted to the role of grain boundaries and their importance in magnet materials and thin film recording media, intermetallics, where boron segregation at grain boundaries is believed to improve ductility, and to ceramic oxides where grain boundary composition and processing are related strongly to properties. Three such areas are grain boundary transport of oxygen in fast ion conductors, inter-

1

faces in PZT thin films and the new high T_c superconducting oxides, where preferred orientation and texture increases greatly the current-carrying capacity. The meeting ended with a consideration of the pitfalls of chemical microanalysis of grain boundaries using the TEM and, very appropriately, Ray Smallman provided the closing remarks.

I R Harris and I P Jones
School of Metallurgy and Materials
The University of Birmingham

The organisers would like to thank the following sponsors:

Assay Office Birmingham
BNFL Magnox Generation
FEI Limited
IRC in Materials for High Performance Applications
School of Metallurgy and Materials
The University of Birmingham

Electronic Aspects of Grain Boundary Strengths

ALAN COTTRELL

Department of Materials Science and Metallurgy,
University of Cambridge, UK

ABSTRACT

Helium and hydrogen, atomically dispersed in metallic grain boundaries, tend to drive the adjoining grains apart so as to reduce their Pauli repulsion. Carbon and boron, by contrast, have partly-filled atomic states at about the Fermi energy level of the metal and in a transition metal these can form strong covalent bonds with the valency states of the adjoining atoms, which resist intergranular fracture. The corresponding behaviour of other foreign atoms in boundaries is briefly discussed.

THE EMBEDDED ATOM THEORY

Traditionally, the free electron theory of metals has dealt with quantum states for the electrons which extend unchanged through macroscopically large regions of uniform potential. Even when the crystal structure is taken into account, as in the Brillouin zone theory, the electrons respond only to thick stacks of crystal planes (by Bragg reflection). Such theory has thus not been suitable for discussing atomically local features such as lattice defects and grain boundaries although Fumi[1] cleverly managed to consider the energies of vacancies in terms of the macroscopic parameters of the theory.

However, a few years ago a new form of the theory was introduced which is ideally suited for discussing local conditions at atomic-sized irregularities. This is the Embedded Atom Theory developed by several people and especially by Nørskov.[2] An indication of its power is shown by the fact that it predicted, in substantial agreement with observation,that, in a metal such as iron, a hydrogen atom, attracted into a vacancy, prefers to sit not at its centre, but at about 0.08 nm to one side of the centre, in the [001] direction.[3]

The EAM goes directly to the heart of the kind of problem of interest here, for it takes a particular single atom, embeds it in a particular site in the host metal, and then evaluates its energy in that position. This is done by exploiting the fact that all such an atom 'knows' about the surrounding metal is what is immediately next to it. And this is the free electron gas of the metal, at the density characteristic of that particular embedding site. It enables a huge simplification to be made. So far as the interaction of the embedded atom with the local electron gas is concerned, this should be the same as if the atom were simply embedded in a completely uniform electron gas everywhere of the same density as that in the actual embedding site. The embedding energy in such a uniform gas is fairly easily calculated. Moreover, once calculated, that same value can be used – whatever the metal, whatever the site – so long as the local electron density is the same there. Generalising, the same calculation for a wide range of densities then provides a ready-made result that can be applied directly to all hosts and sites where the actual local electron density is known. The EAM has thus provided a remarkably simple theoretical tool for opening up many problems of atoms in metals, both in the perfect lattice and also at defects.

The electron density in a given site of the host can of course be calculated from first principles and such calculations have been made, particularly for the free surface and in a vacancy, but surprisingly good estimates can be made by the simple expedient of deducing the density as a superposition of unaltered free atom densities – at the appropriate distance from the nucleus, of course – densities which have been published in detail. For an interstitial lattice site in iron the electron density is, in the usual units used in such work, about 0.03 electrons per bohr cubed (ie about 0.2 per angstrom cubed or 200 per nm cubed). Moving out through the free surface or into the centre of a vacancy the density rapidly diminishes, since the electron gas becomes tenuous away from the centres of electropositive attraction. In iron the density at the centre of a vacancy is expected to be about 0.008. I have made rough estimates for incoherent large-angle grain boundaries on the assumption that in the largest inter-granular sites the atoms on opposite crystal faces are separated by about one-half of an atomic diameter, giving an electron density in these sites of about 0.02 in iron.[4]

HELIUM EMBRITTLEMENT

It is convenient to begin with helium because things are simplest here and also because helium is known to be a potent grain boundary embrittler, of importance in nuclear energy materials. It might be thought that a helium atom with its two electrons very stably housed in their 1s orbital would, through consequential chemical inertness, sit equally well in any electronic surroundings. Not so. There is a strong repulsion, which stems from the Pauli principle. A free electron of the metal, when it passes through the helium atom, cannot locally take on the form of a 1s wavefunction because the 1s state of the helium atom is already filled by the two atomic electrons. It is thus forced to make its traverse of the atom in some higher energy state, which has nodes that make it orthogonal to the 1s state and thereby acceptable to the Pauli principle. The extra oscillation of the free electron wavefunction, due to this, implies a higher energy and it is this energy which is responsible for the repulsion between the helium atom and its free electron surroundings. This repulsive energy rises almost linearly with the density of the electron gas, reaching 149 eV at the density of 1 electron per bohr cubed. It follows that the embedding energy in an interstitial site in iron is approximately 4.5 eV. In practice the value is lower than this because the helium atom can reduce its energy by pushing the iron neighbours apart, so reducing the local host electron density there, admittedly at the cost of some elastic expansion energy. As a result, the total embedding energy is reduced to about 3.5 eV. This is still large, however, so that there is an incentive for the atom to seek a site where the local density is smaller. A vacancy is an obvious candidate and the atom could become trapped at its centre with an embedding energy of about 1.2 eV. Alternatively, the atom could become trapped in a large grain boundary site, with an energy of about 3 eV, so gaining about 0.5 eV relative to the lattice interstitial site, which at room temperature could give a thermodynamic concentration factor, exp $(\Delta H/kT) \approx 10^9$.

Although trapped in such a boundary site, the helium atom could reduce its embedding energy further, eventually to zero, if the two crystals adjoining this boundary were pulled right apart. In considering the energy changes involved in fracturing the metal along this boundary, the 3 eV which the helium atom could thus release is a negative contribution, which weakens the boundary cohesion, tending to embrittle the metal. It is a large contribution so that a boundary which is greatly infected with helium atoms can be seriously weakened and embrittled. If we assume 1 helium atom per 4 iron atoms of one face of the boundary and 1.6×10^{19} such iron atoms per m^2, this 3 eV per helium atom corresponds to a negative surface energy of 1.9 J m^2. The ideal work of intergranular fracture in iron, estimated from measured surface and grain boundary energies, is only about 4 J m^2, so that this concentration of helium would severely weaken the boundary. If helium-impregnated metal is held at temperatures where vacancies are mobile, then helium atoms trapped in vacancies can also migrate to boundaries. Such a helium atom would release only a small embedding energy when the boundary is fractured, or when its vacancy joins others to form a gas bubble. However, the replacement of a bonding iron atom in the boundary by a non-bonding helium-containing vacancy must significantly reduce the cohesion of the boundary.

HYDROGEN

The Pauli repulsion shown by helium reappears in all other atomic species, at an intensity that depends on the electronic structure of the atom concerned. For a hydrogen atom in a high-density site, such as an interstitial one in iron, this repulsion raises the embedding energy by about 80 eV per electron per bohr cubed. The total embedding energy of hydrogen is, however, considerably different from that of helium because of a strong attraction to low-density electron gas. The story here begins with the 1s state of the free neutral hydrogen atom, which contains a single electron and so, quantum-mechanically, has room for a second. However, this free atom only weakly attracts the second electron because of the intense electrostatic repulsion between two electrons confined in the small 1s orbital of this atom. Thus, although 13.6 eV is released when the first electron joins the hydrogen nucleus, only 0.75 more comes out when the second one joins and converts the atom into a negative ion, so that there is an electrical repulsion between them amounting to 12.85 eV. It is this repulsion which provides the opportunity for an attractive embedding energy in low-density electron gas. For the electrons of this, by redistributing themselves near the embedded hydrogen atom, in its negative ionic state, can screen the charges of the two 1s electrons and so reduce their electrical repulsive energy. Although more complicated, the electron gas here behaves somewhat as if it were a medium of high dielectric constant.

The 1s orbital of the embedded ion is no longer tightly confined to the nucleus but spreads out well into the surrounding electron gas, so that the two 1s electrons are more widely separated.

In the limit of an infinitely dilute electron gas, ie a near vacuum, the favourable embedding energy is easily given. It is simply -0.75 eV, ie the energy released by the attachment of a second electron to a free hydrogen atom. As the density of the gas is increased, the screening effect causes this energy to drop sharply to yet more favourable values, before the growing Pauli repulsion eventually reverses the trend at high densities. As a result, the embedding energy reaches a favourable optimum of -2.45 eV at a density of 0.01 electrons per bohr cubed. Both of these numbers are interesting. The first falls close to the binding energy of the hydrogen molecule, -2.4 eV per atom, so that the ability of even the most favourable metallic sites to decompose molecular hydrogen into embedded atoms is marginal. In transition metals, however, hydrogen enjoys a small covalent bonding which, particularly in the earlier ones of the long periods, enables them to dissociate hydrogen to form hydrides. The other number, 0.01, is interesting because it is larger than that at the centre of a vacancy (0.008) in iron but smaller than that at the centre of a grain boundary site (0.02). The first of these has the effect that a hydrogen atom, trapped in a vacancy, sits not at its centre but a little to one side, where it can find and enjoy the optimum value of 0.01. By contrast, in the grain boundary it will sit at the centre of its site but retains there the ability to lower its energy further if this site can be expanded, as for example when the metal is fractured along the boundary. Its embedding energy in a site of density 0.02 is, however, about -2.3 eV, only 0.15 above that of the optimum. The gain from fracturing the boundary is thus small, but the cumulative effect when large numbers of these atoms segregate in a boundary can lead to severe embrittlement. Other aspects of hydrogen embrittlement are mainly due to further effects, such as migration to expanded crack fronts, gas bubble formation and interference with the motion of dislocations.

LARGER ATOMS

When the embedded atom is large enough to prefer substitutional sites, an additional source of grain boundary mechanical change results from a very simple effect. An atom of the host has to be removed from the boundary to make room for the substitutional impurity or alloy atom and so its contribution to the boundary cohesion is lost. If the latter is weaker in its cohesive properties, as for example tin in iron, then the boundary is weakened, so that such an atom is a grain boundary embrittler. Conversely if its cohesion is strong, then the boundary is strengthened, as appears to happen with the boundary segregation of nickel (often in association with boron) in Ni_3Al.

STRONG COVALENT BONDING

Carbon and its neighbours, boron and nitrogen, are unique in having atoms small enough to enter metals interstitially, but also able to form strong covalent bonds, particularly with transition metals. The cohesion of an 'early' transition metal such as titanium can be boosted considerably by filling its interstices with small atoms such as carbon which make strong cohesive bonds to their neighbouring metal atoms. It is as if the coordination number of the metal has, for the purposes of covalent cohesion, been substantially increased by the presence of these interstitial atoms. Considering in particular carbon, the reason for its strong bonding with such a metal is that its free neutral atom has four electrons in the $2s2p$ shell which has capacity for eight. Inevitably, this outer valency shell of the atom remains partially empty when the atom enters the metal. As a result, the energy levels of both the filled and empty $2s2p$ states must then straddle the Fermi level of the metal, which of course is also where the numerous partially filled d states of the metal are located. Conditions are thus ideal for intense hybridisation of these d states with, particularly, the $2p$ states of the interstitial atoms, leading to the formation of strong covalent bonds between them.

All this is very different from the picture of embedding in a free electron gas, so that the elementary EAM theory, outlined above, is no longer applicable. The theory has been elaborated to take strong covalent interaction between an embedded atom and its host into account,[5] but it then loses much of the simplicity which makes it attractive. A better strategy is to switch completely away from the free electron type of theories to those that envisage the entire metallic system as a giant covalent molecule, a strategy that has proved most successful for discussing the cohesion and structures of transition metals.

This has the advantage that the most elementary versions of such theories are also extremely simple, yet give a good semi-quantitative account of transition metals But the question should first be asked, does nature justify this switch? What has happened to the free electron effects in this covalent picture? This has been discussed.[6] If the carbon atoms were to attract electrons into the unfilled parts of their p states, these extra electrons would then – as with the other p electrons – hybridise with the d electrons of the adjoining metal atoms, making more covalent bonds. The net effect would be to form, in essentially the same covalent structure, a slightly increased electron density on the carbon atoms. Detailed calculations confirm this, which is all that remains of the negative ion formation, characteristic of the EAM theory, in these alloys.

This covalent bond theory has been greatly helped by two simplifying effects. The first is that of the *bond order,* ie the difference in the numbers of bonding and anti-bonding orbitals in the interatomic interaction, by which Friedel[7] explained the variation of cohesion along the transition series. The second is the theorem of Cyrot-Lackmann[8] that, to a good approximation, the cohesion of an atom in such a system increases as the square root of the coordination number, z, of the bond partners to this atom. This increase can be interpreted as due to the freedom of a valency electron of the atom to enter any of the z bond orbitals which connect this atom to its neighbours. Refined versions of this simple result have been developed which can take account of variations in the individual bond strengths due to differences in atomic composition and spacing and also take in higher order terms in the interaction, which can describe angular dependencies of these bond strengths.[9]

I have used the $z^{1/2}$ principle to explain the strengthening of boundaries in iron by atomically dispersed carbon atoms, segregated in them. Measurements show that the energy released when a carbon atom, initially free and neutral, is embedded in a favourable boundary site, is 6.95 eV,which is the same as in an interstitial site in FCC iron and – not surprisingly – some 0.65 eV better than in a site in BCC iron. The observed value for chemisorbed carbon on iron is 7.2 eV which at first sight is surprising since, according to the $z^{1/2}$ principle, we might expect the 6.95 to drop by the factor $(0.5)^{1/2}$ to about 5 eV on the assumption that the surface atom has only about half the number of near neighbours that it has inside the metal. However, chemisorption measurements are made under conditions which allow surface atoms such as carbon to bury themselves just under the surface, where they can enjoy a full coordination number.

Conditions are different from this when a carbon-containing boundary is broken in a fast fracture. There is not time for the carbon atoms to bury themselves under the newly created free surfaces while the crack is passing through. Hence in this case the $z^{1/2}$ effect should apply and increase their energy by about 2 eV. This is a large energy change which,in a boundary densely populated with such atoms, will increase its resistance to fracture substantially, ie by about 40%. In this way, segregated carbon atoms in grain boundaries play an important role in suppressing intergranular brittleness in carbon steel.

BORON AND CARBON

Boron and carbon segregate strongly to boundaries in nickel, Ni_3Al and iron. Since boron thereby strengthens Ni_3Al boundaries, as does carbon in iron, it is intriguing that carbon intergranularly embrittles Ni_3Al. The explanation probably depends on the amounts of such segregates, escaping from supersaturated solution, reaching levels at which they can form a second phase in the boundaries. In iron, of course, this is a film of iron carbide, Fe_3C, the sides of which cohere strongly to the iron crystals so that the intergranular strengthening is maintained But the corresponding Ni_3C

is less stable than Fe_3C and in fact nickel is known to promote graphitisation in steel. Thus the expected boundary second phase in Ni_3Al is graphite, which weakly adheres to its neighbouring metal, giving embrittlement. The difference between iron and nickel in their carbide forming tendencies probably results from the larger d electron content in nickel which, when interstitial carbon is present, can take the total pd hybridisation content up into the range where antibonding orbitals begin to be occupied, so electronically destabilising the system.[10]

OXYGEN, SULPHUR AND PHOSPHORUS

The deduction that the energy of a carbon atom is raised when the atom becomes exposed on a newly created free surface, during fast fracture, depends of course on its broken covalent bonds not rotating round to join this face, but remaining instead as 'dangling' bonds. This is consistent with the general behaviour of covalent orbitals, which require large angular separations

Conditions are different, however, when an atom such as oxygen or sulphur is similarly exposed on a new fracture surface. Their valency shells as free atoms are nearly full and their remaining electron holes lead to ionic bonding, which can concentrate itself fully on the underlying substrate, so giving strong bonding and hence a low surface energy. Hence, unlike carbon in iron, these atoms do not resist fast separation of their host grain boundary faces, but in fact encourage it. Phosphorus, with its three valency electron holes in the free atom, is an intermediate case between its covalent and ionic neighbours in the Periodic Table. The observed intergranular brittleness of steels, attributed to phosphorus, may in fact be due to its interactions with other segregated species.

ACKNOWLEDGEMENTS

I am grateful to Professors Windle and Humphreys for making available to me the facilities of the Department of Materials Science and Metallurgy, University of Cambridge, during the course of this work.

REFERENCES

1. F. G. Fumi, *Phil. Mag.,*1955, **46**, 1007.
2. J. K.Nørskov, *Phys. Rev. B,* 1982, **26**, 2875
3. F. Besenbacher, S. M. Myers and J. K. Nørskov, *Nuclear Inst. Meth. Phys. Res.,* B7–8, 55, North-Holland, 1985.
4. A. H. Cottrell, *Mater. Sci. Technol.,* 1990, **6**, 325.
5. K. W. Jacobsen, J. K. Nørskov and M. J. Puska: *Phys. Rev. B* 1987, **35**, 7423.

6. A. H. Cottrell, *Mater. Sci. Technol.,* 1990, **6**, 807.
7. J. Friedel in *'The Physics of Metals 1 – Electrons*, J. M. Ziman, ed., Cambridge University Press, 1969.
8. F. Cyrot-Lackmann, *Adv. in Phys.,*1967, **16**, 393.
9. D. G. Pettifor, *Phys. Rev. Lett.,* 1989, **63**, 2480.
10. A. H. Cottrell, *Mater. Sci, Technol.,* 1991, **7**, 585.

The Effect of Dislocation Density and Structure on the Grain Boundary Segregation of Phosphorus and Carbon in a Model Fe–P–C Alloy

J. R. Cowan,[a] H. E. Evans,[b] R. B. Jones[a] and P. Bowen[b]

[a]Technology and Central Engineering Division, BNFL Magnox Generation, Berkeley Centre, Berkeley, Gloucestershire, GL13 9PB, UK

[b]School of Metallurgy and Materials, University of Birmingham, Edgbaston, Birmingham, B15 2TT, UK

ABSTRACT

Grain boundary segregation in the Fe–P–C system has been examined using Auger Electron Spectroscopy (AES) on specimens subjected to tensile pre-strains of 0 and 10%, followed by ageing at 500°C for times up to 1800 hours. The aim was to monitor the effect of microstructure on the kinetics of grain boundary phosphorus and carbon segregation, with a view to highlighting the important factors which may control grain boundary segregation in commercial steels used in the nuclear power industry. The results revealed a complex pattern of segregation, with phosphorus segregation being affected by the presence of dislocations which provided rapid diffusion paths along dislocation cores during short term ageing, and preferential segregation/precipitation sites in the bulk after longer term ageing. It was inferred from segregation measurements and hardness results that carbon precipitation occurred on dislocations after long term ageing, causing a shift in the phosphorus–carbon site-competition/chemical interaction, resulting in increased grain boundary phosphorus segregation.

INTRODUCTION

The segregation of solute atoms to alloy grain boundaries during high-temperature exposure, and the effect on mechanical properties has been the subject of considerable study over many years,[1–8] with particular interest being paid to the behaviour of phosphorus, and more recently carbon in ferritic steels.[9–11] Isothermal segregation measurement of both phosphorus and carbon in Fe–P–C alloys by Grabke and co-workers[5,7] showed that increasing the alloy's bulk carbon content increased the quantity of its grain boundary segregation at 600°C at the expense of the grain-boundary phosphorus content. This apparent competition between phosphorus and carbon was explained in terms of terms of site competition between the solutes at the grain boundary. However, Guttmann[12] preferred to account for competition between phosphorus and carbon in terms of a repulsive interaction between the two solutes. Recent work looking at the interactions between phosphorus and carbon during slow

11

cooling[13] has highlighted new complexities in the segregation process, showing that, while competitive segregation between phosphorus and carbon may occur, the grain boundary segregation of one solute need not result in displacement of the other.

While a great deal of work has been carried out to determine the effects of time, temperature, and composition on the segregation of phosphorus and carbon in iron, very little has been done concerning the possible effects of microstructure and/or applied stress during thermal ageing. This is perhaps surprising given that variations in both microstructure and applied stress are often present in industrial applications where segregation may be an issue. Two possible effects of stress on segregation have been proposed by Shinoda and Nakamura;[14, 15] the first, an effect on the driving force for segregation is often termed a 'thermodynamic' effect while the second, manifest through changes in bulk diffusion rates of the solute to the grain boundary is often termed a 'kinetic effect'. Shinoda and Nakamura[14, 15] showed a complex variation of grain boundary phosphorus level with ageing time under applied stress, with increased phosphorus segregation during short ageing times attributed to enhanced diffusion. A retardation of phosphorus segregation with stress, seen after 30 hours, was explained in terms of phosphorus segregating to dislocations where it is trapped, as opposed to grain boundaries. However, possible thermodynamic effects confused the understanding of the role of microstructure. Similar effects were seen by Mackenbrock and Grabke[16] where dislocation networks stabilised by fine carbide precipitates act as traps for phosphorus atoms. In the present work the effect of changing the microstructure on the segregation of phosphorus and carbon in an Fe–P–C alloy has been investigated by utilising tensile pre-strains to alter the dislocation density and structure prior to ageing.

EXPERIMENTAL PROCEDURE

The material composition was nominally Fe–0.06wt%P–0.002wt%C with full composition as given in Table 1. The alloy was produced from vacuum-melted iron stock, which had been subjected to the following thermo-mechanical treatment. Firstly all faces of the ingot were machined flat. The ingot was up-ended, giving a 50% reduction to break up the as cast structure, forged and hot-rolled at 1180°C to the final size of 30 × 100 × 700 mm. The alloy was heat-treated at 850°C for 1 hour, followed by a rapid water quench to obtain a structure suitable for grain boundary composition analysis using Auger Electron Spectroscopy (AES), with an initial grain size of approximately 90 μm.

Table 1 Plate composition.

Element	C	P	Al	S	Ti	N	O	Ni
Wt.%	0.002	0.058 ±0.002	0.005 ±0.001	<0.003	0.003 ±0.001	<0.001	0.017 ±0.004	0.032 ±0.001

All other elements <0.005 wt%. Balance pure iron.

Standard tensile specimens were machined from heat-treated alloy, with specimen gauge lengths parallel to the rolling direction. Specimens were subjected to a 10% tensile pre-strain at room temperature using a screw driven Instron tensile testing machine with an initial strain rate of $3.78 \times 10^{-5}\,\text{s}^{-1}$, in accordance with ASTM E21M – *Standard Practice for Elevated Temperature Tensile Tests of Metallic Materials*. Material was also retained in the un-strained condition.

Sections from as-received (0% pre-strain) and initial 10% pre-strained materials were aged at 500°C for 4, 16, 100 and 1800 hours in an air-circulating furnace, followed by water quenching to room temperature. Sections were also retained in the un-aged condition. After ageing Auger 'matchsticks', approximately 10 mm in length, with a 3 mm diameter were machined, the length of the Auger matchsticks being parallel to the gauge length of the tensile specimens. The Auger matchsticks were notched to facilitate brittle fracture. Grain boundary segregation measurements were carried out in a Fisons Instruments 310–F Scanning Auger Microprobe under the standard conditions given in Table 2. Prior to fractures the specimen chamber was subjected to a bake-out anneal of approximately 1 hour at 80°C, to ensure a ultra-high vacuum and limit carbon contamination. Specimens were fractured within the microprobe using an impact fracture stage, where the specimens are held at temperatures below the ductile-to-brittle transition temperature, facilitating brittle fracture. Auger spectra were obtained in a direct form from approximately 20 intergranular facets and 6 cleavage facets, per specimen. In each case the type of facet was determined by inspection. The spectra were differentiated using a 9 step Savatsky and Golay cubic numerical algorithm. The atomic percent of phosphorus and carbon on each facet was determined using the following relation:

$$\text{at.\%}\ X_i = \frac{I_i/I_i^\infty}{\Sigma I_i/I_i^\infty} \tag{1}$$

where i = phosphorus, carbon or iron, I_i is the peak-to-background height and I_i^∞ is the relative sensitivity factor for element i (Table 2). Atomic percent grain boundary coverage were converted to monolayer coverage using the procedure given by Seah.[17] It was assumed that each side of the grain boundary contained 50% of the total grain boundary coverage of solute. Due to the inherent scatter of segregation results obtained using AES on a polycrystalline material, statistical analysis was carried out to determine whether any differences between sets of AES results could be considered significant. Results were analysed using a standard t-test procedure, taking into account the difference in the variance of the populations (using a folded F-test) using SAS® (statistical analytical systems) computer software. A 5% level of significance was used in all cases.

Material in the aged and unaged conditions were examined using a transmission electron microscope (TEM). Slices approximately 0.5 mm thick were cut from each tensile specimen and un-strained material. Each slice was thinned to 0.1 mm by careful grinding. TEM discs with a 3 mm diameter were punched from the thinned slices. These were then electropolished to perforation using a solution of 95% methanol and 5% perchloric

13

Table 2 Standard operating conditions for the Fisons 310-F Auger microprobe.

Sensitivity Factor Fe (I_{Fe}^{oc})	0.22
Sensitivity Factor P (I_{Fe}^{oc})	0.7
Sensitivity Factor C (I_{c}^{oc})	0.19
Incident beam energy (keV)	5
Analyser pass energy (eV)	200
Recorded spectra energy range (ev)	15–1000
Probe size (nm)	~20
Angle of surface normal to incident beam	30°
Constant retard ratio (CRR)	4
Dwell time (ms)	43
Vacuum (mbar)	$< 2 \times 10^{-9}$
Fracture temperature (°C)	< -120

acid. Polishing was carried out at $-60°C$ with a polishing potential of 27 volts D.C. Each foil was examined using a JEOL 3010 TEM operating at 300 kV. The dislocation density was calculated using a linear intercept technique (A.S.T.M E112-188, *Standard Test Methods for Determining Average Grain Size*) and the following equation given by Ham:[18]

$$\rho_D = \frac{2N}{LW} \qquad (2)$$

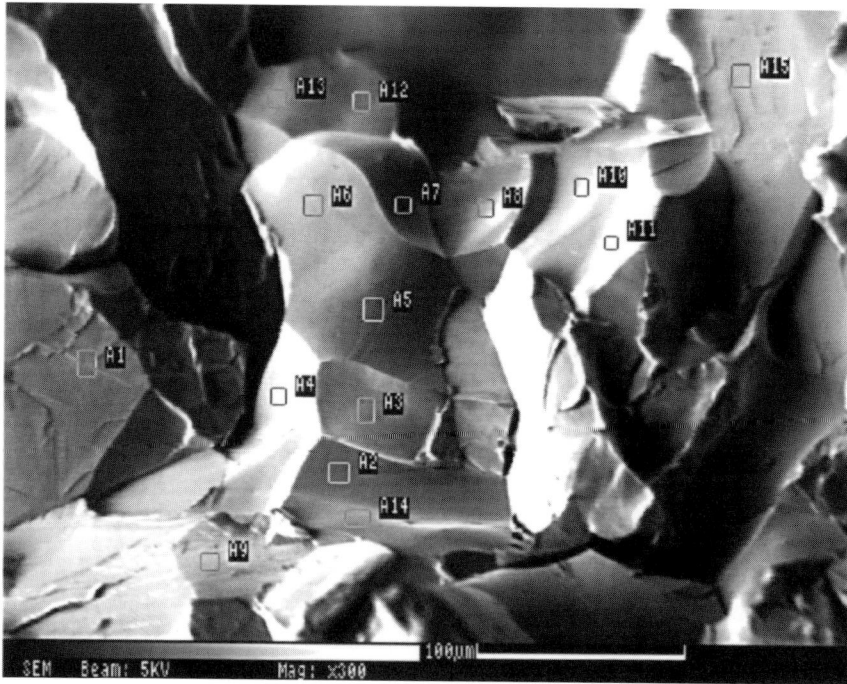

Fig. 1 Typical Auger fracture surface as imaged in the 310-F Auger microprobe showing a brittle fracture surface consisting smooth intergranular and cleavage facets. The latter are readily identified by the presence of river lines.

14

where ρ_p is the dislocation density (number m^{-2}), L is the total line length, N is the number of intersections and W is the foil thickness. The foil thickness was calculated from a convergent beam diffraction image at a moderately low Bragg position, utilising the dynamical diffraction theory of Kelly *et al.*[19] to relate dark field disc fringe spacing to foil thickness. Chemical identification of precipitates and particles was carried out using a Link Analytical Energy Dispersive X-ray (EDX) system within a Vacuum Generators HP501 scanning transmission electron microscope fitted with a field emission gun (FEGSTEM) operating with a beam energy of 100 kV.

RESULTS

Auger Electron Spectroscopy (AES)

After thermal ageing and low-temperature fracture, each specimen was analysed in the Auger microprobe. Figure 1 shows a typical fracture surface as imaged in the microprobe with areas for Auger analysis marked electronically. The fracture surface

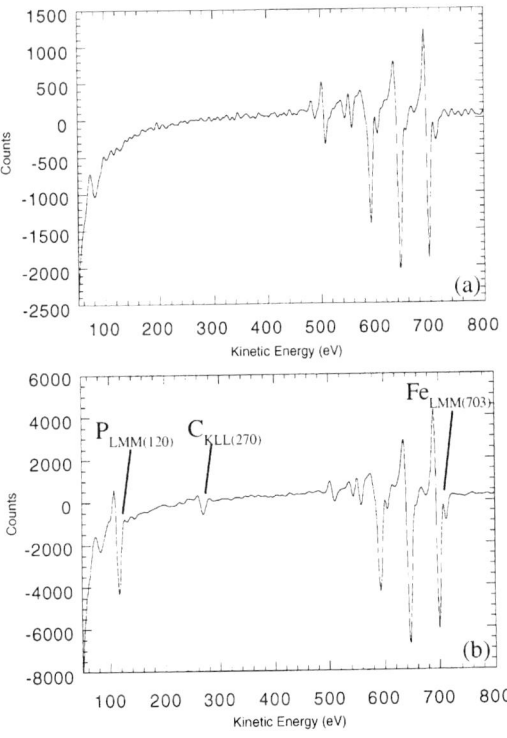

Fig. 2 Differential Auger spectra from (a) cleavage and (b) intergranular facets, showing enrichment of phosphorus and carbon at the grain boundaries.

Table 3 AES results for all ageing treatments.

Pre-strain (%)	500°C 4 hours		500°C 16 hours		500°C 100 hours		500°C 1800 hours	
	at.% $P_{g.b}$ ±95%C.I	at.% $C_{g.b}$ ±95%C.I	at.% $P_{g.b}$ ±95%C.I	at.%$C_{g.b}$ ±95%C.I	at.% $P_{g.b}$ ±95%C.I	at.% $C_{g.b}$ ±95%C.I	at.% $P_{g.b}$ ±95%C.I	at.% $C_{g.b}$ ±95% C.I
0	13.79 ±1.2	14.18 ±1.08	16.00 ±1.56	15.62 ±1.24	18.18 ±1.72	13.94 ±1.32	24.96 ±2.22	12.12 ±0.84
10	15.82 ±1.14	14.56 ±1.44	20.00 ±1.40	15.62 ±1.44	15.26 ±1.30	12.14 ±1.66	30.28 ±2.34	3.08 ±1.50

95% C.I – double sided 95% confidence interval on the mean

consisted mainly of cleavage and intergranular facets, the former readily identified from the presence of river lines.

Auger peaks associated with phosphorus and carbon segregation were clearly visible on spectra obtained from intergranular facets (Fig. 2b). Analysis of cleavage facets (Fig. 2a) showed no evidence of segregation or carbon contamination arising from the machine environment.

Table 3 gives the AES measurements for all experiments. The effect of an initial 10% pre-strain on the grain boundary segregation of both phosphorus and carbon is shown in Fig. 3. For comparatively short ageing times (4 and 16 hours, Regime 1) initial

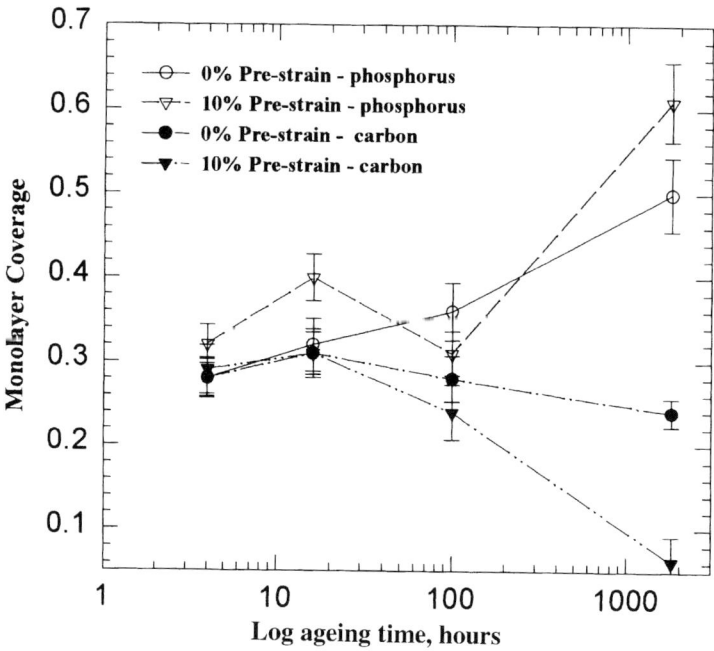

Fig. 3 Plot showing the effect of an initial 10% tensile pre-strain on grain boundary segregation of phosphorus and carbon during ageing at 500°C.

pre-strain increases the level of phosphorus segregation, with no statistically significant variation in carbon segregation with pre-strain during this period. During the next period of ageing (16 to 100 hours, Regime 2) the effect of pre-strain changes, with a significant reduction in phosphorus and carbon segregation for 10% pre-strain compared with the un-strained specimen. The total level of phosphorus and carbon segregation for 10% pre-strain after 100 hours is significantly lower than that after 16 hours, indicating that de-segregation from the grain boundary of both phosphorus, and to a lesser extent carbon has occurred. During significant extended ageing for times approximating to those required to obtain equilibrium phosphorus segregation (1800 hours, Regime 3) the level of phosphorus segregation for 10% pre-strain compared with the unstrained sample increases. During this final period carbon desegregates for the 10% pre-strained specimen, such that the final grain boundary carbon level is almost zero.

HARDNESS TESTS

Vickers hardness results are shown in Fig. 4. A 10% pre-strain increased the material hardness from 84 Hv to approximately 147 Hv. The hardness of the material decreased with increasing ageing time, consistent with recovery of the microstructure. However, during the ageing period between 100 and 1800 hours the hardness of the initial 10% pre-strained material was found to increase dramatically from approximately 104 Hv to 125 Hv. Material in the as-received condition (0% pre-strain) showed a small reduction in hardness after 16 hours, with no further reduction during subsequent ageing.

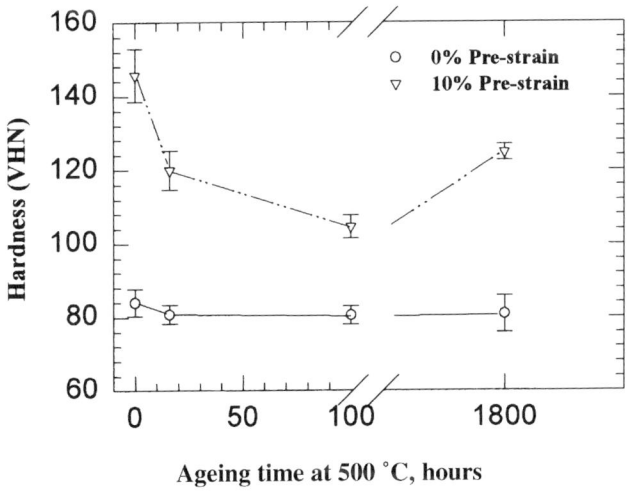

Fig. 4 Plot showing the variation in material hardness during ageing at 500°C for material in the as-received (0% pre-strain) and initial 10% pre-strained conditions.

17

MICROSTRUCTURAL EXAMINATION

In order to determine how the microstructure might affect segregation, TEM examination was carried out. Figs 5a and b show, respectively, the dislocation structures for 0 and 10% tensile pre-strain prior to ageing. Material in the as-received condition consisted of individual dislocations, with bulk and grain boundary dislocation densities of the order of 1×10^{13} dislocations per m^2. Initial tensile 10% pre-strain resulted in the formation of a distinct cellular sub-structure, where sub-cells of dislocations are separated by dislocation walls. The formation of a substructure made determination of the dislocation density impossible, however, Fig. 5 indicates that the dislocation density was significantly greater for 10% pre-strain than as-received material, taking into account the relative thickness of each specimen.

As ageing occurs changes in both the dislocation densities and structures takes place. With no initial pre-strain the bulk and near grain boundary dislocation density appears to remain constant up to 100 hours, after which it is seen to decrease slightly from the prior ageing values of 1.36×10^{13} and 1.20×10^{13} dislocations per m^2 (bulk and near grain boundary) to 6.40×10^{12} and 8.80×10^{12} dislocations per m^2 (bulk and near grain boundary) after 1800 hours. The dislocation structure for the specimen with an initial 10% pre-strain refined during ageing, with the sub-cell walls becoming distinct as ageing continues, resulting in a refined structure with a sub-cell size of approximately 1–2 μm. Large numbers of dislocations were observed threading into the grain boundaries from the sub-structure in the near grain boundary region. Although the structure was too complex to allow quantitative measurement, the

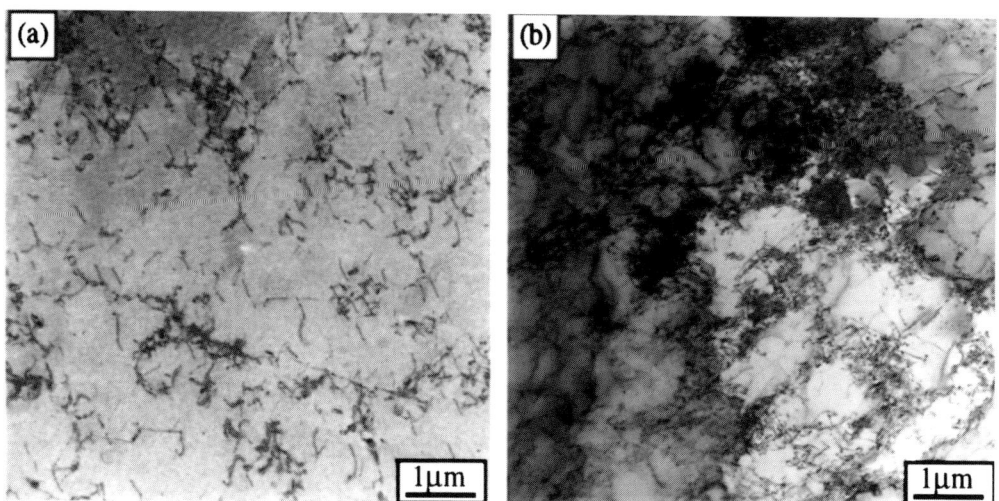

Fig. 5 Bright field TEM micrographs showing dislocation structures in (a) as-received (0% pre-strain) and (b) 10% pre-strained material.

dislocation density in the near grain boundary region remained significantly higher in the 10% pre-strained specimens than the zero pre-strain specimen during ageing. There was no evidence of particles or precipitates after ageing for 4 or 16 hours. However, after 100 and 1800 hours precipitates were observed on dislocations within the sub-cell structure of the 10% pre-strained specimen (Fig. 6). Precipitate analysis of as-received and 10% pre-strained material after 1800 hours ageing showed the presence of large particles (containing S, Si, Mn and Cu). A population of smaller precipitates was also observed in the 10% pre-strain specimen, but analysis of precipitates associated with dislocations proved to be difficult, mainly due to the small size of the precipitates and the thickness of the foil specimen. Where analysis was possible, precipitates were found to contain copper, with both sulphur and phosphorus also present. A number of precipitates near to or at dislocations were examined where no distinct EDX peaks (except iron) were found.

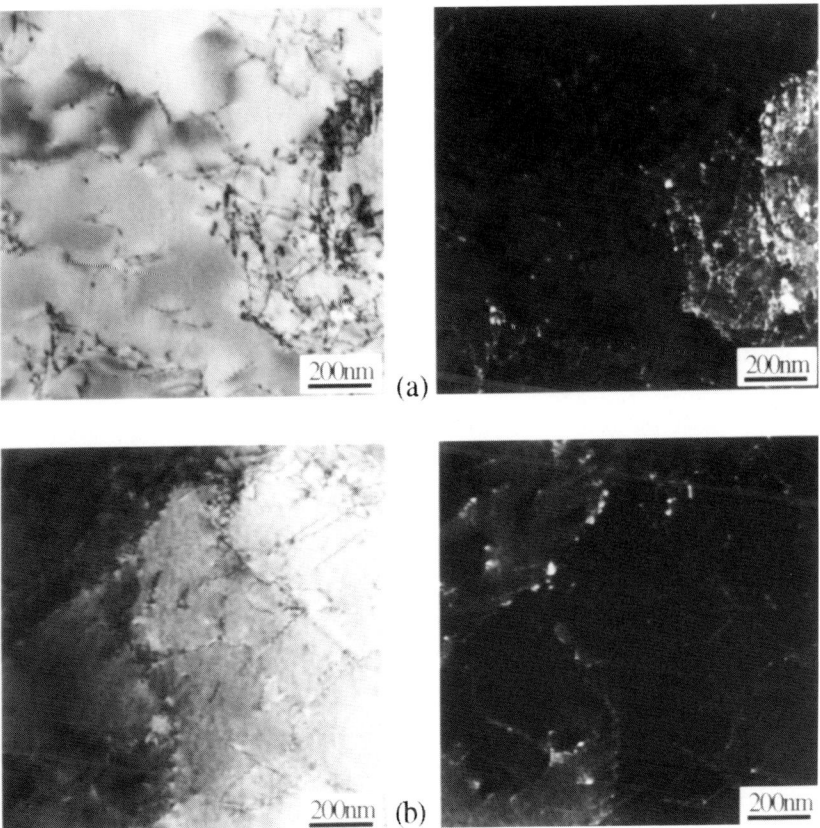

Fig. 6 Bright field and weak beam dark field TEM micrographs showing precipitates associated with dislocations in material subjected to an initial 10% pre-strain after ageing for (a) 100 and (b) 1800 hours at 500°C.

DISCUSSION

The results clearly show a complex relation between microstructure, ageing time and the degree of phosphorus and carbon segregation, with three distinct regimes of behaviour being identified. In Regime 1 (short ageing times) phosphorus segregation increases for specimens with an initial pre-strain compared to the specimen with zero pre-strain; there is no effect of pre-strain on carbon segregation. As ageing continues (Regime 2, 16 to 100 hours) an initial 10% pre-strain appears to cause desegregation of both phosphorus and carbon from the grain boundaries. Extended ageing for long times (Regime 3, up to 1800 hours) results in almost complete desegregation of carbon for pre-strained specimens and a relative increase in phosphorus segregation compared to the specimen with zero pre-strain.

In the as-received (0% pre-strain) condition phosphorus segregation increases with ageing time while carbon segregation decreases, consistent with the McLean type segregation theories. The behaviour of the pre-strained specimens in Regime 3 is consistent with the usual models of segregation which envisage a competition for grain boundary sites between phosphorus and carbon, either through a direct site-competition[7] or repulsive chemical interaction[12] mechanism. However, during Regime 1 phosphorus segregation increases with no variation in carbon segregation, while in Regime 2 both solutes are seen to desegregate from the grain boundary. While the classical antagonistic behaviour of phosphorus and carbon at grain boundaries in iron exists, the overall behaviour is much more complex.

The effect of an initial 10% pre-strain is most likely to be manifest through variations in both the dislocation density and structure during ageing at 500°C, and depend on whether kinetic or thermodynamic factors limit the degree of grain boundary segregation. Segregation data are often described using the classical models of McLean[1], with the kinetics given by:

$$\frac{X_I^{\phi}(t) - X_I^{\phi}(0)}{X_I^{\phi*}(t) - X_I^{\phi}(0)} = 1 - \exp\left(\frac{FDt}{a^2_2 d^2}\right) erfc\left(\frac{FDt}{a^2_2 d^2}\right)^{\frac{1}{2}} \tag{3}$$

where F is taken as 4 for grain boundaries and 1 for a free surface, $X_I^{\phi}(t)$ is the boundary content at time t, $X_I^{\phi}(0)$ is the boundary concentration at time $t = 0$ (taken as 0.1 monolayers), D is the diffusion rate of phosphorus in iron (taken as $2.5 \times 10^{-5}\exp(-200 \text{ kJ}/RT)$, m^2 s^{-1}, Druce et al.[20]), d is the grain boundary thickness (taken as 8×10^{-8} cm ~ 3 atomic layers), $X_I^{*\phi}$ is the equilibrium grain boundary concentration at the ageing temperature, α_2 is given by $X_I^{\phi*}/X_I$ and $erfc = 1$ – error function(x). The equilibrium coverage can be calculated using the McLean–Langmuir[1] isotherm for binary segregation (equation 4) and the value for free energy of phosphorus segregation in iron calculated from the equilibrium segregation values for the Fe–P–C alloy[9] of $\Delta G_P = -35700 - 17.37T$ (J mol^{-1}):

$$\frac{X_P^{\phi*}}{} = \frac{X_P}{1 - X_P} \exp\left(\frac{\Delta G_P}{RT}\right) \tag{4}$$

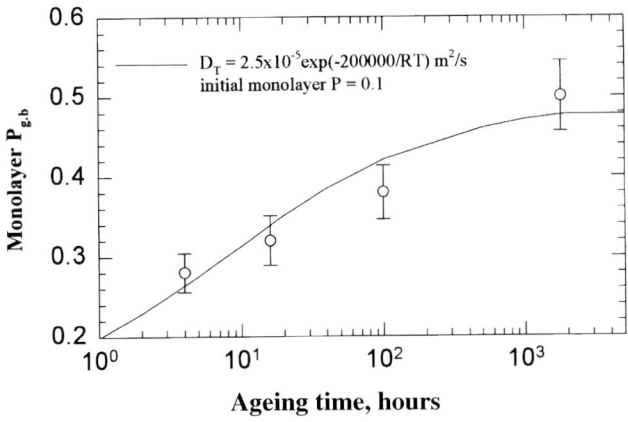

Fig. 7 Plot showing the experimental and theoretical monolayer phosphorus grain boundary levels during ageing at 500°C for material in the as-received condition.

where X_p is the bulk solute atomic fraction, R is the gas constant and T the absolute temperature. These theoretical models fit the experimental data for zero pre-strain over the entire range of ageing times (Fig. 7).

During initial ageing (Regime 1) changes in the level of grain boundary segregation will be dominated by the rate of solute transfer to the grain boundary. The effect of pre-strain will be to increase the dislocation density in the near grain boundary region, resulting in increased phosphorus diffusion by the provision of pipe diffusion paths. The apparent diffusion rate, D_T, combining lattice and pipe diffusion according to Hart[22], under the condition that $2(D_L t)^{1/2} < \rho^{-1/2}$ (clogged pipe model), where t is the ageing time, D_L is the lattice diffusion coefficient and ρ is the dislocation density is given by:

$$D_T = D_L + f(D_p \times D_L)^{1/2} \qquad (5)$$

where D_p is the pipe diffusion coefficient and the term f is given by $f = A_p \rho$, where A_p is the pipe-diffusion cross-section given by $A_p \approx 4a^2$, where a is the lattice parameter of iron, taken as 0.287 nm. For ease of calculation the dislocation densities for 0% pre-strain and 10% pre-strain in the near grain boundary regions can be taken as approximately 1×10^{13} and 1×10^{14} dislocations per m^2 respectively. Values of f, determined for 0 and 10% tensile pre-strain were 3.3×10^{-6} and 3.3×10^{-5} respectively, with lattice diffusion of phosphorus in iron given by:[23]

$$D_L = 13.8 \exp(-332000/RT) \qquad (6)$$

Values for phosphorus pipe-diffusion in iron are not readily available in the literature. However, the ratio of $\Delta H_{lattice}/\Delta H_{pipe}$ for iron self diffusion is 1.85[23], and assuming this ratio holds for phosphorus diffusion ΔH_{pipe} for phosphorus pipe-diffusion in iron can be taken as approximately 180 kJ mol^{-1}. Using the phosphorus lattice diffusion pre-exponential constant value of 13.8 as an upper limit (the same assumption made for sulphur pipe diffusion in iron by Arabczyk *et al.*[24]), the phosphorus pipe-diffusion coefficient in iron can be given by:

$$D_P = 13.8\exp(-180000/RT) \tag{7}$$

Using equations 5, 6 and 7, and the values of f given above for 0% and 10% pre-strain, the maximum apparent diffusion rate for pre-strained specimens can be calculated as 3.8 times greater than that for the un-strained specimens. Experimental segregation results are plotted, along with theoretical fits based on the McLean kinetic model (equation 3) in Fig. 8, where it is assumed that an initial 10% pre-strain increases the diffusion rate of phosphorus in iron by 3.8 times. Good agreement is shown between experimental results and theoretical fits for both zero and 10% pre-strained specimens. It is clear from Fig. 8 that increased grain boundary phosphorus segregation for material with an initial 10% pre-strain, over the initial 16 hours of ageing at 500°C, can be adequately explained and modelled according to the 'clogged pipe' theory of enhanced diffusion via dislocation-cores. There is no apparent effect of increased dislocation density on carbon segregation during this period due to the significantly higher diffusivity of carbon in iron compared with phosphorus, such that the effect of enhanced pipe diffusion is negligible.

As ageing continues (Regime 2) and the limiting parameter changes from kinetic to thermodynamic, both phosphorus and carbon desegregated from the grain boundaries in specimens with an initial pre-strain. Evidence for a reduction in segregation to grain boundaries due to the presence of dislocations was noted by Mackenbrock

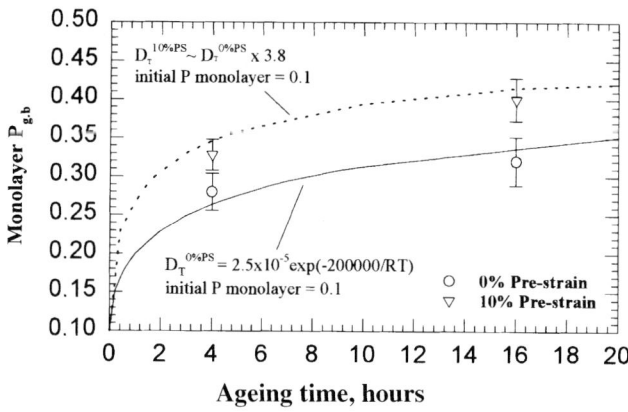

Fig. 8 Plot showing the effect of tensile pre-strain on the segregation of phosphorus during initial ageing for up to 16 hours.

and Grabke,[16] who found a reduction in isothermal segregation level of phosphorus for segregation in a 12Cr–Mo–V steel. This was attributed to segregation of phosphorus to dislocations which had been stabilised by the presence of carbides in the structure, reducing the amount of available solute to segregate to the grain boundaries. In the current work limited evidence of precipitation after ageing for 100 hours with an initial 10% pre-strain was observed (Fig. 6). Chemical analysis (EDX) of the precipitates in material aged for 1800 hours with an initial 10% pre-strain showed a range of precipitate types, including those containing P. In order for desegregation to occur the level of phosphorus must change over the ageing period of 16 to 100 hours such that the ratio of grain boundary to transfer zone solute content is greater than the equilibrium level. Assuming equilibrium segregation (un-strained condition) of approximately 25at% (1800 hours), and an unchanged bulk content of 0.1046at% (0.058wt%) then the equilibrium enrichment ratio is approximately 240 (a typical value for phosphorus equilibrium segregation in iron[25]). Taking the level of grain boundary phosphorus after 16 hours with an initial 10% pre-strain to be approximately 20at%, then in order for desegregation to occur after 16 hours ageing, the level of bulk phosphorus must be less than 20/240 = 0.0833at%, giving a reduction in bulk phosphorus content of 0.022at%. An order of magnitude calculation for the number of precipitates required to produce a reduction to 0.022at% shows a requirement for approximately 1.65×10^{21} precipitates per m^3; approximately 80 precipitates in the field of view shown in Fig. 6. This level of precipitation is unlikely, however, difficulties in accurately measuring the total population of precipitation exist. Desegregation of carbon indicates that similar segregation to/precipitation on dislocations is also likely to occur for carbon during this period. However, the lack of increase in the hardness values (Fig. 6) for times from 4 to 100 hours indicates no significant carbon precipitation, suggesting that the incubation or nucleation period for carbon precipitation often observed before detectable hardness changes occur is greater than 100 hours.[26]

Ageing for 1800 hours (Regime 3) results in increases in phosphorus segregation, with desegregation of carbon for the 10% pre-strained specimens, consistent with competition between phosphorus and carbon. Large increases in hardness and observations of possible carbon precipitation after 1800 hours in the 10% pre-strained specimen suggest that the competition has arisen due to a reduction in available free carbon in the matrix. What is unclear is whether the level of phosphorus segregation for pre-strained specimens would have continued to decrease were it not for the shift in the phosphorus-carbon competition caused by the reduced bulk carbon content.

The effect of dislocation density and structure, combined with competition between phosphorus and carbon and changes in controlling parameters cause complex variations in segregation during ageing. The results are consistent with enhanced phosphorus diffusion due to increased numbers of dislocations during short ageing periods (kinetic dominated). As ageing continues segregation and precipitation of both phosphorus and carbon on the increased number of dislocations results in desegregation due to a reduction in bulk free solute levels. As precipitation continues the level of grain boundary segregation changes to reflect shifts in the phosphorus-carbon grain boundary competition reaction caused by increasing reductions in bulk solute level.

Work is continuing involving further pre-strains prior to ageing and more detailed TEM examination of the interactions between solute and microstructure. This further work will be reported at a later date.

CONCLUSIONS

The effect of dislocation density and structure on the grain boundary segregation of phosphorus and carbon in a Fe–P–C alloy has been investigated by isothermal ageing at 500°C of specimens which had received a prior 10% tensile pre-strains for times up to 1800 hours. The development of grain boundary segregation has been followed using Auger spectroscopy, while changes in dislocation density and structure were observed using standard transmission electron microscope (TEM) techniques. Both phosphorus and carbon segregation varied in a complex manner dependent on both dislocation density and structure and ageing time.

During short periods of ageing (up to 16 hours) phosphorus segregation was shown to be enhanced by the presence of a higher dislocation density in the near grain boundary region. During this period no effect of dislocation density on carbon segregation was observed. As ageing time increased the parameters controlling the extent of segregation changed from kinetic to thermodynamic, with desegregation of both phosphorus and carbon from the grain boundaries occurring for pre-strained specimens during the time period of 16 to 100 hours. Microstructural examination suggested the presence of precipitation on dislocations in pre-strained specimens which could result in reductions in the bulk solute levels. Ageing for times where phosphorus segregation would be expected to approach equilibrium levels (1800 hours) resulted in almost complete desegregation of carbon from grain boundaries in all pre-strained specimens, with a resultant increase in phosphorus segregation caused by the shift in the phosphorus-carbon grain boundary competition reaction.

ACKNOWLEDGEMENTS

The authors would like to thank BNFL Magnox Generation for provision of facilities and for permission to publish this paper. One of us (JRC) would also like to thank the company for financial support. Thanks are also expressed to the members of the Materials Group and Electron Optics Team at Magnox Generation, Berkeley, for many helpful discussions.

REFERENCES

1. D. McLean, *Grain Boundaries in Metals*, Clarendon Press, Oxford, 1957.
2. E. D. Hondros and M. P. Seah, *Proc. Roy. Soc. London,* 1973, **335A**, 191.
3. M. Guttmann, *Surface Sci.,* 1975, **53**, 213.
4. L. Marchutt and C. J. McMahon Jr., *Metall. Trans.,* 1981, **12A**, 1135.
5. H. Erhart and H. J. Grabke, *Metal Sci.,* 1981, **15**, 401.
6. M. Hasimoto, Y. Ishida, R. Yamamoto and M. Doyama, *Acta Metall.,* 1984, **32**, 13.
7. H. Hansel and H. J. Grabke, *Scripta Metall.,* 1986, **20**, 1641.
8. C. M. Liu, K. Abiko and H. Kimura, *Metall. Trans.,* 1992, **23A**, 1515.
9. C. J. McMahon Jr and W. Yu-quig, *Mater. Sci. Technol.,* 1987, **3**, 207.
10. C. J. McMahon Jr, *Acta Metall.,* 1966, **14**, 839.
11. H. Matsui, S. Moriya, S. Takaki and H. Kimura, *Trans. Japan Inst. Metals,* 1978, **19**, 163.
12. M. Guttmann, in *Proceedings of the Second Conference on Ultra High Purity Base Metals,* J. LeCoze ed., *J. de Physique IV,* 1995, C7–85.
13. J. R. Cowan, H. E. Evans, R. B. Jones and P. Bowen, *Acta Mater.,* 1998, **46**, 6565.
14. T. Shinoda and T. Nakamura, *Trans. Japan Inst. Metals,* **21**, 1980, 753.
15. T. Shinoda and T. Nakamura, *Trans. Japan Inst. Metals,* **21**, 1980, 781.
16. M. Mackenbrock and H. J. Grabke, *Mat. Sci. Tech.,* **8**, 1992, 541.
17. M. P. Seah, in *Practical Surface Analysis,* 2nd edn, D. Briggs and M. P. Seah eds, John Wiley and Sons, 1990, 201.
18. R. K. Ham, *Phil. Mag.,* 1961, **6**, 1183.
19. P. M. Kelly, A. Jostons, R. G. Blake and J. G. Naiper, *Phys. Stat. Sol. (a),* 1975, **31,** 771.
20. S. G. Druce, G. Gage and G. Jordan, *Acta Metall.,* 1986, **34**, 641.
21. J. R. Cowan, Ph.D Thesis, 'The Role of Stress, Microstructure and Cooling Rate on the Grain Boundary Segregation Of P and C in an Fe-P-C Alloy', The University of Birmingham, Birmingham, UK, 1998.
22. E. W. Hart, *Acta. Metall.,* 1957, **5,** 597.
23. *Smithells Metals Reference Book,* 7th Edition, E. A. Brandes and G. B. Brook eds, 1992.
24. W. Arabczyk, M. Militzer, H.-J. Mussig and J. Weiting, *Surface Sci.,* 1988, **198**, 167.
25. E. D. Hondros and M. P. Seah, *Int. Met. Rev.,* 1977, **22,** 262.
26. A. C. Damask, G. C. Danialson and G. J. Dienes, *Acta. Metall.,* 1965, **13**, 973.

Segregation to Grain Boundaries in Steels

R. G. FAULKNER

Institute of Polymer Technology and Materials Engineering, Loughborough University,
Loughborough, Leicestershire LE11 3TU, UK

ABSTRACT

The kinetics of equilibrium and non-equilibrium impurity intergranular segregation are discussed. New developments in the understanding of both quench induced and radiation-induced segregation are highlighted. Examples of applications of the theory to practical situations in steels are given. Some discussion is made of the links between intergranular segregation and mechanical ductility and of the various ultrahigh resolution microanalytical techniques required to study segregation.

INTRODUCTION

Segregation to boundaries occurs by either equilibrium and/or non-equilibrium mechanisms. There have been several useful reviews of intergranular segregation (for example, Lejcek and Hoffman,[1] Seah and Hondros[2] and Faulkner[3]). In this paper the details of the segregation mechanisms will be considered under the conditions appropriate to various driving forces, notably thermally and irradiation induced. The quantitative models used for these mechanisms will be described and the best experimental techniques for studying the phenomenon will be discussed. Finally the industrial implications of boundary segregation effects in steels will be covered.

EQUILIBRIUM SEGREGATION

This segregation mechanism was proposed by Gibbs early in the 20th century. The basis to the segregation lies in the provision of a disordered interface or boundary like a grain boundary or a precipitate matrix interface where misfitting impurity atoms can find low energy sites. This reduction in the energy provides the driving force to transfer impurity atoms from regions of perfect crystallinity to the boundary. Since the attractiveness of the boundary reduces as the temperature rises because of the greater disorder introduced by greater atomic vibrational amplitudes, this effect reduces in magnitude with increasing temperature. The kinetics of this mechanism were treated initially by Hondros and Seah.[2] The impurities diffuse less rapidly as temperature reduces and so less segregation is expected from a kinetic viewpoint at lower temperatures. The overall effect of temperature is to produce a 'C' curve distribution in the segregation intensity variation with temperature (Fig. 1). This illustrates the temperature dependence of phosphorus grain boundary segregation in a plain

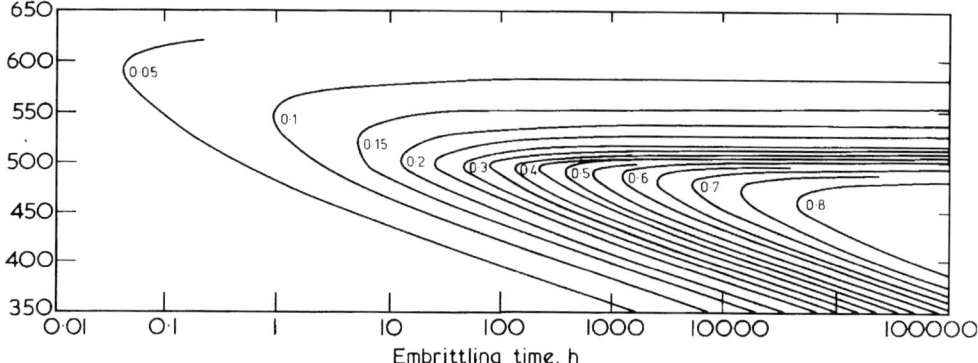

Fig. 1 Kinetics of intergranular segregation of phosphorus in SAE3140 steel as a function of temperature, showing a 'nose'. Numbers on curves show segregation increase in mono-layers (after Hondros and Seah.)[2]

carbon steel. Confirmation of the validity of this analysis is given by the observation that plain carbon steels become temper embrittled in the 500–550°C regime and the embrittlement occurs in an intergranular mode. Phosphorus has been seen on the intergranular facets. It is important to realise that this is an equilibrium process, and so the impurity atoms once they reach the boundary see no driving force to remove them if the system is kept at a fixed temperature. The impurity segregates in a mono-layer fashion on the boundary.

NON-EQUILIBRIUM SEGREGATION

This mechanism was first proposed by Aust and Westbrook[4] in 1967 and independently by Anthony[5] in 1969. It relies on the creation of a non-uniform distribution of point defects around boundaries by either quenching the material or irradiating it with energetic particles like neutrons. Under these circumstances a quenched in supersaturation of point defects is produced in regions of the material where there are locally no good sinks for the defects. In defect regions like boundaries, a good sink efficiency exists and the thermal equilibrium point defect concentration is maintained irrespective of whether the material has been quenched or irradiated. In these non-equilibrium conditions a concentration gradient for the point defects is set up with the net result that the point defects want to diffuse down the concentration gradients towards the lower concentration regions on the boundary plane. Now if there are misfitting impurity atoms present in the lattice these will undergo a small but important attraction with the point defects so that impurity–point defect complexes form. A proportion of these complexes will be dragged with the point defects which are already moving towards the boundary because of the non-equilibrium concentration gradient set up during the quenching or irradiation treatment. This complex flux has the effect of increasing the amount of impurity in the vicinity of the boundary (see Fig. 2). The impurity concentration distribution is spread out over several nanometres on either side of the boundary and this is to be contrasted with the monolayer impurity distribution found in the equilibrium segregation case (Fig. 2). The kinetics of the process are accounted for in the mechanism description. The magnitude of the enrichment depends on the starting temperature in the case of thermal quenching-induced segregation or on irradiation dose and dose rate in the case of irradiation-induced segregation. The binding energy of the impurity to the

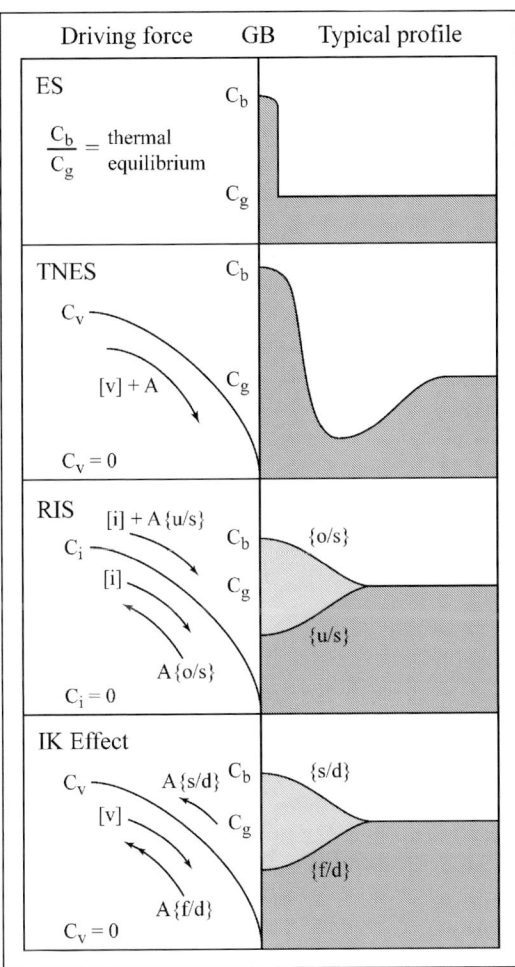

Fig. 2 Summary of Equilibrium Segregation (ES), thermal non-equilibrium segregation (TNES), and radiation-induced segregation (RIS). A stands for solute atoms; *u/s* means under-size and *o/s* means oversize, *v* for vacancies, and *i* for interstitials. The IK effect refers to Inverse Kirkendall which is a non-equilibrium process involving uranium, fast diffusing solute (f/d) and slow diffusing solute (s/d).

point defect and the point defect formation energy are also important. Because of the non-equilibrium nature of the quench-induced mechanism, if the system is left for great lengths of time at a fixed temperature the enrichment will finally die away; the critical times required are determined by the diffusion coefficients for the impurity atom in the lattice and by the inter-sink spacing. Typically in the case of intergranular segregation this spacing will be the grain radius. Quenching relies on vacancies as the major transport point defect. Irradiation damage creates large quantities of self-interstitials and these point defects dominate in assisting the transport mechanism in this case. The critical time situation also occurs with irradiation except that after this time a steady state is reached which remains until the radiation conditions are altered or turned off.

MODELLING OF SEGREGATION

EQUILIBRIUM SEGREGATION

Prediction of the amount of impurity segregation to boundaries is based on the models of McLean[6] and Hondros and Seah.[2] Essentially the segregation intensity or ratio of boundary concentration, c_b to that of the matrix concentration, c_g is given by

$$\frac{c_b}{c_g} = \exp\left(\frac{E_b}{kT}\right) \tag{1}$$

where E_b is the binding energy of the misfitting impurity atom to the boundary, k is Boltzmann's constant and T is the absolute temperature. The kinetics are introduced into the prediction by assuming that the concentration on the boundary builds up and is monitored as a function of time by the concentration c_x, where

$$\frac{c_x - c_g}{c_b - c_g} = 1 - \exp\left(\frac{4D_l t}{\alpha^2 d_1^{\,2}}\right) erfc\left(\frac{4D_l t}{\alpha^2 d_1^{\,2}}\right) \tag{2}$$

where D_l is the lattice diffusion coefficient, α is the enrichment ratio given in (1), t is time and d_1 is the grain boundary width.

Using these equations a graph of temperature plotted against time to produce a boundary with a given enrichment ratio yields a characteristic 'C' curve, an example of which is shown in Fig. 1.

NON-EQUILIBRIUM SEGREGATION (NES)

There are basically two main methods employed to predict the magnitude of NES. The first is the more rigorous rate theory or Inverse Kirkendall model, which uses a multiplicity of parameters many of which are unknown. The second is the solute drag model, which is a simplified version of the rate theory. It has the advantage that it requires fewer unknown input parameters and can be easily used in a truly predictive sense to assist in engineering design calculations of plant integrity in relation to impurity boundary segregation.

RATE THEORY

Rate theory (see for example, English)[7] attempts to assess the time dependence of the concentration at the boundary for all the solute atoms and point defects (vacancies and self-interstitals) using partial differential equations based on non-steady state diffusion theory. These equations are then solved simultaneously to give the concentration variation for the impurity in question as a function of the heat treatment. The boundary conditions are determined by this heat treatment. Typical equations that have to be solved, for vacancies, interstitials and impurity are as follows.

$$\frac{\partial C_i}{\partial t} = G_i - Rc_iC_v - divJ_i$$

$$\frac{\partial C_v}{\partial t} = G_v - Rc_iC_v - divJ_v$$

$$\frac{\partial C_s}{\partial t} = -divJ_s \tag{3}$$

where

$$J_i = -D_{ii}\nabla C_i - D_{iv}\nabla C_v - D_{is}\nabla C_s$$
$$J_v = -D_{vi}\nabla C_i - D_{vv}\nabla C_v - D_{vs}\nabla C_s$$
$$J_s = -D_{si}\nabla C_i - D_{sv}\nabla C_v - D_{ss}\nabla C_s$$

where the subscripts i, v, and s refer to self-interstitials, vacancies and solute respectively. C is concentration. G is the point defect generation rate, and R is the recombination rate between vacancies and self-interstitials: these terms only apply to irradiation-induced segregation. The diffusion coefficients, D, refer to transport by exchange with the two subscripted species. The interstitial equations only apply to irradiation-induced segregation.

SOLUTE DRAG

The solute drag model (see, for example, Faulkner)[8] determines the boundary conditions set up by the NES treatment (quenching or irradiation) and assumes that all the solute is moved to the boundary by binary point defect-solute complexes moving down the point defect concentration gradients defined by the boundary conditions. This leads to a maximum solute enrichment ratio given by the following equation for the quenching type of NES, where vacancies are the dominant point defect.

$$\frac{c_b}{c_g} = \exp\left(\frac{E_{b1} - E_f}{kT_i} - \frac{E_{b1} - E_f}{kT_{0.5Tm}}\right)\frac{E_{b1}}{E_f} \tag{4}$$

Where E_{b1} is the binding energy of the solute atom to the point defect and E_f is the formation energy of the point defect responsible for transporting the solute to the boundary (vacancy or self-interstitial). T_I is the initial temperature before the quench and $T_{0.5Tm}$ is the temperature below which all diffusion is assumed to be minimal, usually half the absolute melting point. The segregation kinetics to give the boundary concentration, c_{bt}, at any given time, t, are defined by the following equation.

$$\frac{c_{bt} - c_g}{c_b - c_g} = 1 - \exp\left(\frac{4D_c t}{\alpha^2 d_1^2}\right)erfc\left(\frac{2(D_c t)}{\alpha d_1}\right)^{\frac{1}{2}} \tag{5}$$

Where D_c is the diffusion coefficient for the binary solute-point defect complex and α is the avoidment ratio given in (4). After a critical time t_c desegregation begins to occur. Another equation must be used to quantify the desegregation phase. The critical time is given by

$$t_c = \frac{d^2 \ln\left(\dfrac{D_c}{D_I}\right)}{48(D_c - D_I)} \tag{6}$$

Where d is the grain size and δ is a numerical constant (usually 0.5), and the desegregation equation is as follows.

$$\frac{c_x - c_g}{c_b - c_g} = \left(\frac{D_c t_c}{D_I t}\right)^{\frac{1}{2}} \exp\left(\frac{-x^2}{4D_I t}\right) \tag{7}$$

where x is the distance from the grain boundary.

For irradiation, self-interstitials are the dominant point defect species. The equations controlling the process predict a time dependence similar to that for quenching but where radiation dose rate and defect production and annihilation bias have to be considered.[9] The relevant equations for a binary situation are as follows. Site competition is also included in these equations.

$$C^m_{br\,(Sj)} = C^{Sj}_g \frac{E^{ip}_{b(Sj)}}{E^p_f} \left[\frac{C^{Sj}_g \exp\left(\frac{E^{ip}_{b(Sj)}}{kT}\right)}{\sum_j C^{Sj}_g \exp\left(\frac{E^{ip}_{b(Sj)}}{kT}\right)} \right] \left[1 + \frac{BGF(\eta)}{A_p D_p k^2_{dp}} \exp\left(\frac{E^p_f}{kT}\right) \right] (j = 1,2) \qquad (8)$$

and

$$\frac{C^{Sj}_{br}(t) - C^{Sj}_g}{C^m_{br(Sj)} - C^{Sj}_g} = 1 - \exp\left(\frac{4D_{c(Sj)}t}{\alpha^2_{Sj} d^2_1}\right) erfc\left(\frac{2\sqrt{D_{c(Sj)}t}}{\alpha_{Sj} d_1}\right) (j = 1,2) \qquad (9)$$

where $C^m_{br(Sj)}$ is the maximum time-independent concentration of the Sjth element on the boundary, C^{Sj}_g is the grain concentration of the Sjth element, $E^{ip}_{b(Sj)}$ is the interstitial-impurity (Sj) binding energy, E^p_f is the interstitial formation energy, k is Boltzmann's constant, T is the absolute temperature, B is the proportion of freely migrating defects (assumed to be 1%), G is the point defect generation rate, $F(\eta)$ is the recombination rate, discussed more fully in Faulkner *et al.*,[9] A_p is the pre-exponential term in the equation describing the interstitial concentration, D_p is the interstitial diffusion coefficient, k^2_{dp} is the sink efficiency of the matrix for the interstitial point defect, taking into account the dislocation density and grain size, $C^{Sj}_{br}(t)$ is the boundary concentration of the Sjth element after time t, $D_{c(Sj)}$ is the complex diffusion coefficient for the Sjth element, $\alpha_{(Sj)}$ is the maximum enrichment ratio for the Sjth element, ie, $C^m_{br(Sj)}/C^{Sj}_g$ from equation (8), and d_1 is the grain boundary width (assumed to be 1 nm).

Site competition is totally accounted for by assuming that $C^{Sj}_{br}(t)$ is modified to $C_{br}(t)^*$, according to the relative binding energies which the segregating species have to the grain boundary itself, Q_{Sj}.

$$C^{S1}_{br}(t)^* = C^{S1}_{br}(t) \left(\frac{C^{S1}_g \exp\left(\frac{Q_{S1}}{kT}\right)}{C^{S1}_g \exp\left(\frac{Q_{S1}}{kT}\right) + C^{S2}_g \exp\left(\frac{Q_{S2}}{kT}\right)} \right) \qquad (10)$$

After the critical time a steady state is reached with the flux of solute caused by the irradiation equalling the reverse segregation attempting to return the system to equilibrium, ie, there is no desegregation phase.

DATA REQUIRED FOR MODELLING RADIATION-INDUCED SEGREGATION

The most inscrutable data required for modelling radiation-induced segregation are those connected with the diffusion characteristics of the interstitial-impurity complexes and the impurity-interstitial binding energy. The complex diffusion can take place in bcc lattices by a variety of mechanisms. Barbu and Lidiard[10] have treated these and it seems that the most likely mechanism is the interstitial migration plus 60° rotation (RT_1). Under these circumstances we have assumed that the total energy required to move the complex involves binding a self-interstitial/impurity atom complex (energy required, E_b^{ip}), followed by migration of the interstitial (energy required, E_m^p). The activation energy for diffusion of the complex ($D_{c(Si)}$) is given by the sum of the two above energies.

The solute-interstitial binding energy is derived from a continuum elasticity analysis of the lattice surrounding the impurity atom in the interstitial-impurity split (110) dumbbell configuration. Although Hardy and Bullough[11] have shown that lattice statics based on the discrete nature of the lattice is a better approach to use, we feel justified in using continuum elasticity because the continuum approach only breaks down in the long range component of the energies involved. For nearest neighbour distances it seems a good approximation, unless, of course, strong electronic effects are important, as is the case if non-metallic bonding applies, for example with ceramics or polymers. It may be that in future molecular dynamics calculations will provide answers to these binding energy questions (eg, Calder and Bacon).[12] For the time being, the best information available comes from the continuum elasticity results and these are shown for austenitic steel matrices in Fig. 3.[3]

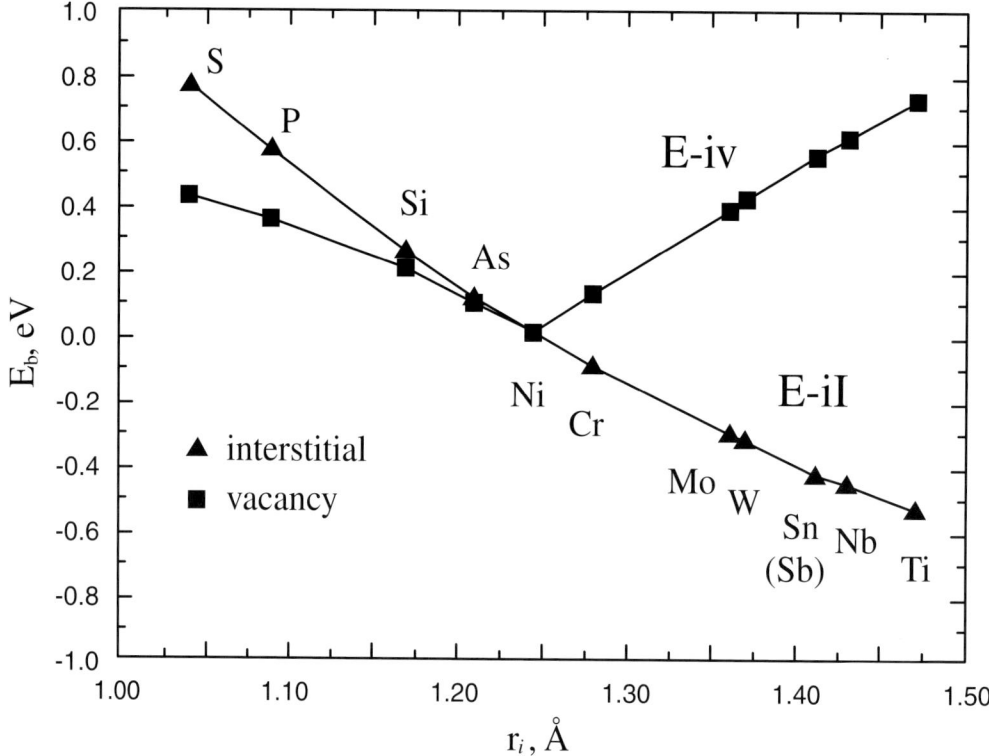

Fig. 3 Impurity-point defect binding energies, E_b, as a function of impurity atom radius, r_i, in a ferritic steel matrix. Note that the interstitial binding energies become negative for positive misfits.

It is seen that for vacancy-impurity interactions the binding energy is always positive (attractive). But for interstitial-impurity complexes the binding energy goes negative for positive misfit situations. This implies that, for positive misfits, the impurities lock the dumbbells in stable positions, and no further impurity drag towards the boundary occurs.

EXPERIMENTAL TECHNIQUES TO STUDY INTERGRANULAR SEGREGATION

One of the difficulties encountered in quantifying segregation is the very small scale of the composition accumulation. Usually microanalytical spatial resolutions of around 1 nm are necessary. The only techniques that satisfy this requirement are field emission gun transmission electron microscopy (FEGTEM), Auger electron spectroscopy (AES), and atom probe field ion microscopy (AFIM). FEGTEM illuminates the specimen with a very bright beam that is capable of producing large quantities of X-rays from very small areas. This method has already shown clear examples of sub-nanometre segregated layers of phosphorus, Ni, and Mo in steels. The benefit of the method lies in its capability to see and position segregated layers in the global environment of the grain boundary. AES is a surface analytical technique and an intergranular fracture must first be made in order to study segregation effects on the exposed fracture surfaces. For this reason the precise position of the segregated layer can never be ascertained by this method. It does, however, unlike the FEGTEM, have very high sensitivity and can detect light elements. Successful studies of P, S and C segregation in ferritic steels have been made by this technique. AFIM is an atomic resolution microanalytical technique requiring painstaking specimen preparation and ultra-high vacuum technology with time-of-flight mass spectrometers. It has been used to study Mo and P intergranular segregation in ferritic steels.

INDUSTRIAL IMPLICATIONS OF INTERGRANULAR SEGREGATION

The importance of intergranular segregation in austenitic and ferritic steels is manifested in the premature intergranular cracking of pressure vessels and superheater components, particularly in welds. Very high resolution analytical electron microscopy indicates that phosphorus and sometimes sulphur is observed in thin layers a few nanometres thick on the grain boundaries in susceptible materials. An example of a FEGTEM result of this kind for quench induced NES of phosphorus in a 2.25%Cr–1Mo ferritic steel is given in Fig. 4 (after Song).[13] The kinetics of the accumulation of these elements can be predicted by the basic models for quench-induced segregation described above. The desegregation effect is illustrated in Charpy measurements made on Fe–12Cr steels during quenching and tempering (Fig. 5).[8] During a fixed tempering treatment of 2 hours at 600°C, the silicon segregation reaches a maximum at a quench rate of $50\,\mathrm{K\,s^{-1}}$. The

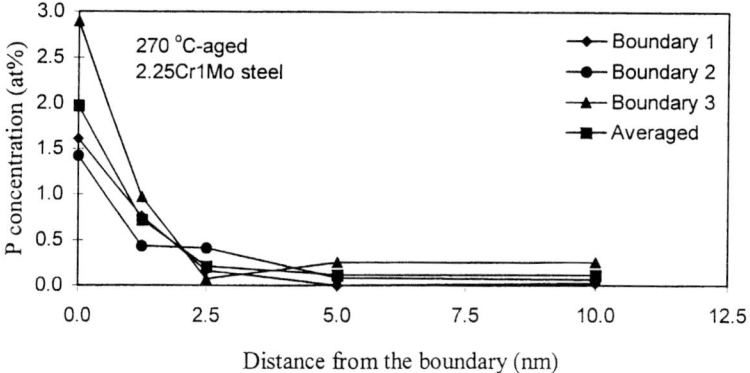

Fig. 4 Phosphorus concentration profiles from the grain boundary to the matrix for 2.25 Cr–1.0Mo steel aged at 270°C for 3000 hours (data points away from the boundary are produced by averaging the measured values from both sides of the boundary).

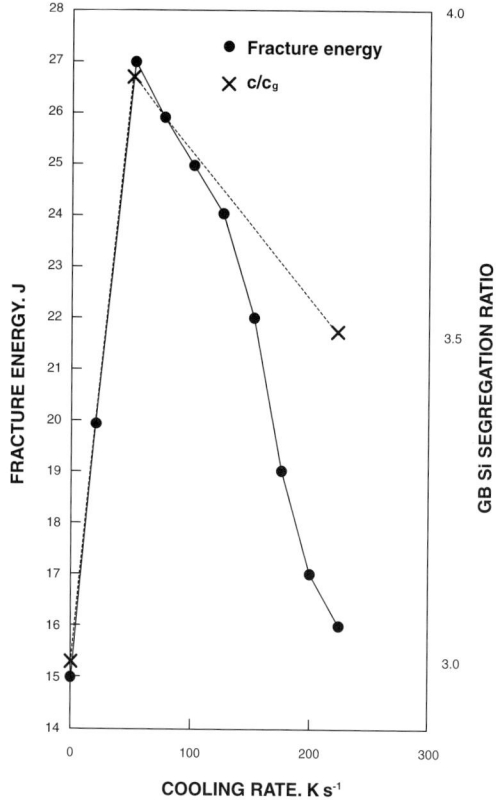

Fig. 5 Impact fracture energy and predicted Si intergranular segregation (c/cg) as a function of cooling rate for fixed 600°C: 2 hours tempering treatments for an iron–12 Cr ferritic steel.

corresponding Charpy energy reaches a maximum after the same time, additionally confirming the relation between intergranular segregation and mechanical properties.

Probably the most work done in recent years has been on irradiation-induced inter-granular segregation in steels, and phosphorus has been the major element of interest.

A combined survey picture of the experimental results obtained for P grain boundary segregation as a function of temperature for a variety of steels and iron-based binary alloys is shown in Fig. 6. It should be emphasised that these results come from experiments with widely-differing conditions of dose, dose rate, composition and microstructure and should therefore only be reviewed in a qualitative sense.

Several general observations can be made about phosphorus segregation behaviour in steels. The first is that, taken in conjunction with the modelling predictions to be discussed later, segregation increases with increasing temperature as seen in Fig. 6, but the temperature dependence appears to have two peaks. The one at low temperature (250°C–300°C) is due to non-equilibrium segregation (RIS), caused by streams of self-interstitial-impurity complexes being drawn to the boundary by the irradiation. The second, high temperature peak (>400°C) is due to thermal equilibrium segregation (ES). Secondly, there is evidence for a site competition effect, particularly from carbon. The effect of increasing carbon causing reduced phosphorus segregation is seen in the work of Beere.[14] Kameda and Bevolo[15] have observed similar trends. Thirdly, increasing dose increases the amount of phosphorus segregation, but there does seem to be a saturation limit.

There are strong correlations of these effects with mechanical properties: higher phosphorus coverage leads to larger amounts of intergranular fracture. It should be realised

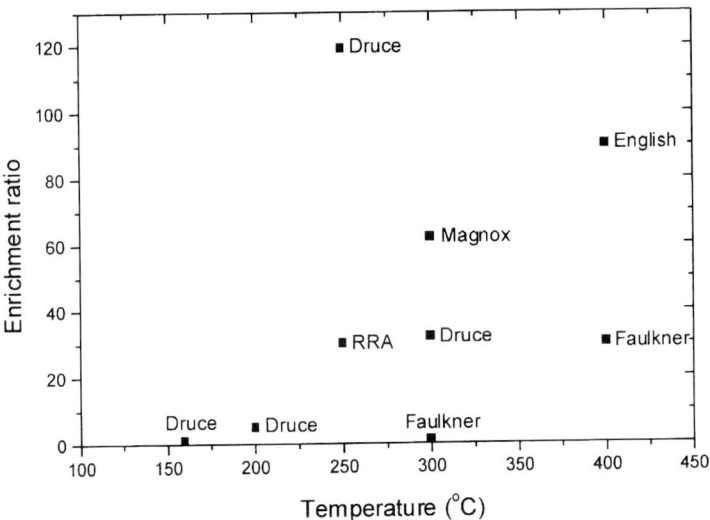

Fig. 6 Magnitude of radiation induced phosphorus intergranular segregation in pressure vessel steels as a function of temperature. Reference Key: Druce *et al.*;[17] English, private communication, after Kameda;[15] RRA – Meade;[20] Magnox – Jones *et al.*,[22] 1998; Faulkner *et al.*[9]

that the yield properties also increase with dose due to matrix hardening, and, in the case of some older pressure vessel steels, copper-rich precipitation due to irradiation.

There have been even higher resolution studies of segregation than those from FEGTEM or AES. Atom probe field ion microscopy (AFIM) has confirmed the presence of P and Mo on grain boundaries in VVER irradiated pressure vessel steels.[16]

Figure 7 provides the rate theory and solute drag theory (with and without carbon) modelling predictions of the temperature dependence of phosphorus grain boundary segregation in ferritic steel, and a summary the experimental data from Fig. 6 is given for comparison/validation purposes.

The solute drag model is very sensitive to the site competition effect when carbon is considered. The rate theory predictions[17] give quite good fits with observed data, but they do not include site competition, and, in any case, have been evaluated by curve-fitting to the data in the first place. Two carbon concentrations are shown for the solute drag model. The best fit lies with a curve with between 50 and 500 appm carbon. This represents the free carbon content, after the carbon required for the carbides has been subtracted from the alloy composition. An independent metallographic study of a 0.4 at.%C–0.14at.%P–2.25%Cr–1.0%Mo steel has shown that the free carbon is around 0.11 at.% (1100 appm) with a slightly lower stress relieving treatment temperature (400°C) than typical for pressure vessel steels (650°C). This indicates that a good estimate of the free carbon content for the steels in Fig. 7 is somewhere between 50 and 500 appm. It is seen that this complies well with the experimental curve.

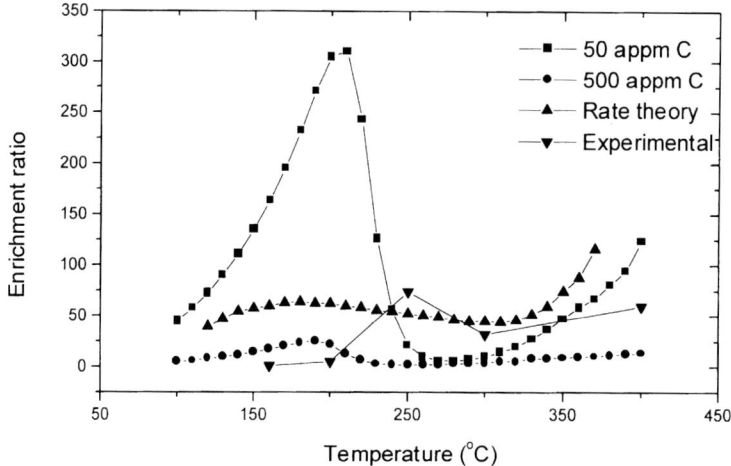

Fig. 7 Theoretical predictions of phosphorus intergranular segregation in ferritic steels as a function of temperature. A summary of experimental data from Fig. 6 is also included. Two carbon contents are included in the solute drag theory predictions. The rate theory predictions are taken from Druce.[17] Dose: 1 dpa. Dose rate: 10^{-8} dpa s^{-1}. Grain size: 10 μm. Dislocation density: 10^{16} m^{-2}.

The observed parabolic dose-dependence is supported by both rate theory and NES RIS models.

An example of irradiation-induced segregation in a positive misfit situation is seen in intergranular stress corrosion cracking in austenitic steels used in modern power plant. Here the reverse of segregation, chromium depletion is responsible for accelerated oxidation and cracking in the intergranular regions. This is explained on the basis of depletion caused by combined reverse segregation and precipitation effects. This observation is the conclusion drawn from several experimental studies completed in the USA, Japan and the UK (a typical example of these results is taken from the work of Bruemmer,[18] Fig. 8). No studies have so far been undertaken to separate the segregation and precipitation components of the observations and this must be an urgent and necessary component of future work. The results described next insist on the fact that no chromium carbide precipitation was present on the boundaries from which the depletion measurements were taken. The interesting point to note is that there is a reversal of the effect on moving from the heat treated condition prior to irradiation, when the Cr is segregating, to depletion when the irradiation is turned on. This can be explained by solute drag theory on the basis that thermal effects cause solute drag mechanisms to operate based on the vacancy-impurity complex. The binding energy of Cr to vacancies in austenitic matrices is positive (Fig. 3). For irradiation, we have indicated that self-interstitials play a more important role than vacancies. In this case, Fig. 3 shows that the slight positive misfit of Cr leads to a negative binding energy with the interstitial (repulsion, implying zero drag). Under these circumstances it is assumed that all the segregation calculations will be reversed so that enrichment predictions now become equal but opposite depletions. Therefore it is expected that

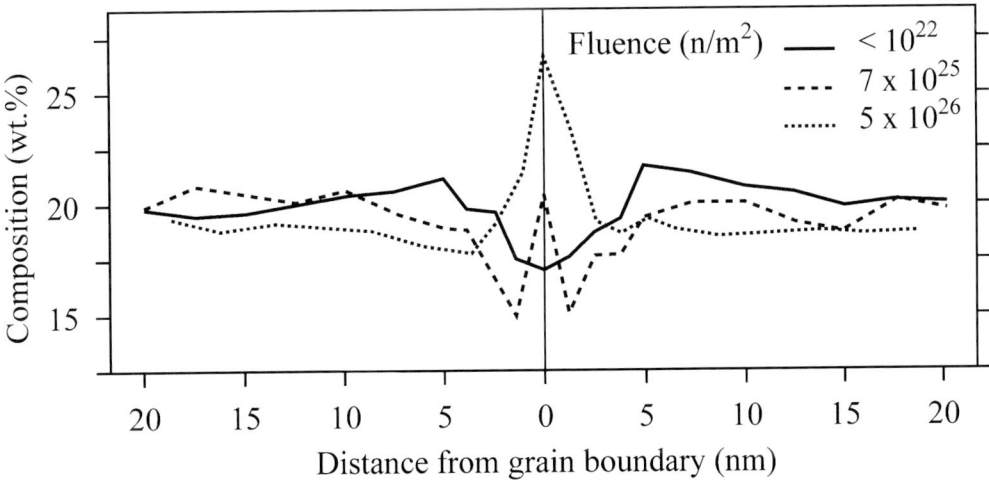

Fig. 8 Irradiation effects on Cr composition profiles measured by FEGTEM across grain boundaries in 316 stainless steel (after Bruemmer).[18]

Cr depletion will occur during irradiation and that the depletion will increase with dose, and this is supported by the observations (Fig. 8).

It is important to remember that solute drag theory has been developed for dilute impurity scenarios. Clearly Cr in austenitic steels is a concentrated alloy situation and therefore the predictions do not have so much validity as those for the P in ferritic steel. It is also noteworthy that rate theory predictions concerning the Cr depletion effect have recently been made by, for example, Was *et al.*[19]

Probably some of the most commercially-significant work is that of Kato *et al.*[21] They show that the addition of Zr and Hf reduce the Cr depletion effect to almost zero. They also have explored a range of strong carbide forming additions and they show that the effect is not simply due to size factor effects. Nb and V have very large positive misfits but they do not have any effect on irradiation-induced Cr depletion.

CONCLUDING STATEMENT

A review of the major types of intergranular segregation is made. The key features of the equilibrium (ES) and non-equilibrium (NES), solute drag mechanisms are highlighted. Modelling and experimental approaches to the problem are surveyed, and examples are given of where intergranular segregation is important in ferrous industrial applications.

REFERENCES

1. P. Lejcek and S. Hoffman, *Crit. Rev. Solid State Mater. Sci.,* 1995, **20**, 1.
2. E. D. Hondros and M. P. Seah, *Int. Met. Rev.,* 1977, **22**, 262.
3. R. G. Faulkner, *International Materials Reviews,* 1996, **41**, 198.
4. K. T. Aust, S. J. Armijo, E. F. Koch and J. H.Westbrook, *Trans ASM,* 1967, **60**, 360.
5. T. R. Anthony, *Acta. Metall.,* 1969, **17**, 603.
6. D. McLean, *Grain Boundaries in Metals,* Oxford University Press, 1957, 131.
7. C. A. English, S. M. Murphy and J. M. Perks, *J. Chem. Soc., Faraday Trans.,* 1990, **86**, 1263.
8. R. G. Faulkner, *Acta. Metall.,* 1987, **35**, 2905.
9. R. G. Faulkner, S. Song, P. E. J. Flewitt, M. Victoria and P. Marmy, *J. Nuclear Materials,* 1998, **255**, 189.
10. A. Barbu and A. B. Lidiard, *Phil. Mag.,* 1996, **A74**, 709.
11. J. R. Hardy and R. Bullough, *Phil. Mag.,* 1967, **15**, 237.
12. A. F. Calder and D. J. Bacon, *J. Nuclear Materials,* 1993, **207**, 25.
13. S. Song, R. G. Faulkner, P. E. J. Flewitt, P. Marmy, and M. Victoria, ASTM STP1366, M. L. Hamilton, A. S. Kumar, S. T. Rosinski and M. L. Grossbeck eds, ASTM, 2000, p. 343.
14. W. Beere, Rep. No. TE/GEN/REP/0152/97, Berkeley Centre, BNFL-Magnox, UK, 1997.
15. J. Kameda and A. J. Bevolo, *Acta. Metall.,* 1989, **37**, 3283.
16. M. K. Miller and M. G. Burke, *J. Nuc. Mat.,* 1992, **195**, 68.
17. S. G. Druce, C.A. English, A. J. E. Foreman, R. J. McElroy, I. A. Vatter, C. J. Bolton, J. T. Buswell and R. B. Jones, *Proc 17th Symposium on Radiation Effects in Materials,* ASTM STP 1270, D.S. Gelles, R. K. Nanstead, A. S. Kumar and E. A. Little eds, ASTM, 1996, p. 119.
18. S. M. Bruemmer, Materials Science Forum, 1999, **294–296**, 75.

19. G. S. Was, T. R. Allen, J.T. Busby, J. Gan, D. Damcott, D. Carter, M. Atzmon and E. A. Kenik, *J. Nuc. Mat.*, 1999, **270**, 96.
20. D. Meade, PhD Thesis, Loughborough University, 1998, 238.
21. T. Kato, H. Takahashi and M. Izumia, *J. Nuc. Mat.*, 1992, **189**, 167.
22. R. B. Jones, J. R. Cowan, R. C. Corcoran and J. C. Walmsley, BNFL-Magnox Generation, Private communication, 1998.

The Role of Austenite–Ferrite Interface in the Decomposition of Super Duplex Stainless Steels

M. STRANGWOOD

The University of Birmingham, School of Metallurgy and Materials/IRC in Materials,
Elms Road, Edgbaston, Birmingham B15 2TT, UK

X. LI

Thermtech Ltd, Surrey Technology Centre, Guildford, Surrey GU2 5YG, UK

ABSTRACT

The ageing behaviour of Zeron 100 TIG weldments at 850°C has been characterised in terms of the phases formed in ferrite and their rate of formation both microstructurally using quantitative SEM and TEM and through their effects on hardness. The weldment decomposition has been compared with the basemetal and separate rolled plate material and a commercial thermodynamics package has been used to rationalise the behaviour. Hardening has been related to the formation of σ-phase in ferrite, but this is preceded by interfacial formation of χ-phase. Equilibrium thermodynamics can explain the behaviour and its dependence on composition qualitatively, but not quantitatively.

INTRODUCTION

The extremely high corrosion resistance of super duplex stainless steels (SDSSs), defined as having a pitting resistance equivalent (PRE) value greater than 40, is achieved through extensive alloying, principally with Cr, Mo, Ni and N. The highly alloyed nature of SDSSs imparts a high degree of metastability on the ferrite phase (SDSSs are usually balanced at the 50:50 mix of austenite and ferrite),[1] which gives rise to a complex series of precipitate phases. Precipitation is possible over the temperature regime 250 to 950°C[2,3] with each of the precipitate types formed leading to a reduction in the corrosion resistance and, generally, toughness of the alloy. These ferrite decomposition processes limit the service temperature range over which these alloys can operate. In addition to thermal exposure in service, processing of SDSSs will also involve elevated temperatures and the risk of precipitation, particularly in the weldbead and heat affected zones (HAZs) of fusion welds. Weldmetal is often more highly alloyed with respect to Ni in order to maintain the 50:50 phase balance, which has been reported to accelerate the decomposition of ferrite on thermal exposure.[4] Further development of SDSSs will involve increased alloying additions and so the processing and service windows will be reduced, especially where fusion

43

welding is concerned. In order to optimise future composition–process combinations, it will be necessary to establish accurate models for the development of microstructure in these systems capable of expansion to proposed future alloy compositions. As part of this process, the thermal decomposition of existing SDSS products are being quantified and a comparison between wrought plate and TIG weldments is reported here.

EXPERIMENTAL

Samples were machined from a cross-rolled Zeron 100 plate and two TIG weldments, Tables 1 and 2. These coupons were sealed in silica tube under a partial pressure of argon and aged at 850°C for times up to 1 hour followed by water quenching. After heat treatment the samples were sectioned normal to the major rolling direction/welding direction, mounted, polished to a 0.25 µm diamond paste finish and electrolytically etched in an aqueous 10% oxalic acid solution (5–30s at 3–6 V). These samples were used for optical and scanning electron microscopy, the latter carried out in a JEOL 6300 SEM operating at 20 kV and equipped with a Link QX2000 energy dispersive X-ray spectroscopy (EDS) system using an ultrathin window detector.

0.5 mm thick slices were cut from selected specimens using a slow speed diamond saw to minimise deformation. 3 mm diameter discs were punched from these slices and mechanically ground to 100–150 µm thickness prior to perforation using a Tenupol III twin jet electropolisher in a solution of 2% perchloric acid in 2-Butoxy ethanol at −15°C and 60 V. The resulting foils were examined in JEOL 4000fx and Philips CM20 transmission electron microscopes (TEMs) operating at 200 kV and equipped with Link QX2000/AN10000 EDS systems. Samples were taken from basemetal and weldmetal but not from any HAZ regions.

Both microhardness (10 g load) and macrohardness (10 kg load) values have been measured.

RESULTS AND DISCUSSION

As-Received Material

The as-rolled and as-welded specimens all contained an austenite volume fraction of 0.5 ± 0.05 although the scale and morphology changed with finer, more acicular austenite seen in the weldmetal. Despite the increased amount of γ/δ interfacial area

Table 1 Chemical composition (wt%) of as-received Zeron 100 plate.

C	Si	Mn	P	S	Cr	Mo	Ni	Cu	W	N	Fe
0.017	0.3	0.46	0.021	0.001	25.3	3.7	7.2	0.59	0.57	0.24	Rem

Table 2 Compositions (wt%) of as-received Zeron 100 DSS TIG welds.

	C	Si	Mn	P	S	Cr	Mo	Ni	Cu	W
TIG weld 1	0.025	0.41	0.75	0.05	<0.01	24.62	3.36	9.04	0.71	0.91
TIG weld 2	0.017	0.33	0.66	0.04	<0.01	23.32	3.04	8.93	0.65	0.5

present in the weldmetal, TEM of this region did not reveal any precipitation at these sites or elsewhere. The plate and basemetal were also free of precipitation.

AGEING RESPONSE OF PLATE

The austenite phase of the rolled plate was unaffected by thermal exposure at 850°C, whereas microhardness testing revealed that the ferrite phase started to harden after between five and fifteen minutes exposure with hardness increasing up to 60 minutes, Fig. 1. A similar trend was observed in macrohardness values. Microstructurally, precipitation was observed, by SEM and TEM, to occur after 5 minutes ageing. Although this did not result in increased hardness, the etching behaviour of the ferrite was altered indicating a change in the corrosion resistance; this aspect requires further study. SEM backscattered electron images revealed that the γ/δ interfaces were decorated by a series of bright (increased heavy alloying element content) precipitates. The fine scale of these precipitates precluded EDS analysis in the SEM but TEM examination revealed them to be χ-phase (Im3m, $a = 0.892$ nm) enriched in Mo and W whilst depleted in Ni, Table 3.

The χ-phase exhibited a cube–cube orientation relationship with the remaining ferrite into which it grew; the precipitate did not appear to develop into the austenite phase, Fig. 2. Also observed at the γ/δ interfaces was a small amount of Cr_2N, which appeared to form with or prior to the χ-phase and grows into both phases. The volume fraction of these nitrides was not determined due to the small amount and their fine scale. Extended thermal exposure gave an increase in the volume fraction of

Table 3 EDS analyses (wt%) for Zeron 100 DSS plate and weldmetal aged at 850°C for 1 hour (plate) or 5 minutes (weld).

		Cr	Ni	Mo	Mn	Si	W
δ (as-rec)	Plate	27.85	5.62	4.22	0.64	0.42	0.64
	Weldmetal I	27.45	7.42	3.93	0.62	0.68	0.98
	Basemetal 1	27.24	5.33	4.05	0.71	0.64	0.85
χ-phase	Plate	25.8	2.6	17.4	0.4	0.5	3.5
	Weldmetal 1	25.6	3.2	22.2	0.4	0.7	3.1
σ-phase	Plate	31.2	3.3	7.6	0.2	0.5	1.2
	Weldmetal 1	32.1	4.5	7.6	0.5	0.4	1.5
Eutectoid	Plate	21.2	9.3	2.1	0.4	0.3	0.3
	Weldmetal 1	20.1	10.1	1.2	0.9	0.3	0.4

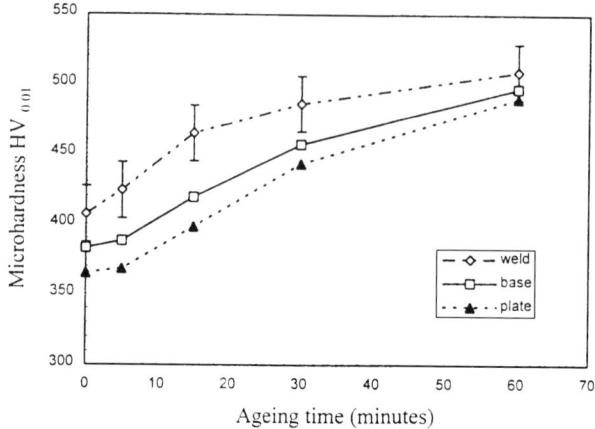

Fig. 1 Microhardness of ferrite $HV_{0.01}$ for ageing at 850°C.

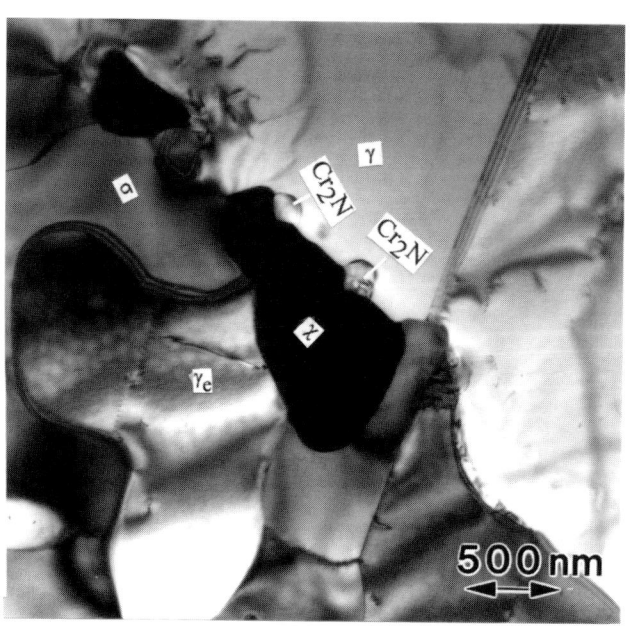

Fig. 2 Zeron 100 rolled plate aged at 850°C for 1 hour.

χ-phase up to 1.9% for 30 minutes after which the amount decreased, falling to 1.6% after 60 minutes. Thus the χ-phase appears to be a metastable transient phase and is replaced by the σ + γ eutectoid, which first appeared after 15 minutes ageing and so corresponds to the increase in hardness noted earlier. The eutectoid colonies appear to nucleate at γ/δ or χ/δ interfaces and grow into the remaining ferrite consistent with the reaction: δ σ + γ. The σ-phase itself is enriched in Cr and Mo and depleted with respect to Ni, the latter being taken up in the eutectoid austenite. During the earlier stages of the formation of the eutectoid it forms in interfacial regions where the development of χ-phase along the γ/δ interfaces is less complete, consistent with the local loss of Mo from ferrite coupled with increased Ni levels rejected from the interfacial phase.

AGEING RESPONSE OF THE BASEMETAL

The hardening response of ferrite in the basemetal follows the same trend as that for the plate material although the initial, as-rolled hardness exceeds that of the plate material. The phases observed on thermal exposure are the same and form in the same order as for the plate material. Thus χ-phase is observed to decorate γ/δ interfaces and δ/δ grain boundaries after 5 minutes exposure with a small amount of the σ + γ eutectoid; the latter increases in extent as exposure time increases, in line with the hardening curve for ferrite. The two basemetals differ in the extent of ageing and also differ from that shown by the plate material in line with their composition (SEM characterisation was carried out for both but only weldment 1 was studied using TEM). γ/δ interfaces in the basemetals of both welds showed greater decoration by χ-phase than the plate material with the basemetal of weld 2 having the greatest fraction of χ-phase formed, Fig. 3. In terms of the σ + γ eutectoid, the formation of this phase again correlated with the change in mechanical properties and, after 60 minutes exposure, resulted in almost complete transformation of the ferrite. In comparison with the eutectoid formed in the plate material, that in the basemetals was generally coarser and more divorced in nature, although the transformation rate was similar.

AGEING RESPONSE IN THE WELDMETAL

Macrohardness tests, Fig. 4, reveal that the weldmetal hardens much more rapidly than either basemetal or plate material up to 15 minutes exposure when the hardness values plateau out. The ferrite microhardness values do exhibit such a dramatic rise initially and also appear to continue hardening up to 60 minutes exposure. At all times the weldmetal ferrite hardness exceeded that of both basemetal and plate. No evidence was found for alteration in the austenite characteristics on thermal exposure and so the discrepancy may arise from the smaller resolution of the macrohardness

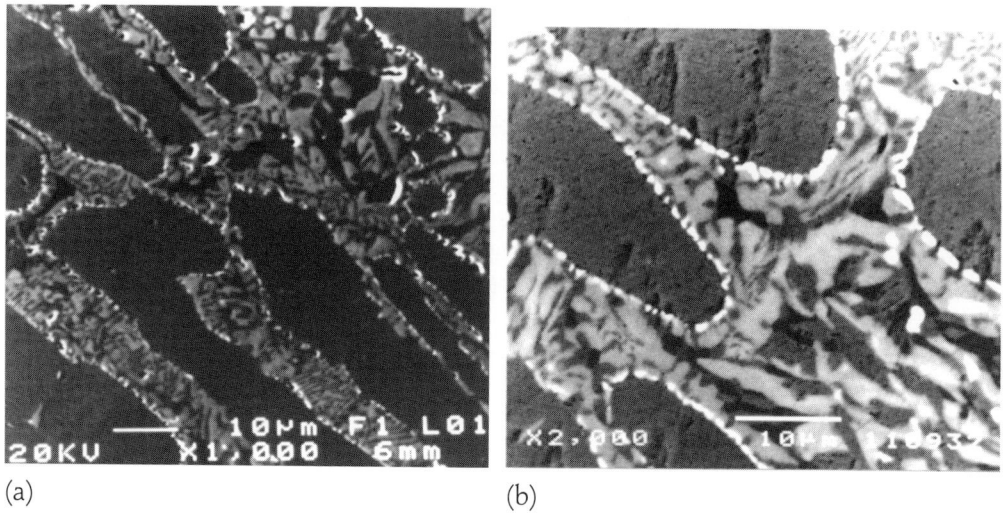

(a)   (b)

Fig. 3 Zeron 100 TIG weld basemetal aged at 850°C for 1 hour (a) weld 1, (b) weld 2. Brightness decreases from 1996, χ-phase (white), through σ + γ eutectoid, then primary γ to residual ferrite (darkest grey).

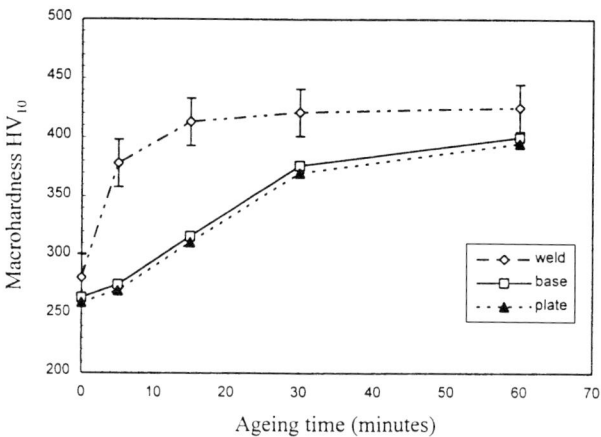

Fig. 4 Macrohardness of Zeron 100 aged at 850°C.

testing. Microstructurally, the ferrite phase decomposes at a much more rapid rate than that in plate and basemetal so that even after the shortest exposure time – five minutes – significant precipitation has occurred, Fig. 5. In contrast to the plate and basemetal ferrite, decomposition of ferrite proceeds to the $\sigma + \gamma$ eutectoid, confirmed by TEM, with a much smaller proportion of the γ/δ interfaces decorated by Mo-rich χ-phase. Thus, the increase in hardness again correlates with the formation of the eutectoid. The volume fraction of χ-phase formed is very low in the weldmetal of weld 1 with only isolated precipitates observed. Decomposition of the ferrite continues up to 60 minutes exposure, in line with the microhardness tests, when only 6% ferrite remains in the structure. The decomposition of ferrite in weld 2 proceeds with more χ-phase, Fig. 6, and at a slower overall rate, so that after 60 minutes exposure 9% ferrite remains, Fig. 7.

COMPOSITION EFFECTS

Initial rationalisation of the microstructural and hardening behaviour for the plate, basemetals and weldmetals has been attempted using the Thermo-Calc thermodynamics package (version L), which has previously modelled these steels successfully.[2] Concerning the overall equilibrium structure at 850°C (Table 4), close agreement is noted between the prediction and the observed phases for the plate material, where the transient χ-phase has started to re-dissolve and around 12.5% ferrite remains after 60 minutes exposure.

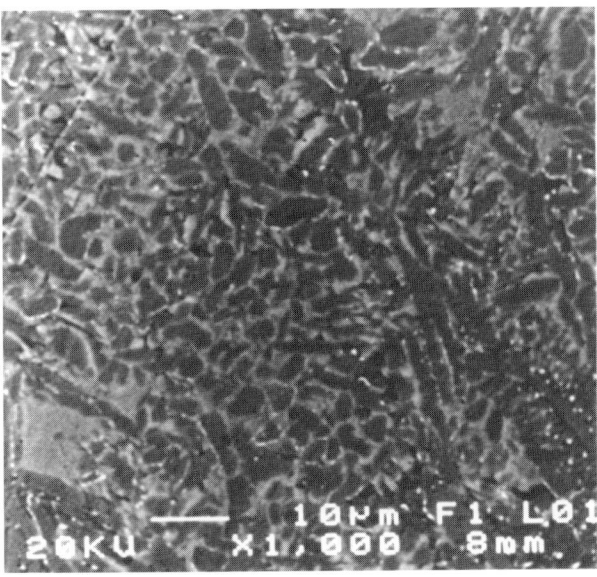

Fig. 5 Zeron 100 TIG weld 1 weldmetal aged at 850°C for 5 minutes. Brightness as for Fig. 3.

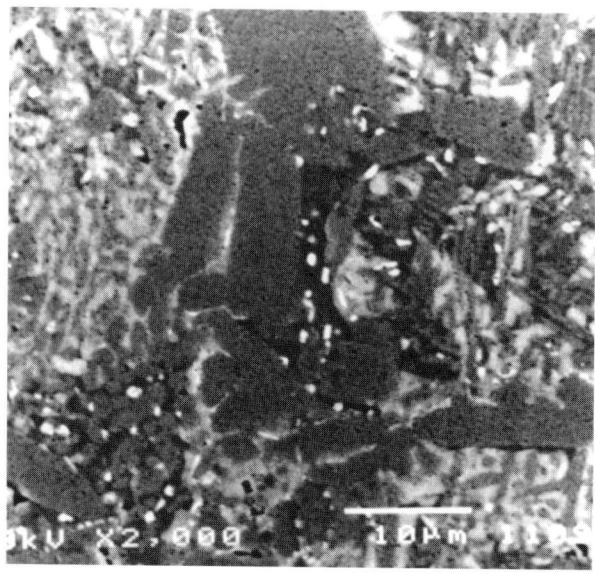

Fig. 6 Zeron 100 TIG weld 2 weldmetal aged at 850°C for 5 minutes. Brightness as for Fig. 3.

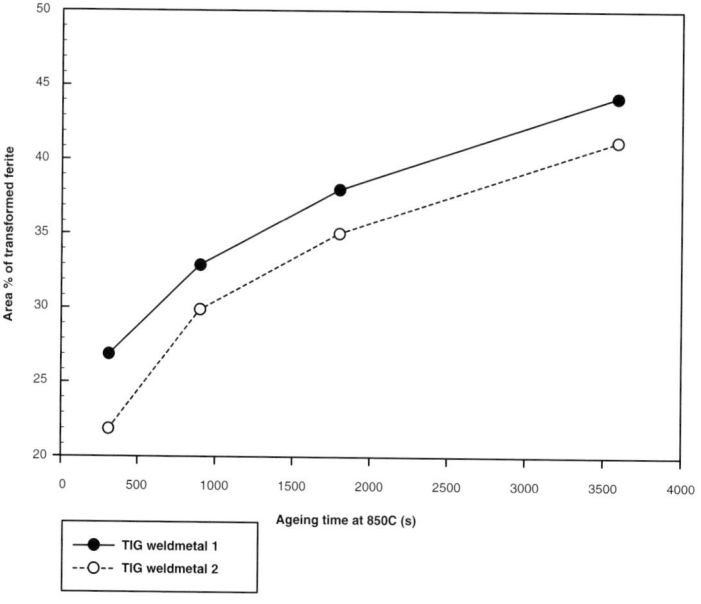

Fig. 7 Decomposition of ferrite in Zeron 100 weldments at 850°C.

Table 4 Thermo-Calc equilibrium predictions (%) at 850°C.

	δ	γ	σ
Plate	11.69	74.68	13.62
Weld 1	–	87.31	12.7
Weld 2	0.02	91.4	6.4

The observed trends for the weldmetals also agree with those predicted with greater stability of ferrite in weld 2, however the absolute values are at variance as the levels of retained ferrite predicted are negligible. This discrepancy may be related to uncertainty in the nitrogen levels actually present in the weldments. The σ + γ eutectoid in the weldmetals is much finer than that in plate and basemetals but still contains much more σ-phase than predicted. These aspects are being further investigated to improve agreement. However, in semi-quantitative terms, the model already accounts for the effects of Ni, W and Mo, which are the major compositional variations between these samples. Prediction of the equilibrium structure covers the variation in ferrite, austenite and σ-phase, which relates to the hardening of the steel, but does not address the formation of chromium nitride and the χ-phase, which has an influence on the chemical behaviour and corrosion resistance. Restricting the allowable phases and utilising the locally determined ferrite composition allows the initial driving force for the δ → χ transformation to be estimated. This driving force is needed for future time-temperature-transformation (TTT) and continuous cooling transformation (CCT) curve generation. The values determined for plate and weld 1 are presented in Table 5 utilising TEM-EDS compositions for substitutional alloying elements and published partition coefficients[5] for the interstitial contents.

The driving force values, which are zero for stable phases and hence the driving force increases the closer to zero its value, indicate that the tendency to form χ-phase decreases in the order:

basemetal weld 1 > plate > weldmetal weld 1

agreeing with the extent of precipitation observed. Thus the composition variations between the samples are covered in a (currently) qualitative manner.

Table 5 Theoretical driving forces ($\Delta G/IRT$) for δ → χ transformation at 850°C.

	Plate	Weld 1 Basemetal	Weld 1 Weldmetal
Driving Force	-3.346×10^{-2}	-2.7962×10^{-2}	-4.096×10^{-2}

OVERALL TRANSFORMATION RATE

The final measurements needed prior to TTT and CCT prediction refer to the overall transformation rates, which have been dealt with at this stage by an Avrami-type plot in order to highlight the differences between wrought material and weldmetal before a more mechanistic approach is adopted. The Avrami coefficients are presented in Table 6 for plate and weldmetals and indicate significantly higher k values for the weldmetals consistent with faster nucleation and growth kinetics.

Measurements of the mean eutectoid colony radius as a function of time for the plate material indicate parabolic growth ($r \propto \sqrt{t}$) in the plane of the SEM images consistent with a diffusional mechanism. Hence, the low value of n suggests decreasing nucleation rates (confirmed by number density measurements) associated with site saturation at γ/δ interfaces. The fine scale of the eutectoid colonies in the weldmetals precludes the corresponding colony growth measurements. However, the compositions of the parent ferrite, σ-phase and eutectoid austenite in all three samples (plate and two weldmetals, Table 3) are similar suggesting that similar diffusional growth occurs in all three cases. On this basis, the lower n values for the weldmetal would be associated with a greater dependence upon nucleation brought about by (i) greater driving forces due to higher Ni and W levels, and (ii) greater nucleation site density due to the finer scale of austenite. The greater driving forces then result in faster site saturation and a smaller time exponent.

CONCLUSIONS AND FURTHER WORK

The microstructure and hardening behaviour of Zeron 100 SDSS plate and weldments after thermal exposure at 850°C have been quantified and related to composition through thermodynamic modelling. From these studies the following points have been identified:

(i) Decomposition of ferrite occurs along the path:

$$\delta \rightarrow \delta + Cr_2N + \chi \rightarrow \delta + Cr_2N + \chi + \sigma + \gamma \rightarrow \sigma + \gamma$$

(ii) Nucleation occurs on γ/δ interfaces with locally higher concentrations of χ-phase retarding later eutectoid formation.

(iii) χ-phase formation is associated with a decrease in corrosion resistance but little change in hardness. The latter follows the same trend as eutectoid development.

Table 6 Avrami coefficients for Zeron 100 DSS rolled plate and TIG welds aged at 850°C.

	Rolled plate	TIG weld 1	TIG weld 2
n	0.45	0.25	0.25
k	0.053	0.33	0.30

(iv) Precipitation in weldmetal is accelerated compared with wrought product forms due to increased driving force and nucleation site density; nucleation effects having a more dominant role in the overall transformation behaviour in the weldmetals.

(v) Thermodynamic modelling is able to predict the behaviour in a qualitative or semi-quantitative manner.

In providing a quantitative, predictive model for the behaviour of SDSSs and their weldments, this work is being expanded to:

(i) Provide more quantitative agreement between thermodynamic models and experimental observations.

(ii) Develop a more mechanistic model for overall transformation behaviour.

(iii) Relate precipitation processes to corrosion behaviour.

(iv) Provide a more quantitative link to hardness and other mechanical properties.

ACKNOWLEDGEMENTS

The authors would like to thank Professors M. H. Loretto and I. R. Harris for the provision of laboratory facilities and financial support, as well as the British Council and Weir Materials Ltd for further financial support (XL).

REFERENCES

1. J. C. Prouheze and G. Martin, *Proc. Conf. Duplex Stainless Steels '91,* J. Charles and S. Bernhardsson eds, Les Editions de Physiques, 1991, 632.
2. J. O. Nilsson, *Mat. Sci. Tech.,* 1992, **8**, 685.
3. A. Redjaïma, G. Metauer and M. Gantois, *Proc. Conf. Duplex Stainless Steels '91,* J. Charles and S. Bernhardsson eds, Les Editions de Physiques, 1991, 119.
4. J. O. Nilsson, T. Huhtala, P. Jonsson, L. Karlsson and A. Wilson, *Metall. Mat. Trans,* 1996, **27A**, 2196.
5. V. J. Gadgil and E. G. Keim, *Proc. 4th International Conf. Duplex Stainless Steels '94,* TWI, 1994, Paper 5.

Grain Boundary Fracture Processes in Quenched-and-Tempered Structural Steels

A. ISLAM, P. BOWEN and J. F. KNOTT

The University of Birmingham, Edgbaston, Birmingham B15 2TT

ABSTRACT

The paper describes a number of effects produced by the segregation to grain boundaries of the impurity elements phosphorus and sulphur. It is shown that 'classical' (reversible) temper-embrittlement is exhibited in fracture toughness tests carried out at both $-120°C$ and $-196°C$, as a result of phosphorus segregation. As-quenched embrittlement and '350°C' embrittlement are affected by both pre-segregation in austenite and the development of carbides during tempering, concentrating phosphorus by the 'carbide rejection' mechanism. Stress-relief cracking and crack growth at approx. 500°C are associated with the stress-driven segregation of sulphur to the region of high triaxial stress ahead of a stressed crack tip. Reconsideration of earlier results suggests that phosphorus segregation in an initially as-quenched steel may also be stress-assisted at this temperature. Intergranular facets are observed during fatigue crack propagation in air at room temperature. It is suggested that this is a result of hydrogen uptake from the water vapour present in air, but that entry of hydrogen is facilitated by the pre-segregation of impurity elements to grain boundaries, most plausibly sulphur, and possibly phosphorus.

INTRODUCTION

Electrical power-plant and chemical plant contain assemblages of structures, such as boilers, pipe-work, or reactor pressure vessels, which are fabricated by welding together components made in quenched-and-tempered steels. Typical steel compositions are 2¼Cr1Mo for boilers, piping or chemical reaction vessels and A533B (plate) or A508 (forging) versions of NiMnMo steels for nuclear reactor pressure vessels. The welding of such steels usually requires pre-heat to be applied to avoid hydrogen cracking and the welded joints are subjected to stress-relief heat treatments after welding. These involve rather slow heating and cooling rates to and from a maximum temperature of, typically, 600–650°C. Light-water-reactor pressure vessels are often clad with a 5–7 mm thick layer of austenitic steel to minimise aqueous corrosion in service (at temperatures up to 300°C). The clad is deposited by a fusion process and when the vessel is stress-relieved after cladding, extra stress in the clad and sub-clad regions may be generated as a result of the difference in thermal expansion coefficients between austenitic and ferritic steel.

Such heat-treatment sequences can set up conditions in which it is possible for trace impurities in the steel to segregate to grain-boundaries and by this segregation embrittle the boundaries such that the fracture path at low temperatures changes from transgranular to intergranular, with an associated reduction in the steel's fracture toughness. The main elements producing such effects, and those with which this paper is predominantly concerned, are phosphorus and sulphur, but there is good evidence, from previous studies on higher-carbon, alloy steels, that other elements from the B-subgroups of Groups IV and V in the Periodic Table, notably Sn, As, Sb and Bi, are also detrimental. This can provide a problem for 'cold-hearth' all-electric steel-making processes, because the scrap steel input must be sorted extremely carefully to avoid pick-up of Sn (from tin-plate) or of Sb (from white-metal bearings).

In this paper, we address three aspects of intergranular segregation. The first is that of 'classical', or 'reversible' temper-embrittlement. This is exhibited in steels, which are austenitised and cooled rapidly to a martensitic microstructure, such that prior austenite grain-boundaries are preserved, and then tempered/stress-relieved at 600–650°C. Traditionally, embrittlement has been assessed in terms of the shift (ΔTT) of the ductile–brittle transition temperature in Charpy notched-impact tests. It has been observed that if a sample is cooled rapidly from 650°C, it exhibits a low transition temperature and a predominantly transgranular fracture appearance when fractured at low temperatures.[1] If a susceptible steel is cooled slowly from 650°C (allowing impurity elements to segregate to ferrite grain boundaries which are coincident with the original prior austenite grain boundaries), it is found that the transition temperature is increased and that the low-temperature fracture surfaces exhibit a significant proportion of intergranular facets.

Even if a steel has been cooled rapidly from 650 °C, subsequent service operation may expose it to temperatures in the temper-embrittlement range (450–565°C): this would be typical for some chemical reaction vessels fabricated from 2¼Cr1Mo steel.[2] Recently, there have been indications of intergranular fracture in nuclear pressure-vessel steels at temperatures as low as 300°C. This is attributed to the enhancement of the diffusivities of the segregants by the presence of point defects produced by irradiation-damage.[3] In this paper, we explore the effects of different hold-times at 520°C on the fracture toughness of a 2¼Cr1Mo steel tested at −120°C and −196 °C.

A second set of intergranular embrittling phenomena may be observed in steels in an as-quenched condition, after tempering at approximately 300°C, or during exposure to stress in stress-relieving operations at approximately 500°C. Attention is drawn to the importance of the original austenitising heat-treatment, which can produce different amounts of grain-boundary segregation in the as-quenched condition and to the mechanism of 'carbide rejection', which contributes to a reduction of fracture toughness after low-temperature tempering ('500°F', '350°C', or 'one-step' temper embrittlement). Carbide rejection causes impurities to concentrate at boundaries, but the effect is much enhanced if these boundaries are already enriched as a result of segregation in austenite.[4] In our studies, predominantly on A533B nuclear pressure-vessel steel, the impurity of interest is phosphorus, but it is found that, in 2¼Cr1Mo steel at 500°C, sulphur is sufficiently mobile to migrate to the region of high triaxiality ahead of a stressed notch or crack. This 'stress-induced segregation' allows a crack to propagate as a function of time under constant stress at 500°C.[5,6]

The final observations on intergranular fracture relate to fatigue-crack growth in 2¼Cr1Mo steel at room temperature. Previous work on 9Cr1Mo suggested that the occurrence of intergranular facets could be related to pick-up of hydrogen (from water vapour), the effect of which was to promote an intergranular path, rather than markedly affect the growth-rate.[7] The presence or absence of facets was also a function of the steel's heat-treatment and it was suggested that hydrogen might enter the steel preferentially along grain boundaries if these had been pre-segregated with P or S: elements which could depolarise the hydrogen evolution reaction at a cathode. The presence of a precipitate-free-zone (PFZ) in the grain-boundary region (combined with any denudation of Cr when grain-boundary chromium carbides are formed) could promote grain-boundary uptake of hydrogen by providing a locally soft region in which plastic flow could penetrate an oxide film (which might, in any case, be less protective than elsewhere as a result of the Cr denudation). We describe experiments carried out to explore any effects of a similar nature in 2¼Cr1Mo.

EFFECTS OF HOLD TIME AT 520°C ON LOW-TEMPERATURE FRACTURE TOUGHNESS

The steel studied had a chemical composition of 0.15C 2.27Cr 0.9Mo 0.013P 0.023S. It was austenitised for 2 h at 1100°C and oil-quenched to give an auto-tempered martensite with a prior austenite grain size of 120 μm. The as-quenched hardness was 390 HVN. It was then tempered for 2 h at 650°C and either oil-quenched after tempering (unembrittled, QT) or subjected to various embrittling times at 520°C: 24 h (light embrittlement QTLE), 96 h (medium embrittlement QTME) or 210 h (heavy embrittlement QTHE). The hardness values after treatments up to 96 h were in the range 250–260 HVN, but dropped to 242 HVN after 210 h. Figure 1 shows a micrograph of a specimen embrittled for 210 h and etched heavily in picric acid. It can be seen that there is segregation at both grain boundaries and carbide/matrix interfaces.

Fracture toughness specimens were prepared as SEN bars with B = 10 mm, W = 15 mm, and were fatigue pre-cracked following BS7448. Sets of 10 specimens in each condition were tested at temperatures of −196°C (in liquid nitrogen) and −120°C (in a liquid nitrogen/ethanol mixture). After testing, specimens were warmed-up to room temperature in alcohol and were subsequently examined in a JEOL 5410 or 6300 scanning electron microscope (SEM) operated at 20 kV and 0° tilt. Auger spectra for selected specimens were obtained, using an Auger spectrometer with a spatial resolution of order 2 μm.

Table 1 gives values for K_{Ic} at −196°C and −120°C for the QT, QTLE and QTHE conditions, together with indications of the area fractions of intergranular facets.

Table 2 gives typical values for the phosphorus peak-height-ratio (PHR = height of P peak/height of Fe_{703} peak) where measured.

Figures 2(a) and 2(b) show the effect of temperature on fracture toughness for the QT and QTHE conditions. In Figs 3a and 3b the fracture toughness values for QT,

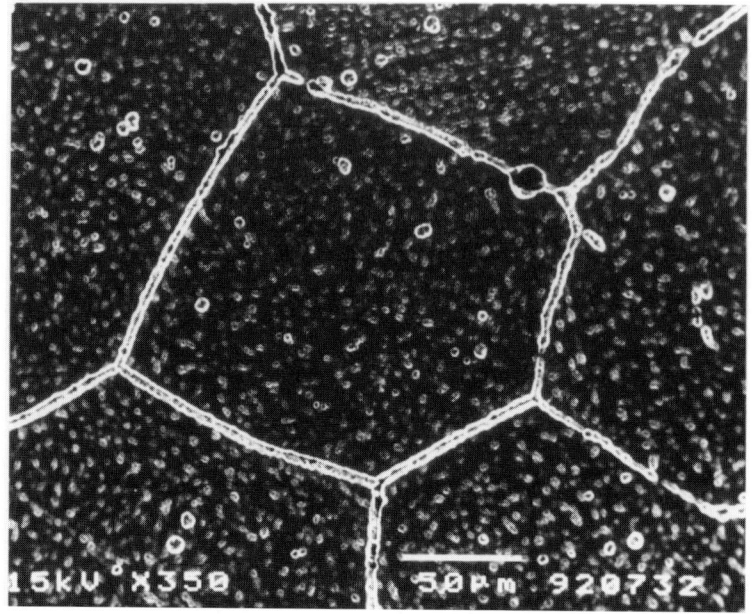

Fig. 1 Micrograph of QTHE specimen after heavily etching in picric acid.

QTLE and QTHE conditions at $-120°C$ and $-196°C$ respectively are shown as cumulative distribution functions (CDFs). Here, median ranking has been used and the points are plotted on normal probability paper, which is scaled about a 50% mean in a manner such that an error function (erf), which is the integral of a Gaussian probability density function (pdf), plots as a straight line. The standard deviation (s.d.) about the mean corresponds to CDF values of 16% and 84%. At each temperature, the fracture toughness values for the QTHE set are significantly lower than those for the QT set. Some typical fracture surfaces are shown in Fig. 4.

A number of interesting points arise. First, it is clear, by examining Table 1 and Figs 2(a) and 2(b), that the area fraction of intergranular facets on the brittle fracture surfaces relates to the segregation heat-treatment and does not depend on test tempera-

Table 1 Fracture Toughness Values and Percentage Intergranular Fracture.

Condition	Test Temp°C	Mean K_{lc} MPa m$^{0.5}$	s.d. MPa m$^{0.5}$	Mean i.g .%	s.d.%
QT	−120	71.2	4.5	0	–
	−196	45.5	3.4	0	–
QTLE	−120	66.4	7.4	26	12.8
	−196	42.1	5.4	18	13.4
QTHE	−120	53.2	4.0	71.3	3.9
	−196	32.1	2.6	63.6	6.1

Table 2 Typical values for the phosphorus peak-height-ratio.

Condition	%i.g.	Phosphorus PHR
QTLE	20–25	11.4
QTME	40–45	15.6
QTHE	65–70	18.8

ture. The area fractions for QTHE at $-196°C$ and $-120°C$ are 64 ± 6 and $71 \pm 4\%$ respectively. It might be noted that the QT (unembrittled) set tested at $-196°C$ exhibits 0% area fraction intergranular associated with a K_{Ic} value of 46 (s.d. ±3) MPa $m^{0.5}$ whereas the QTHE (embrittled) set tested at $-120°C$ exhibits $71 \pm 4\%$ area fraction but a K_{Ic} value of 53 (s.d. ±4) MPa $m^{0.5}$. This demonstrates the important role of the temperature dependence of the yield stress with respect to the variation of K_{Ic}, for any given value of local critical fracture stress, whether relatively high (for QT transgranular fracture) or lower (for QTHE intergranular fracture).

A second finding relates to the phosphorus coverage (as measured by PHR) on intergranular facets, see Table 2. This is found to increase with time at 520°C, *in addition to* the increase in the area fraction of intergranular fracture. The carbide microstructure and the hardness are functions primarily of the 650°C temper and do not change substantially during the periods at 520°C. It therefore appears that the segregation behaviour is associated rather directly with grain-boundary structure. In a randomly-oriented polycrystal, there is a range of grain-boundary misorientations, possessing 'good-fit' (low energy) at certain special values, corresponding to the

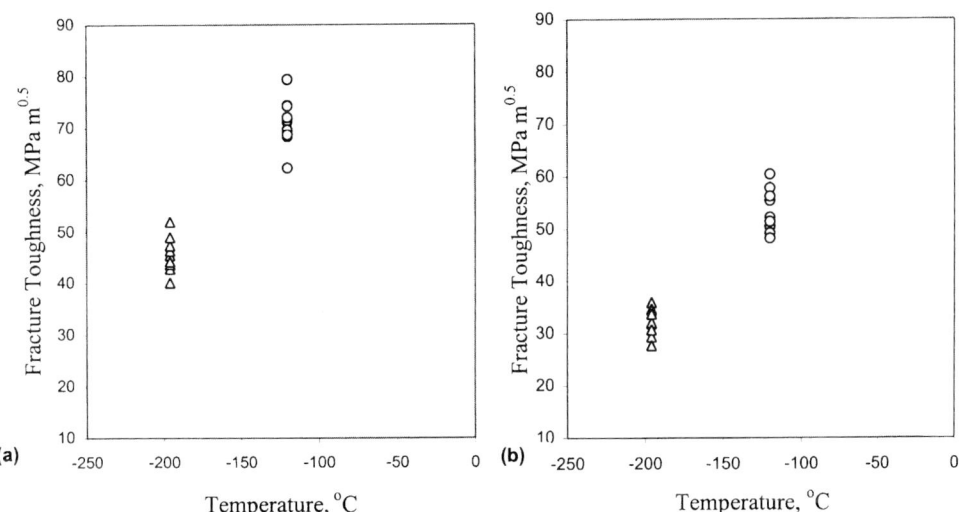

Fig. 2 Plot showing the variations in low-temperature ($-120°C$ and $-196°C$) fracture toughness values (a) QT and (b) QTHE conditions.

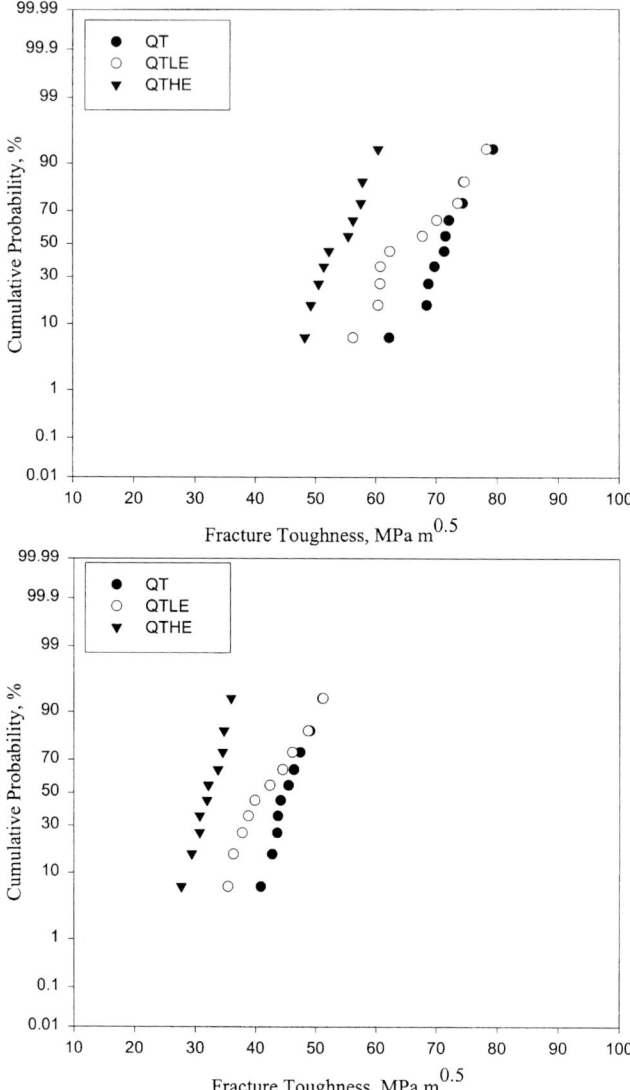

Fig. 3 Cumulative probability distribution functions (CDF) of 2.25Cr–1Mo steel at (a) −120°C and (b) −196°C under different heat-treatment conditions.

coherent twin boundary or to coincident-site lattices, and much more 'open' structures at intermediate misorientations. A useful way of envisaging the degree of 'accommodation' of a particular misorientation is to model a grain-boundary as a set of linked polyhedra.[8]

The QT condition exhibits apparently zero (<2%) intergranular fracture. This

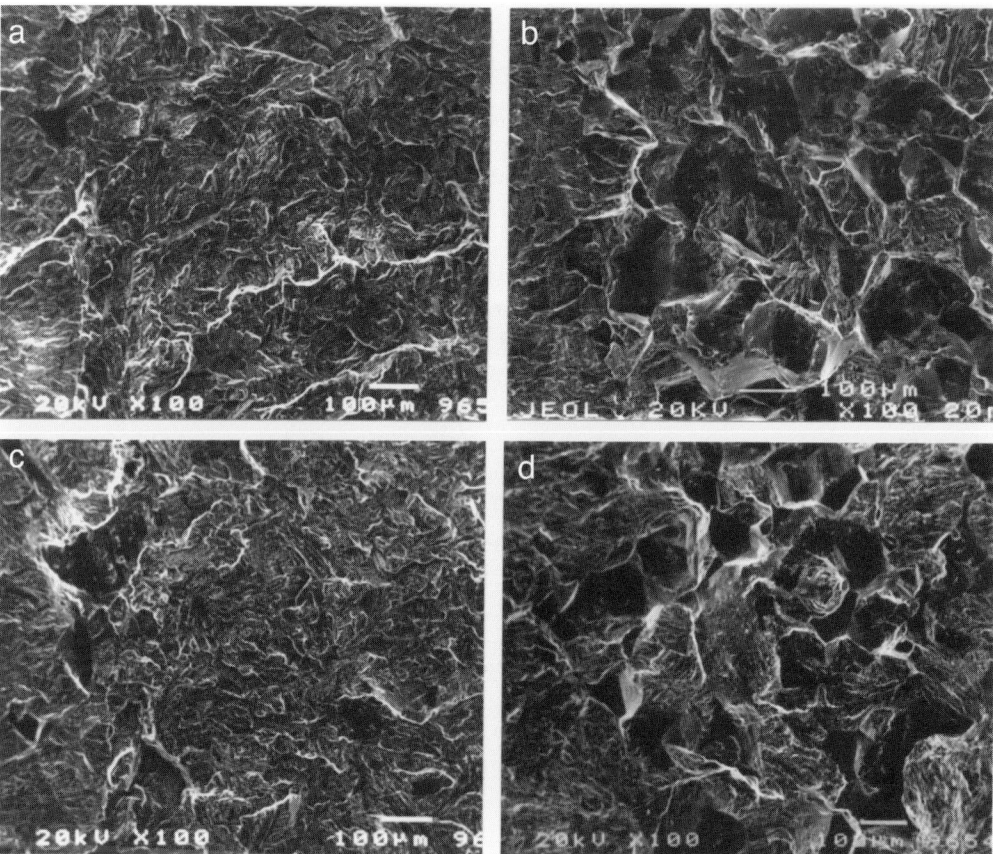

Fig. 4 Fracture surfaces for different embrittling treatments: (a) QT −196°C (zero i.g.); (b) QTHE −196°C (50–60% i.g.); (c) QTLE −196°C (a 'tough specimen' ~zero i.g.); (d) also QTLE −196°C (a 'brittle specimen' ~35% i.g.). See text.

implies, *not* that there is *no* phosphorus segregation to a more substantial fraction of grain boundaries, but that any such segregation is insufficient to lower the local work-of-fracture to a value below that for transgranular fracture. The degree of segregation in QT specimens needs to be established, using other techniques, such as chemical microanalysis of thin foils in the FEG–STEM. During embrittlement, it is presumably the case that the most 'accommodating' boundaries acquire the highest area coverage of phosphorus in the shortest period of time and that they are the first to demonstrate 'embrittlement' ie they possess a work-of-fracture sufficiently low to produce intergranular facets. The implication of the Auger results is that with embrittling time, not only do less 'accommodating' boundaries reach the threshold level required to give embrittlement but that the already-embrittled boundaries become even more embrittled as a result of further phosphorus segregation. The *mean* fracture toughness thus continues to decrease.

The third feature relates to the CDF plots for fracture toughness, as shown in figs 3a and 3b. It will be noted that, at both $-196°C$ and $-120°C$ the distributions for the QT and the QTHE sets are linear, with a standard deviation (s.d.) of around 3–4 MPa $m^{0.5}$. The distributions for QTLE fall in-between the 'tram-lines' set by the QT and QTHE distributions. The mean is an intermediate value, as noted above, but the s.d.s are significantly greater (5.5–7.5 MPa m$^{0.5}$). The reason for this can be attributed plausibly to a degree of 'banding' of alloying elements and impurity elements in the steel, which is deduced to produce spatial variation in the carbide sizes and degree of segregation which exist in the critical 'process zone' ahead of a fatigue crack tip in a fracture toughness test specimen. Sometimes this produces 'tough' behaviour: sometimes, more 'brittle' behaviour.

There are three aspects of interest. First, with respect to 'tough' QTLE specimens whose K_{Ic} values overlap those of QT specimens, it is *only because we know that they have been embrittled for 24h at 520°C* that we can distinguish them as a separate population. If such specimens (the top 5 QTLE at $-196°C$, the top 5 QTLE at $-120°C$) had been mixed with QT specimens and not identified by any special marking, it would have been impossible to establish, *from fracture toughness results alone*, that they were not from a single population. Nevertheless, in terms of further holding time at 520°C, they would have a '24 hour start' on QT specimens and would therefore embrittle to a greater extent in a shorter time. For different batches of material for which the prior heat-treatment history is unknown, it is not therefore possible to rely on start-of-life K_{Ic} values alone to give any consistent, reliable starting-point from which further in-service embrittlement can be assessed.

More information is potentially available from the examination of fracture surfaces. A second aspect of the results is, therefore, the fractography of the QTLE specimens. It is found that, at both $-196°C$ and $-120°C$, the area fractions of intergranular facets in the 'process zone' for specimens exhibiting 'tough' behaviour (indistinguishable from the results for the QT condition) are less than 22% (0–15% at $-196°C$, 10–22% at $-120°C$), whereas the 'brittle' specimens exhibit 25–45% (see Figs 4c and 4d). Recent theoretical analyses by Smith *et al.*[9] predict that, for a randomly oriented b.c.c. array of grains, cleaving only on {001} planes, and with grain boundary strength equal to transgranular cleavage strength, some 20% intergranular fracture would be expected, simply to accommodate geometrical changes implied by cleavage planes in adjacent grains which are tilted and twisted with respect to each other. In the QT condition (~zero% i.g.) the boundaries are presumably stronger than the cleavage planes. The toughness results show that, for less than 22% i.g. the overall resistance to fracture is not impaired. Here, the intergranular work-of-fracture is presumably greater than – in the limit equal to – the transgranular cleavage work-of-fracture. Decreases in fracture toughness occur for 25–45% i.g. indicating presumably that the intergranular work-of-fracture is now less than the transgranular work-of-fracture.

The third aspect concerns 'lower-bound' values of toughness. These are often involved in safety-cases, particularly in the nuclear industry, where a value of fracture toughness at typically 10^{-4} probability may be specified. Such values can only be estimated by extrapolation of points from necessarily rather limited data sets. On referring to Figs 3(a) and 3(b) it may be observed that the CDF distributions for both unembrittled (QT) and heavily embrittled (QTHE) material are linear with values of standard deviation consistent with those for random experimental errors. See also Refs 10, 11. Linear extrapolation to a lower-bound value at 10^{-4} is therefore perceived to be a plausible procedure, although the possibility of outliers occurring must always be recognised. The distributions have low s.d. values because the microstructures sampled in the process zone ahead of a fatigue crack are reproducible from sample to sample. This gives transgranular fracture for *every* QT sample (zero% i.g.) and 50–75% intergranular fracture for *every* QTHE sample.

For the QTLE condition, however, *spatial variation* in the degree of segregation sampled in the process zone may occur, perhaps as a result of 'banding' of composition in the original plate (Figs 4(c) and 4(d)). Some samples in the set are 'more tough': some are 'more brittle'. The s.d. increases, appearing as a decrease in slope of the CDF line on the probability plot. If the effect is large, it is possible that the 10^{-4} extrapolated limit is *lower than* that for QTHE material. This is clearly not physically reasonable and demonstrates the need to establish an extrapolation criterion. From the present and related research,[11, 12] we propose that, if the s.d. is less than 5MPa $m^{0.5}$, the material is classified as 'quasi-homogeneous' and extrapolation is allowed. If the s.d. is greater than 5MPa $m^{0.5}$, the significance of the simply extrapolated value to a safety case should be assessed. If it causes a problem, the data-sets should be re-examined in terms of the behaviour of possible bounding CDF distributions and detailed fractography should be carried out to identify the cause of the more widely-spread distribution.

A final point related to 'classical' temper embrittlement is that, during a hold at, say 520°C, trace impurity elements segregate not only to grain-boundaries but also to carbide/matrix interfaces. In higher carbon, quenched and tempered steels, there is a significant volume fraction of tempered carbides within the grains, which contribute to the steel's strength by providing barriers to dislocation movement. The dislocations tangle around the carbides and exert stress on the carbide/matrix interface. As the overall level of plastic strain is increased, the density of tangles increases and more stress is generated. Eventually, the local stress becomes sufficiently high to decohere the interface and increased shear strain is then transferred to other carbides. Often the overall shear band is localised between voids formed around non-metallic inclusions and *ductile fracture* proceeds by a process of 'shear decohesion': a sequential 'unzipping' of carbide/matrix interfaces. If a hold at 520°C after tempering at higher temperature causes impurities to segregate to the carbide/matrix interfaces (as observed in Fig. 1), these can reduce the local work-of-fracture, such that fewer dislocations are needed to produce decohesion. The 'unzipping' process then occurs at lower matrix strain, so that the crack-tip opening displacement (CTOD) required to initiate *ductile fracture* and the slope of the ductile fracture resistance curve are both decreased. In a 0.4C 4.25Ni steel, the initiation CTOD was decreased from 0.075 mm to 0.03 mm as a result of such segregation.[13] A similar effect can be seen in Figs 5(a) and 5(b) which show ductile fracture surfaces at 0°C for the QT and QTME conditions respectively. Voids are formed around MnS particles but it can be observed that the inter-void regions for QTME show fine-scale micro-voids centred on carbides (Fig.5(b)). These are less pronounced for the QT specimen (Fig. 5(a)). From such results, it may be inferred that the 'upper shelf' energy levels in notched-impact tests would also be decreased as a result of segregation. This has implications with respect to the estimation of any transition temperature shift, ΔTT. Values of ΔTT for rather low energies relate simply to the values of critical fracture stress for transgranular and intergranular fracture. At higher energies, any reduction in upper shelf energy for ductile fracture introduces a further factor. The commonly quoted 50% FATT, of course, assumes 50% ductile fracture.

AS-QUENCHED EMBRITTLEMENT, '350°C' EMBRITTLEMENT, STRESS-RELIEF CRACKING

This section is included to demonstrate effects on fracture toughness that can be produced by segregation of impurity elements in austenite and those attributable to microstructural changes during the early stages of tempering. The mobility of sulphur also gives rise to fascinating behaviour during stress-relief heat-treatments. For clarity, we focus on the elements phosphorus and sulphur in the pressure-vessel steels 2¼Cr1Mo and A533B. The heat-treatments of interest simulate those that might be

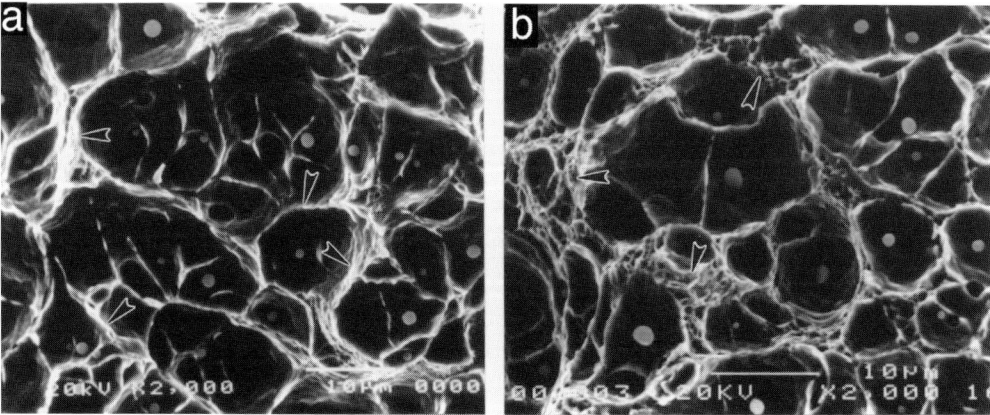

Fig. 5 Ductile microvoid coalescence on the fracture surface at 0°C, (a) QT and (b) QTME conditions. Note that the ridges between large voids are essentially free of microvoids in (a) as shown by arrows, but contain many microvoids in (b).

experienced in the coarse-grained HAZ of (a) a primary weld: high austenitisation temperature, fast cool, stress relief; (b) a repair weld: as (a) no stress relief, but subsequent operation at temperatures in the range 300–500°C; (c) the 'under-clad' region: as (a) but with additional stress arising from the mismatch in austenite/ferrite thermal expansion coefficients.

At high austenitisation temperatures, both P and S are taken into solid solution in austenite. The subsequent cooling-rate determines whether they are retained in supersaturated solid solution in the as-quenched (usually auto-tempered) martensite or whether they precipitate on prior austenite grain-boundaries or other interfaces (e.g. carbides, or, in weld-metals, on oxide/silicate inclusions). The effect of cooling-rate was demonstrated in Ref. 4 where A533B Steel was first austenitised at 1250°C, to establish an austenite grain size of 130 μm and was then either oil-quenched to room temperature (RT) or cooled at 20°C per minute from 1250°C to 900°C, held for 1h at 900°C and then oil-quenched to RT. Fracture toughness tests were then performed at −120 °C. To simulate a repair weld (non stress-relieved before entry into service) some specimens were fatigue precracked (and simultaneously tempered) at 290°C, rather than at RT. A 'control' specimen (fast quench) was tempered at 290°C for a period of time equated to that for the fatigue precracking and was then precracked at RT. The fracture values at −120°C, the area fractions of i.g. fracture and the corresponding coverage of P (in terms of PHRs) are shown in Table 3.

It is clear that, if we compare, first, the fast Q (RT) and the slow Q (RT), the segregation, *in austenite*, of P to grain-boundaries has produced a reduction in K_{Ic} from 58 to 50 MPa m$^{0.5}$ an increase in % i.g from <5% to 23% and an increase in PHR from 1.7% to 2.7%. Tempering at 290°C coarsens inter-packet cementite and particularly inter-granular cemenite. For the fast Q material, where rather little P has segregated to prior austenite grain boundaries, the 290°C temper, for either the specimen precracked at 290°C or the 'control' (tempered at 290°C, fatigue precracked at RT) increases the % i.g. slightly (9% or 6% w.r.t. 5%) and barely effects the PHR% (1.9% or 1.7% w.r.t. 1.7%), but reduces the value of fracture toughness from 58 MPa m$^{0.5}$ to 46 or 47 MPa m$^{0.5}$. It is deduced that this is because micro-crack nuclei (inter-packet, inter-granular cementite) have become larger, whilst the yield stress (dependant on transformation dislocation density, lath width, intra-lath carbides) has not been affected. From the observations on the QTLE condition compared with the QT condition

Table 3 Brittle Fracture Behaviour at −120°C.[4]

Specimen Condition	K_{1c} (MPa m$^{1/2}$)	% i.g.	Auger PHR% (P_{120}/Fe_{703})
Fast Q (RT)	58	<5	1.7
Fast Q (T 290°C)	46	9	1.9
Fast Q (control)	47	6	1.7
Slow Q (RT)	50	23	2.7
Slow Q (T 290°C)	40	67	3.7

in the previous section, variations in % i.g. from <5% to 9% or 6% would not appear to affect the experimentally observed fracture toughness to an extent differentiable from those of normal experimental errors.

A more dramatic effect is seen for the slow Q material subsequently tempered at 290°C. It is clear that, in the as-quenched condition, there is segregation of P to prior austenite grain-boundaries, an associated increase in % i.g. to 23%, and a decrease in fracture toughness from 58 to 50 MPa m$^{0.5}$. After tempering at 290°C, however, the area fraction of i.g. increases to 67% and the PHR to 3.7%. The fracture toughness decreases to 40 MPa m$^{0.5}$, only two-thirds that of the fast Q (RT) condition. The reason for this is two-fold. First the carbides coarsen, just as for the fast Q (T290°C). However, P is not soluble in Fe$_3$C and so, as the carbide coarsens, P is *rejected* to the carbide/matrix interface.[14] The most potent (carbide) nuclei are located at grain-boundaries and it is here that the P is segregated after the 'slow-Q' heat-treatment. There is, therefore, a combination of effects which reduces the fracture toughness: first, the coarse carbide nuclei, giving rise to larger microcrack nuclei; secondly, the accumulation of P, reducing the local work-of-fracture.

It is of interest to compare the K_{Ic} values obtained for A533B tested at −120°C with those for the fully tempered (and subsequently embrittled) samples of 2¼Cr1Mo discussed in the previous section. The exercise helps to distinguish the effects of % i.g. from those of microstructure (micro-crack nucleus size) and yield strength on the overall K_{Ic} values. The average K_{Ic} values for 2¼Cr1Mo (0.15 %C) at −120°C are: QT, 71.2 MPa m$^{0.5}$ (0% i.g.) QTLE 66.4 MPa m$^{0.5}$ (10–45% i.g.) and QTHE 53.2 MPa m$^{0.5}$ (64–75% i.g.). A perfect match with A533B (0.25%C) would not be expected, but it is significant that the higher hardness of the A533B (compared with 2¼Cr1Mo) gives rise to a fracture toughness of only 46–58 MPa m$^{0.5}$ for <10% i.g. compared with 71.2 MPa m$^{0.5}$ for the fully tempered QT condition. For slow QT (290°C) 67% i.g. is observed, but the K_{Ic} value is only 40 MPa m$^{0.5}$ contrasted with 53.2 MPa m$^{0.5}$ for QTHE (64–75% i.g.). At −120°C the yield strength of the A533B steel was 1450 MPa, whereas that for 2¼Cr1Mo was 900 MPa.

A further finding for the fast Q condition was that the fatigue pre-crack, grown in at 290°C, exhibited a very high proportion (70%) of intergranular facets even though the subsequent fracture surface in the −120°C fracture toughness test exhibited only 9% intergranular fracture.[4] This was attributed to the effect, not of phosphorus, but of sulphur. A slow cool in the austenite range permits sulphur to precipitate as solid sulphide particles, but a fast quench retains sulphur in supersaturated solid solution. It is well established that such sulphur can segregate to grain-boundaries in QT structural steels subjected to high triaxial stress, when a notched or precracked bar is subjected to monotonic load at a temperature in the range 450–550°C. Such segregation can give rise to intergranular separation and sulphur has been detected on the i.g. facets, using Auger analysis. It may be noted that the separation is at *constant chemical potential* rather than constant surface excess. There is no general embrittlement remote from the high triaxial stress region ahead of an advancing (intergranular) crack, the rate-of-growth of which is controlled by the rate of arrival of sulphur atoms. Figure 6 shows results of recent calculations for the flow of impurities ahead of a growing crack.[20] The observation that an apparently similar effect can occur during the fatigue of A533B at a temperature of only 290°C is of interest. It may be that, at the frequency used for pre-cracking (25 Hz), the mechanical work put into the specimen causes the local temperature in the reversed plastic zone to rise to a level comparable with that associated with segregation under monotonic load.

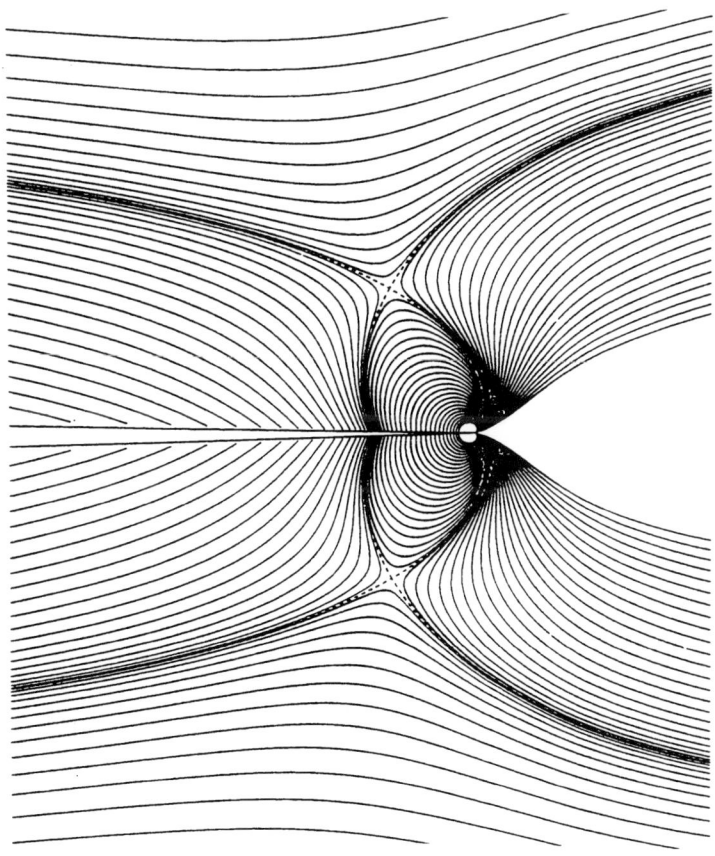

Fig. 6 Illustration of the flow lines for impurities which migrate to the tip of a non-stationary crack (moving from left to right). Only those impurities possessing flow lines which have one end at the crack tip can migrate to the tip. (Prof. P. Streitenberger ref 20)

INTERGRANULAR SEPARATION DURING FATIGUE CRACK GROWTH AT ROOM TEMPERATURE

The final part of the paper relates to observations made first on fatigue crack propagation in 2¼CrlMo steel. The first observations were made on the steel in the QT condition, for fatigue cracks grown in laboratory air at 65 Hz ($R = 0.1$). A maximum of 20% intergranular facets was observed for $\Delta K = 18$ MPa m$^{0.5}$. For the QTHE condition (210 h at 520°C after tempering) the % intergranular at $\Delta K = 18$ MPa m$^{0.5}$ increased to 45% and slightly further, to 50%, when the frequency was reduced to 0.25 Hz (see Fig. 7a). These figures are generally similar to those observed for En24 steel[15] and for 9CrlMo steel tested in room-temperature air.[7]

For QTHE tested at 0.25 Hz in air at 110°C, however, the area percentage of i.g. facets decreased to zero (contrasting with 50% at RT) and for QTHE tested at 65 Hz at −196°C, the % i.g. facets was again zero (contrasting with 45% at RT). Further observations are that when the steel is tested in RT air at 65 Hz in the oil-quenched (OQ) condition (after 2 h at 1100°C), no i.g. facets are observed (20% for the QT condition). If the OQ material is tempered for 96 h at 520°C, and tested similarly, no i.g. facets are observed. If the steel is tempered for 20 mins at 650°C and oil-quenched, approximately 5% i.g. facets are observed.

These various observations can be put into a common framework, similar to that argued in Ref. 7. First, for the QT and QTHE treatments, it is proposed that the basic cause for the presence of intergranular fracture is hydrogen, but that its effect is enhanced by segregation of impurity elements. In the present case, hydrogen is produced by the decomposition of water vapour. If the partial pressure of water vapour at the crack tip is reduced virtually to zero by testing at 110°C (or −196°C), there is no hydrogen available and even the QTHE condition shows no evidence of i.g. facets (see Fig. 7b). Reference 7 contains the results of experiments on 9CrlMo in air, in vacuum, at 90°C and 110°C in air, and at 50°C and 70°C in distilled water.

The QT condition exhibits 20% i.g. facets at RT whereas the QTHE condition exhibits 45% at 6 Hz, 50% at 0.25 Hz. Hippsley and Druce (16) examined fatigue crack growth in 2¼CrlMo in gaseous hydrogen and observed ~40% i.g. facets for an unembrittled condition (yield stress 500 MPa) but 75% i.g. facets for an embrittlement of 1000 h at 500°C. It is clear, therefore, that for the same hydrogen potential (generated by decomposition of water vapour or by gas pressure), the 'classical' temper-embrittling heat-treatment produces a higher % i.g. facets. The process has been modelled[17] in a manner equivalent to that for stress-driven sulphur segregation. Hydrogen is assumed to accumulate ahead of a crack-tip (at a grain-boundary) by stress-driven diffusion and then to lower the intergranular work-of-fracture (in synergy perhaps with pre-segregated impurity) to produce intergranular fracture. Hippsley reports an increase in growth-rate with K_{max} (in hydrogen and in water) but there is other evidence[7, 15] that, in air, the percentage of i.g. facets is a function of ΔK rather than K_{max}.

It is of interest that neither the OQ nor the OQ + 96 h/520°C specimens exhibit

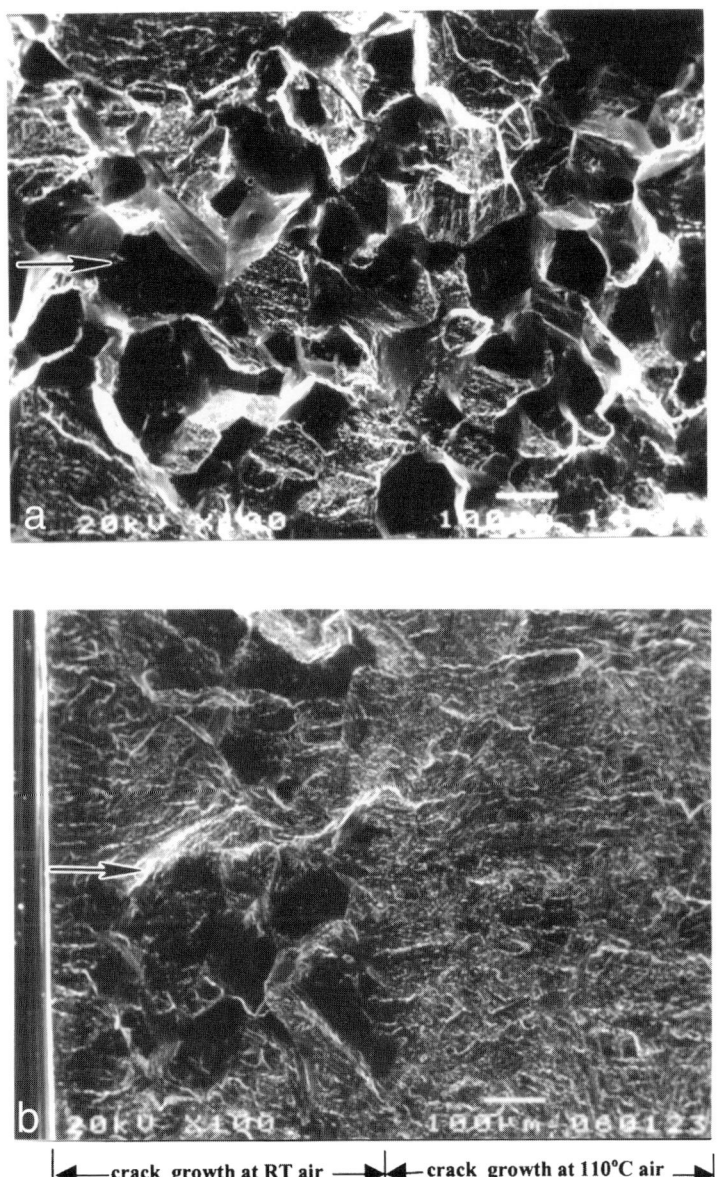

|←—crack growth at RT air —→|←—crack growth at 110°C air —→|

Fig. 7 Fatigue surfaces observed on QTHE specimens. Arrows indicate the direction of crack growth: (a) $\Delta K \sim 18$ Mpa m$^{1/2}$, at room temperature air, 1/4 Hz (~50% i.g.) and (b) $\Delta K \sim 18$ Mpa m$^{1/2}$, at room temperature air and then at 110°C air, 1/4 Hz. Note that specimen (b) shows significant proportions of intergranular facets at room temperature air, but facets are absent when test is continued at 110°C air at the same level of ΔK.

any intergranular facets, despite the fact that the yield strengths (and hence the absolute magnitude of triaxial stress ahead of a loaded crack) are *higher* than those for the QT or QTHE conditions. It could be argued that the OQ gives no segregation of either P or S at prior austenite grain boundaries (here note that QT in RT fatigue gives 20% i.g. facets). Additionally, there are suggestions that the rather high Mo content (0.9%) enables Mo–P clustering to occur, such that less P accumulates at grain boundaries under conditions of light tempering than was observed for A533B. Once the steel has been tempered at 650°C, Mo is removed from solution as molybdenum carbide and P is then free to migrate to grain boundaries during the subsequent embrittlement at 520°C. Against this has to be set the observations[18] of segregation of P to grain-boundaries of 2¼CrlMo during stress-relief heat-treatment which was shown to be responsible for intergranular fracture at temperatures of ~500°C (stress-relief cracking). The segregation was modelled originally as general segregation, but a review of the experimental observations suggests that this model may need revision. In particular, when a specimen containing a stress-relief crack, grown in at ~500°C, is unloaded and fractured at −196°C to examine the fracture surface, it is found (even for a P-doped sample) that the majority of the low-temperature fracture surface is transgranular, not intergranular. This suggests that, like sulphur, the segregation of P is stress-driven.

For hydrogen to enter the metal, either from hydrogen gas or from water/water-vapour, it must be, at least, in atomic form, possibly as a proton. If it remains as a molecule, which becomes part of a hydrogen bubble, it will simply escape the system. From water, a plausible reaction[19] is:

$$Fe + H_2O = FeOH^+ + H^+ + 2e^-$$ (1)

possibly followed by

$$H^+ + e^- = H$$ (2)

The H atoms may combine to give, first, H_2, then, $nH2$ as a bubble, which escapes. The alternative, of hydrogen entry, is effected by elements with electronic structure such that they affect (de-polarise) the hydrogen–electron combination. Sulphur is a well-recognised de-polariser: CS_2 can be added as a 'poison' to sulphuric acid to help facilitate electrolytic hydrogen charging of samples; 'sour gas' (containing H_2S) is a problem with respect to hydrogen embrittlement; and the NACE solution bubbles H_2S through sodium chloride to accelerate corrosion/hydrogen embrittlement. It is not impossible that P could act in a similar way (arsenic, also group V, can be used as a 'poison' to facilitate hydrogen charging), but the role of S is extremely well-established.

The argument put forward in Ref. 7 for 9CrlMo was that the ingress of hydrogen occurred at pre-segregated grain boundaries, because the segregated species de-polarised the H/H₂ recombination. The present results for OQ 96 h/520°C combined with the earlier stress-relief cracking results suggest, *either* that S is more effective than P with respect to hydrogen entry (so that any pre-segregated P does not lead to intergranular facets in RT fatigue) *or* that P is equally effective, but that its segregation at ~500°C is also stress-assisted: the triaxial stress providing sufficient driving force to overcome the strength of the Mo–P attraction, which was invoked to explain the absence of i.g. facets in fracture toughness tests at −120°C for the OQ 96 h/520°C condition. Because the fractures in such tests are 100% transgranular, Auger analysis can give no information on grain-boundary segregation of S or P, and any such segregation needs to be analysed in the FEG/STEM.

A QT specimen (tempered 2 h 650°C OQ) exhibits 20% i.g. facets in fatigue in RT air, but zero i.g. in fracture toughness tests at −120°C or −196°C. After segregating at 520°C, following tempering, a higher % i.g. is observed in fatigue in RT air and in low temperature fracture toughness tests. In fatigue, however, the % i.g. is 45–50% for QTHE, whereas in low temperature fracture it is 64–71% and a significant % P is observed on the i.g. facets. The possibility arises that the two elements P and S are behaving in a different manner: P dominating low-temperature embrittlement; S dominating hydrogen ingress.

The low-temperature embrittlement is attributed to a straightforward lowering of the 'work-of-fracture' by P segregation. This would not, by itself, be sufficient to induce brittle grain-boundary fracture at room temperature: under monotonic loading in the absence of hydrogen, the fracture mode is ductile. With respect to the presence of intergranular facets in fatigue, it is plausible that hydrogen enters preferentially along grain-boundaries pre-segregated with S, producing small increments of i.g. growth each cycle. It is possible that pre-segregation of S may not be fully eliminated even for a 650°C temper: in alloy steel, small sulphide particles at grain boundaries associated with overheating do not re-dissolve at 650°C. It is then possible that there is sufficient sulphur at boundaries in the QT specimen to give 20% i.g. facets in fatigue: phosphorus would be expected to be fully de-segregated at 650°C, and remain desegregated during the oil-quench following tempering. The observation that zero % i.g. facets are observed in RT fatigue for an OQ followed by 96 h at 520°C, whereas sulphur is sufficiently mobile to produce intergranular fracture under 'stress-relief' conditions at this temperature, supports the previous finding that it is necessary to generate a triaxial stress field to produce sulphur segregation. It is now suggested that phosphorus segregation at ~500°C may also be stress-driven. Information on impurity segregation will be enhanced in future studies by our newly acquired facility to carry out fine-scale grain-boundary micro-analyses.

CONCLUSIONS

The paper has described a number of different phenomena associated with grain-boundary fracture in weldable pressure-vessel steels. Although discussion has concentrated on the segregation of impurity elements, such as P or S, to grain-boundaries and the effect that this has on fracture toughness, it is important to incorporate also effects of strength level and microstructure, because carbides provide the micro-crack nuclei. The phenomena described in this paper are attributable (over and above features of heat-treated microstructure) to the effects of three elements: P, S and H. It appears that reductions in fracture toughness at low temperature are mainly a consequence of P concentration at boundaries, whether achieved by segregation or by carbide rejection. Some S is seen on facets. Previous work has shown that S is primarily responsible for stress-driven monotonic crack growth at temperatures of order 500°C (possibly at 290°C in fatigue). The observation of i.g. facets during fatigue in RT air is attributed to a localised crack-tip hydrogen-assisted separation and it is proposed that this is intergranular because H enters preferentially along pre-segregated boundaries. Plausibly, the impurity segregation de-polarises the hydrogen/electron recombination reaction. Although it is not impossible that pre-segregated P could produce such depolarisation, it is suggested that S may be more potent in this respect.

REFERENCES

1. B. C. Woodfine, *Jnl. Iron and Steel Inst.,* 1953, **173**, 229–242.
2. T. Iwadate, J. Watanabe and Y. Tanaka, Trans. ASME *Jn. Pres. Vess. Tech.,* 1985, **107**, 230–238.
3. J. F. Knott, *Proc. Second Griffith Conf.,* Institute of Materials, 1995, 3–14.
4. P. Bowen, C. A. Hippsley and J. F. Knott, *Acta Metall.,* 1984, **32**, 637–647.
5. J. J. Lewandowski, C. A. Hippsley and J. F. Knott, *Acta Metall.,* 1987, **35**, 2081–2090.
6. P. Bowen C. A. Hippsley and J. F. Knott, *Matls. Sci. and Tech.,* 1990, **6**, 562–574.
7. K. Nishioka and J. F. Knott, in *Environment Assisted Fatigue, EGF7,* P. Scott ed., Mech. Eng. Publ., 1990, 241–254.
8. M. F. Ashby, F. Spaepen and S. Williams, *Acta Metall.,* 1978, **26**, 1647–1662.
9. G. Smith, A. G. Crocker, P. E. J. Flewitt and R. Moscovic, in *Damage and Failure of Interfaces,* H.-P. Rossmanith ed., Balkema, 1997, 229–236.
10. X. Z. Zhang and J. F. Knott, *Acta Mater.* 1999, **47**, 3483–3495.
11. X. Z. Zhang and J. F. Knott, 'The Statistical Modelling of Brittle Fracture in Homogeneous and Heterogeneous Steel Microstructures', *Acta Mater.,* in press.
12. J. F. Knott, in *George Irwin Symposium on Cleavage Fracture,* Kwai S.Chan ed., The Minerals, Metals and Materials Society, 1997, 171–182.
13. J. E. King and J. F. Knott, *Metal Science,* 1981, **15**, 1–11
14. J. R. Rellick and C. J. McMahon Jnr., *Met. Trans.,* 1974, **5**, 2439–2446.
15. R. J. Cooke, P. E. Irving, G. S. Booth and C. J. Beevers, *Engng. Fracture Mech.,* 1975, **1**, 69–77.
16. C. A. Hippsley and S. G. Druce, in *Environment Assisted Fatigue, EGF7,* P. Scott ed., *Mech. Eng. Publ.,* 1990, 223–240.
17. C. A. Hippsley and C. L. Briant, *Scripta Met.,* 1985, **19**, 1203–1208.
18. C. A. Hippsley, J. F. Knott and B. C. Edwards, *Acta Met.,* 1980, **28**, 869–885.
19. T. Misawa, N. W. Ringshall and J. F. Knott, *Corrosion Sci.,* 1976, **16**, 805–818.
20. P. Streitenberger 'Stress-driven Solute Migration near a slowly-moving Greek Tip' submitted to *Scripta Mat.,* March 2000.

Influence of Grain Boundaries on Texture Evolution During Deformation of Aluminium

SARAH LILLYWHITE

IRC in Materials for High Performance Applications
University of Birmingham, Birmingham, B15 2TT, UK

ABSTRACT

Texture evolution during thermomechanical processing of aluminium is greatly affected by heterogeneities present in the microstructure such as second phase particles and grain boundaries. Investigations of polycrystalline aluminium alloys with differing initial grain sizes have highlighted the importance of grain boundaries but a systematic study of the grain boundary influence has not been carried out. In the present paper, a combination of experiment and computer simulation was used to investigate the influence of grain boundaries on deformation behaviour of aluminium. Experimental controlled orientation single and bicrystals were prepared and the orientation evolution during channel die compression to $\varepsilon = -0.5$ was studied. The Crystal Plasticity Finite Element Model simulated the deformation behaviour of the experimental specimens. The results showed good agreement between the model and the experiment in terms of the lattice rotations that take place during deformation. However, the deformation banding was predicted to a lesser degree of accuracy, due to the inability of the model to deal with the movement of dislocations which are likely to influence the formation of such microstructural features. The results implied that the mechanical interaction of neighbouring grains, rather than the grain boundary character, altered the deformation behaviour. The change in specimen boundary constraints was considered also to be responsible for this. It was suggested that a more generalised description of the grain boundary to include the orientations of the related grains would better account for the interaction of adjoining grains during deformation.

INTRODUCTION

The preferred orientation (texture) of a component is known to have a strong bearing on the material properties and therefore, the prediction of texture evolution during thermomechanical processing is the aim of industry and academia alike. The textures produced during cold deformation, for example, have proved difficult to predict because of the highly heterogeneous nature of plastic deformation on the microscale. Also, in the case of fcc alloys, the high symmetry results in the possibility of slip occurring on mechanically equivalent sets of slip systems which can give different orientations. Recrystallisation textures depend on the nuclei that form in the deformed microstructure as well as the growth of these nuclei. Since the orientations

73

of new recrystallising grains are present in the deformed microstructure, an understanding of the deformed texture and microstructure is a prerequisite for understanding the recrystallisation textures. Investigations of polycrystalline aluminium alloys with differing initial grain sizes have highlighted the importance of grain boundaries on thermomechanical processing,[1] but a systematic study of their influence has not been completed. Although, for example, grain boundary energies and mobilities can be described as functions of boundary misorientation and character, the influence of grain boundaries on the deformation microstructure is not so readily explained.

As standard personal computers continue to increase in capacity and speed, there is an incentive to attempt the prediction the deformation behaviour through techniques such as Crystal Plasticity Finite Element Modelling (CPFEM).[2] At this stage of development, it is primarily used in situations which are not straight forward to study experimentally, or to reduce the number of specimens. Although the CPFEM has been successful in the prediction of certain features of deformation,[3] at this stage of our understanding, it is often better to compare the CPFEM predictions to experimental results.

Single crystals are frequently used as a method of investigating cold deformation behaviour in terms of the orientation changes that can result[4] or work hardening/softening.[5] However, it can be shown that the orientations of neighbouring grains can have a significant effect on the deformation behaviour.[6] The lattice rotations that cause deformation banding and orientation splitting are dependent on the constraints on each grain.[7] By using controlled orientation bicrystals these phenomena can be investigated in detail – the influence of neighbouring grains and therefore, grain boundaries on deformation can be investigated on specimens where the initial orientations can be well defined. For example, aluminium bicrystals deformed in tension have been used to investigate the influence of grain boundaries on flow stress.[8]

In this paper, a combination of controlled orientation aluminium bi- and single crystals are used to explore the influence of constraints imposed by neighbouring orientations on deformation behaviour. In addition, the ability of the Crystal Plasticity Finite Element Method to predict the deformation microstructures of the bi- and single crystals is assessed. An attempt is made to clarify the role of the grain boundary on orientation evolution during plane strain compression.

EXPERIMENTAL PROCEDURE

Controlled orientation superpurity aluminium bicrystals were produced by diffusion bonding two single crystals.[9] Single crystals of each bicrystal orientation were also produced. The bi- and single crystals were deformed at room temperature by channel die compression under plane strain conditions to $\varepsilon = -0.5$. Friction between the platens and specimen was minimised by using oil lubricant and PTFE film.

The initial orientations are described in Fig. 1. The orientations were chosen such that shearing in the x-y plane was avoided to ensure plane strain conditions held. The bicrystal has a [110]/90° asymmetric tilt grain boundary with (1$\bar{1}$0)‖(011) in the interface. The [110] direction is parallel to transverse direction (y-direction) for all orientations investigated. For the base orientation, the [$\bar{1}$10] direction is parallel to the extension direction (x-direction) whereas for the top grain the [001] direction is parallel to the extension. The deformed microstructures were observed optically using anodized specimens under polarised light and the orientations and banding features

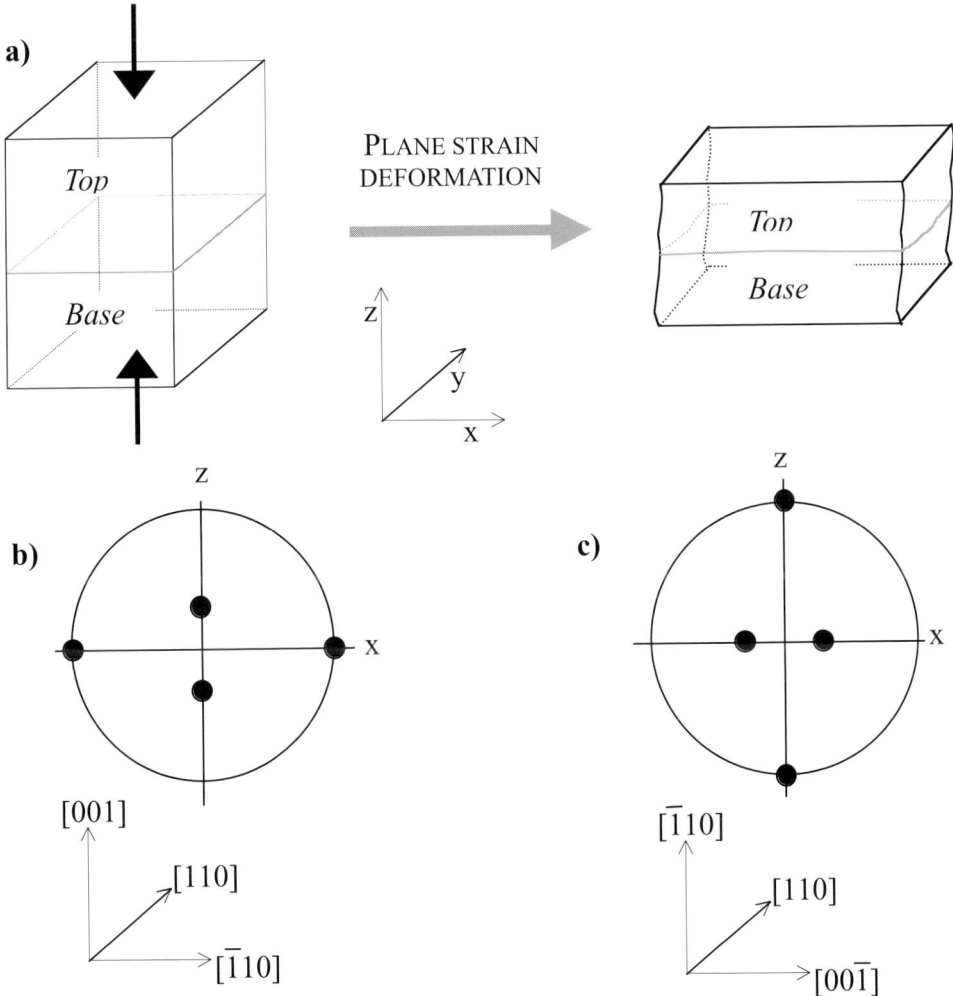

Fig. 1 Schematic diagrams of (a) bicrystal deformation and initial orientations, (b) 'T' orientation and (c) 'B' orientation.

resulting from deformation were investigated by EBSD on the SEM. In the following, each orientation is referred to as a grain if it is part of a bicrystal but as a crystal if it is not. To avoid multiplicity of terms, in all cases, the orientations are referred to as B and T for the 'base' orientation and 'top' orientation, respectively.

EXPERIMENTAL RESULTS

SINGLE CRYSTALS

The stress-stain behaviour of the two types of single crystal showed some differences. The B orientation gave a clearer elastic region before the onset of plastic deformation with a yield point at ~20 MPa whereas the stress-strain curve was smoother for the single crystal of the T orientation (Figs 2(a) and (b)). When the compressive strain reached $\varepsilon = 0.5$ the stress for the T orientation was slightly higher (~80 MPa) than for the T orientation (~75 MPa) although the experimental error may account for this difference.

The differences in deformation behaviour between the two single crystals are highlighted in Figs 3(a) and (b) which show the deformed specimens. Some of the features of deformation were coarse enough to observe with the naked eye. Little gross macroscopic shearing appeared to have taken place of either single crystal orientation during channel die compression. In addition, a slight bulging at the unconstrained boundaries is evident for the T orientation. The deformation banding in this specimen is not easily seen with the exception of some shearing across the diagonal of the specimen. The boundary of this band however is not clear. The B orientation shows coarser deformation banding stretching across the specimen in a 'criss-cross' manner and the deformation band boundaries can clearly be identified. The micrographs of Figs 3(c) and (d) show typical regions of the microstructure. The T orientation has fine, diffuse banding which is probably of the order of one or two subgrains across. Fig. 3(d) shows an example of the coarse banding of the single crystal with the B ori-

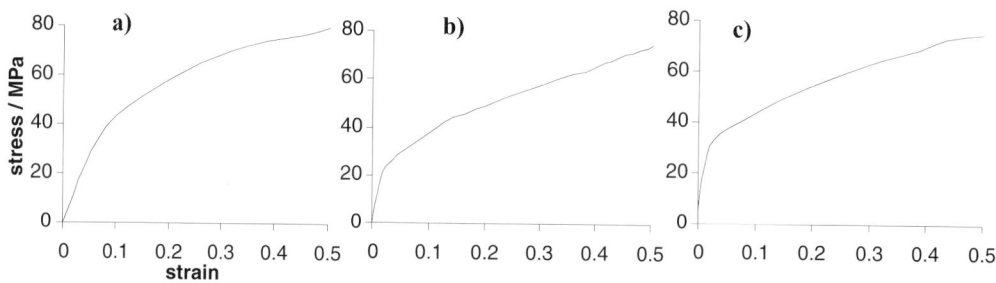

Fig. 2 Stress-strain behaviour during channel die compression for (a) T crystal, (b) B crystal and (c) bicrystal.

entation. Within the coarse deformation features, finer bands can be observed. The substructure of this single crystal appears somewhat coarser than that of the single crystal with the T orientation.

Figure 4 summarises results from EBSD analysis on the SEM. The orientation images show that the clarity of the banding features observed optically is due to the local orientation gradients. For example, the features observed in the single crystal with the T orientation rarely showed local misorientations of greater than ~5° whereas many local misorientations of greater than 10° were present within the microstructure of the B orientation. The pole figure of Fig. 4a shows the lattice as a whole did not rotate significantly for the single crystal with the T orientation, whereas in the single crystal of the B orientation, the orientation has split and rotated in both directions by approximately equal quantities (~±20°).

BICRYSTAL

The stress-strain behaviour during channel die compression is similar to that of the single crystals (Fig. 2(c)). However, observations of the microstructure indicated differences between the deformation behaviour for the two orientations as a bicrystal

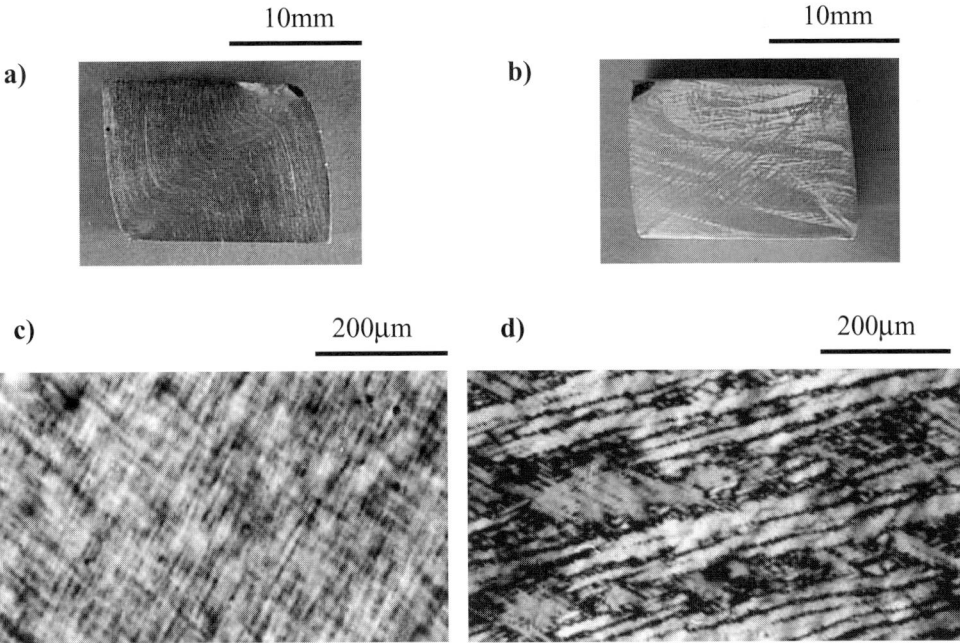

Fig. 3 Deformed single crystal specimens, (a) T crystal, (b) B crystal and optical micrographs of, (c) T crystal and (d) B crystal.

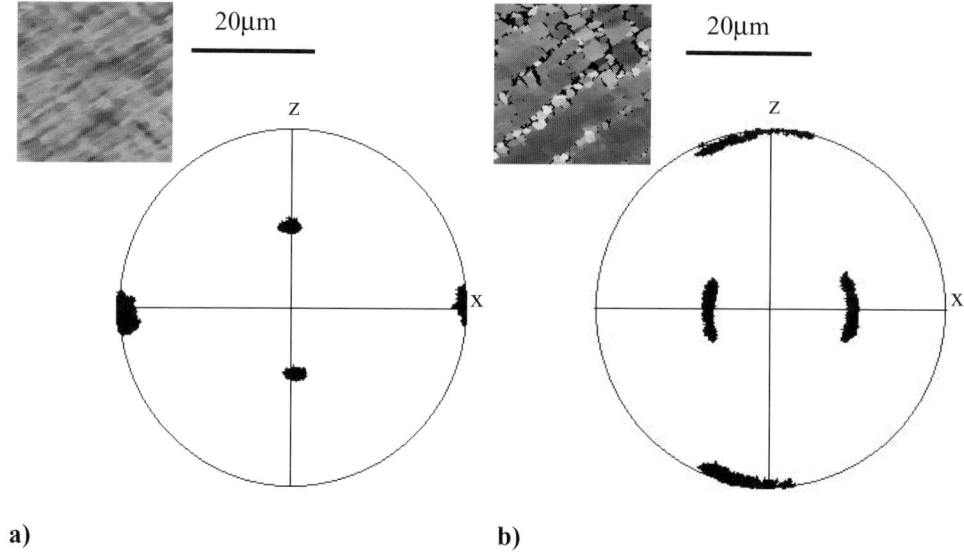

a)   b)

Fig. 4 Orientation image and 100 pole Fig. for (a) T crystal and (b) B crystal.

or as single crystals. Figure 5(a) shows the deformed bicrystal observed using polarised light. The specimen geometry shows some bulging in the B grain whilst the T grain shows some concavity. The T grain has coarse, well defined deformation bands stretching from the unconstrained external boundaries of the specimen whereas very little banding is evident in the B grain. No particular variation in microstructure can be observed in the vicinity of the grain boundary for this bicrystal. Figures 5(b) and (c) show optical micrographs of typical regions from the T grain and the B grain for the bicrystal. Well defined coarse deformation bands were observed in the T grain with more diffuse banding features extending through the coarse bands, whereas only fine, diffuse banding was observed in the B grain of the bicrystal.

Figure 6 shows orientation images and pole figures from EBSD analysis for typical regions of both grains in the bicrystal. In the T grain, deformation resulted in some large lattice rotations which provide deformation bands with misorientations up to ~30°. Although the overall lattice rotated towards the orientation of the B grain, some of the initial orientation in the T grain remained present after deformation. In the B grain, where the microstructure was fine and diffuse, deformation resulted in little orientation spreading and the lattice as a whole appears to have rotated by ~10° away from the initial orientation.

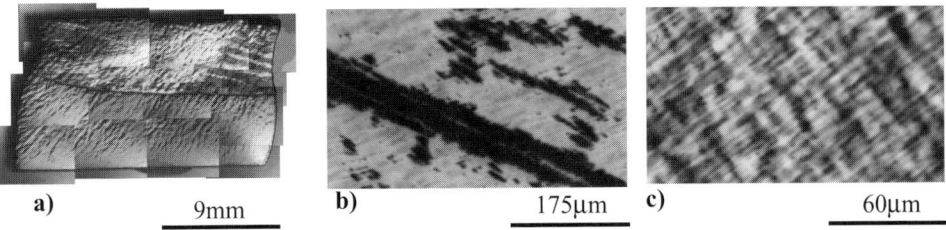

a) 9mm b) 175μm c) 60μm

Fig. 5 (a) Deformed bicrystal, and optical micrographs of (b) T grain and (c) B grain.

CRYSTAL PLASTICITY FINITE ELEMENT MODELLING

Crystal Plasticity Finite Element Modelling incorporates constitutive equations for crystal plasticity into finite element methods. The aim of its use in the present case is to predict orientation evolution during deformation for aluminium single and bicrystals, although it can also be used to predict other features of deformation behaviour such as residual stresses and anisotropy, for example Ref. 10. Discretisation at less than the grain scale may allow prediction of microstructural evolution (deformation banding) from this method.

In order to incorporate constitutive equations of crystal plasticity into finite element simulations, the introduction of slip rate sensitivity is required. This rate sensitivity of slip rounds off the step function which describes whether slip will take place on that system or not; ie, rate dependence overcomes the problem of nonuniqueness in the choice of active slip systems leading to the prediction of lattice rotations that take place in the microstructure. Rate sensitivity rounds off the yield

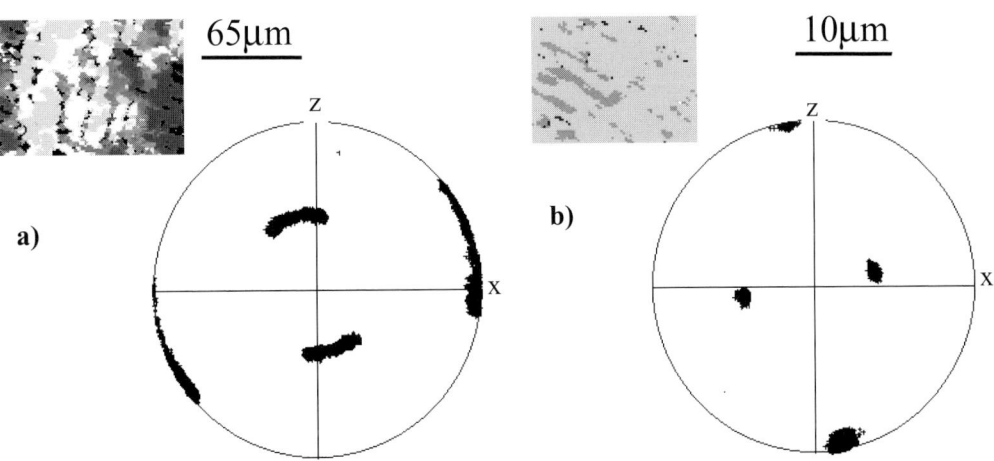

Fig. 6 Orientation image and 100 pole Fig. for bicrystal in (a) T grain and (b) B grain.

surface vertices such that singularities at the vertices of the yield loci can be avoided. The model used for the present work is described in detail elsewhere.[11]

Either 8-node quadrilaterals or 6-node triangles were used as the elements in the initial FE mesh. It was found that the type of mesh used made little difference to the outcome of the model prediction. The overall mesh geometries were chosen to approximate the shape of the experimental specimens so that the external boundary constraints were equivalent. This factor was seen to be important, in that the prediction of deformation behaviour was greatly altered if the initial mesh geometry was altered. The meshes were 'deformed' under relaxed constraint conditions between two rigid dies. The movement of one die simulated the deformation to a strain of $\varepsilon = -0.5$. For all simulations, the friction coefficient used was 0.03.

The predictions were plotted in terms of the variations in effective stress over the mesh and the expected lattice rotations since these parameters are expected to directly relate to the deformation behaviour of the specimens. The effective stress can be thought of as a measure of the stored energy of deformation and as a consequence, changes in effective stress may indicate changes in cell structure (size, misorientation). Therefore, effective stress is particularly important in the context of subsequent annealing since the stored energy provides the driving force for recrystallisation. The lattice rotations that take place during deformation are important since they can cause orientation splitting and other microstructural heterogeneities such as deformation banding. In the results presented below, a positive lattice rotation can be described by an anticlockwise rotation about the y-axis (see Fig. 1(a)).

SINGLE CRYSTALS

The model prediction of variations in effective stress for the single crystals during plane strain compression is plotted in Fig. 7. The distribution of effective stress values across the single crystal with the T orientation is predicted to be significantly more heterogeneous with minimum values at the unconstrained external boundaries of the specimen. Maximum values of effective stress are predicted to occur where significant shearing could be seen on the single crystal specimen (see Fig. 3). The effective stress for the single crystal of B orientation is expected to be relatively homogeneous and is ~90 MPa. The plots of Fig. 7 also indicate that the expected change in specimen geometry involves little gross macroscopic shearing. The external boundaries of B orientation are expected to remain relatively straight while the T orientation is expected to experience some bulging at the unconstrained specimen boundaries.

The variations in lattice rotation that are predicted to take place about the transverse direction (y-direction) in each single crystal are plotted in Figs 7(c) and (d). The single crystal with the T orientation is expected to experience an overall lattice rotation of $\sim -5°$ with a spread of orientations of ~15°. The lattice rotations are predicted to be greatest along a band stretching from opposite corners although this band does not appear to have clear boundaries, ie a sharp orientation gradient is not evident at

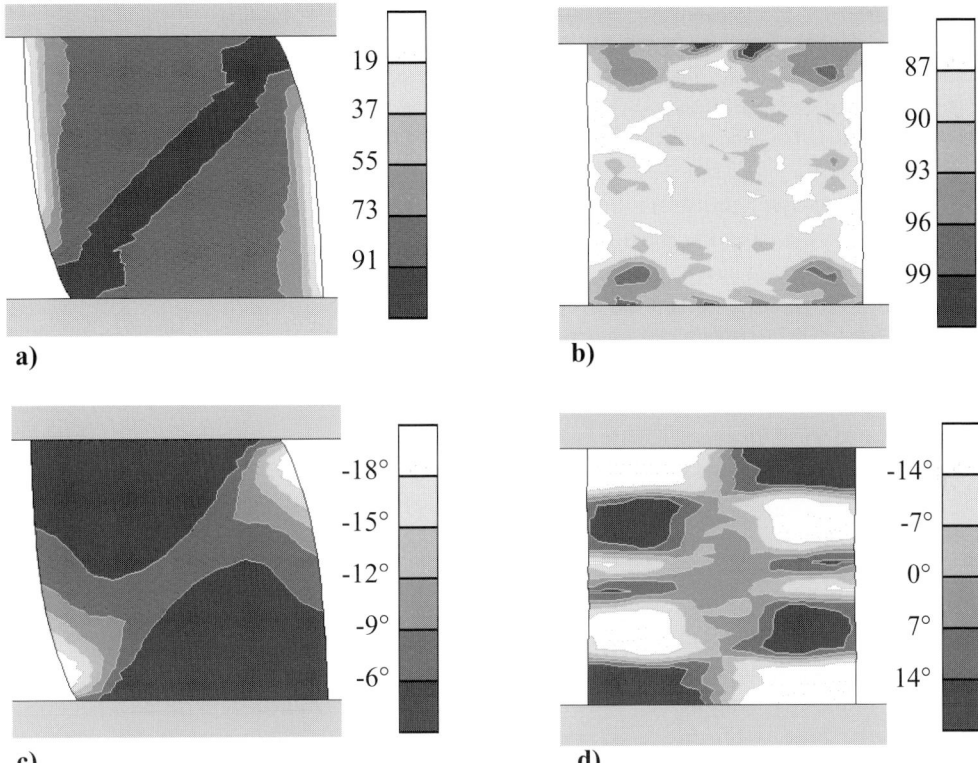

Fig. 7 CPFEM predictions for single crystals variations in effective stress (MPa) for (a) T crystal and (b) B crystal; variations in lattice rotations about the *y*-direction for (c) T crystal and b) B crystal.

the band boundaries. In contrast, for the single crystal with the B orientation, sharp orientation gradients imply that strong banding features are expected. The lattice rotations that are expected during plane strain compression in this case result in the lattice splitting into two complimentary orientations about the original orientation.

BICRYSTALS

The expected variations in effective stress and the lattice rotations about the transverse direction (*y*-direction) for the bicrystal are plotted in Fig. 8. The effective stress is predicted to vary from ~80 MPa to ~120 MPa over the two grains of the bicrystal. The B grain shows that the effective stress is expected to be more heterogeneous than in the T grain. The lattice in the B grain is expected to rotate by between −35° and −15° from the original orientation. The plot of Fig. 8(b) indicates that no large orientation gradients are expected. This implies that deformation banding within the microstructure in this region is diffuse. The lattice in the T grain is predicted to rotate in the same direction as the lattice in the B grain. An exception is the central region where no lattice rotation is expected.

Modifications to the original bicrystal were also modelled in an attempt to investigate the influence of grain boundary character on deformation in aluminium bicrystals. In the first case, the grain boundary of the bicrystal was tilted to 30° with the initial orientations of each grain remaining the same as those of the original bicrystal (Fig. 9(b)). This resulted in a grain boundary relationship that was different to that of the original bicrystal. The results are plotted in Fig. 10 in terms of the expected variations in effective stress and lattice rotations about the transverse direction. The values and distributions of effective stress and lattice rotation are seen to be remarkably similar to those of the original bicrystal.

The second modification to the bicrystal also had the grain boundary tilted to 30° but in addition, the orientations of each grain were tilted by 30°. This gave the same grain boundary relationship that was present in the original bicrystal but the deformation geometry can be considered to be tilted. This configuration is described in the schematic diagram Fig. 9(c). The expected variations in effective stress and the variations in lattice rotations about the transverse direction (y-direction) are plotted in Fig. 11. These plots indicate that the specimen geometry was expected to dramatically change during deformation. The specimen can be seen to expect to have undergone significant shearing in the x–z plane while the grain boundary rotates to become parallel to the extension direction (x-direction). Figure 11(a) shows that the expected effective stress variation is highly heterogeneous ranging from ~30 MPa in some regions of both grains up to ~120 MPa in the base grain. The lattice in each grain is expected to rotate in the same direction but to a higher angle in some regions of the top grain.

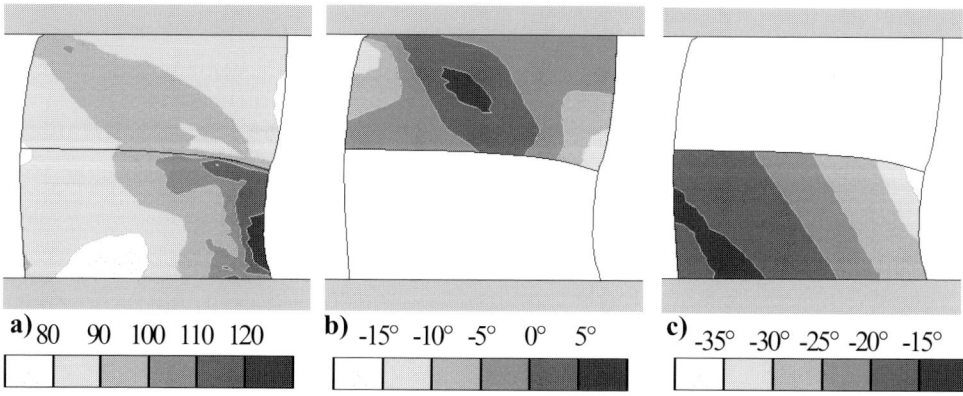

a) 80 90 100 110 120

b) -15° -10° -5° 0° 5°

c) -35° -30° -25° -20° -15°

Fig. 8 CPFEM predictions for bicrystal, (a) variations in effective stress (MPa) and variations in lattice rotations about the y-direction for (b) T grain and (c) B grain.

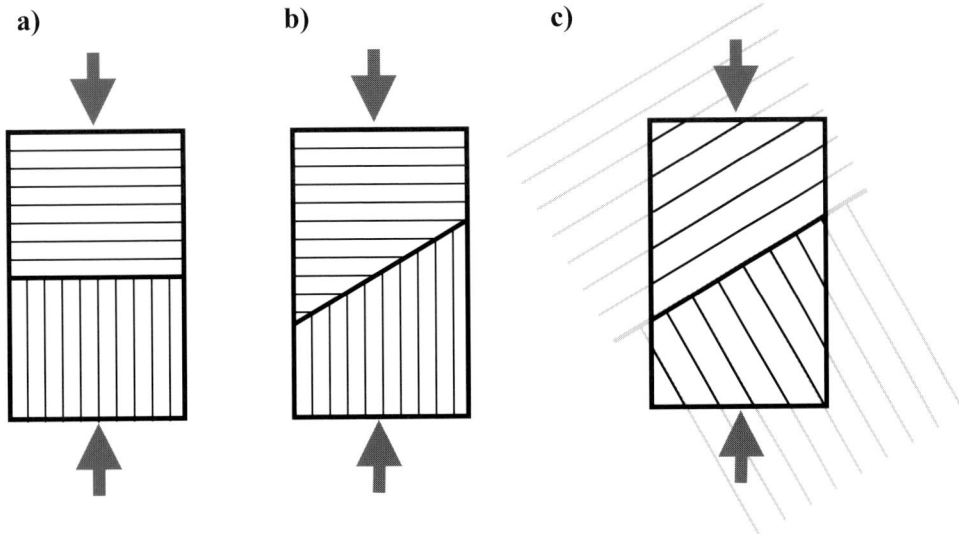

Fig. 9 Schematic diagrams of bicrystal configurations, (a) original bicrystal, (b) modified bicrystal, first case and (c) modified bicrystal, second case.

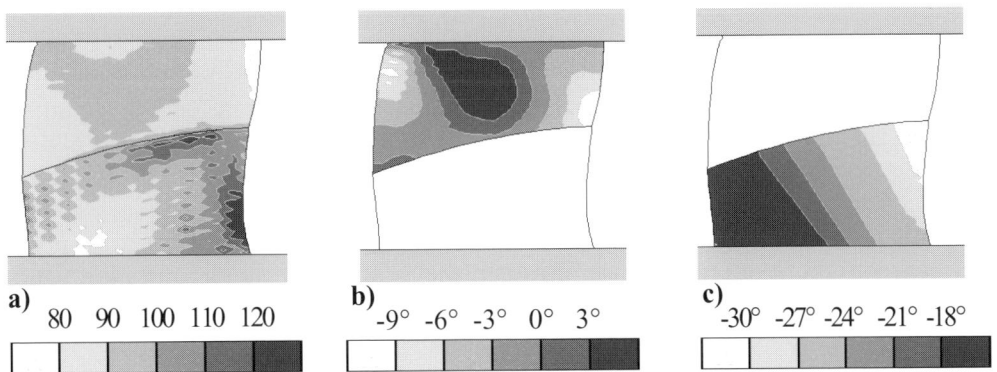

Fig. 10 CPFEM predictions for first case of modified bicrystal (Fig. 9b), a) variations in effective stress (MPa) and variations in lattice rotation about the *y*-direction for, b) top grain and c) base grain.

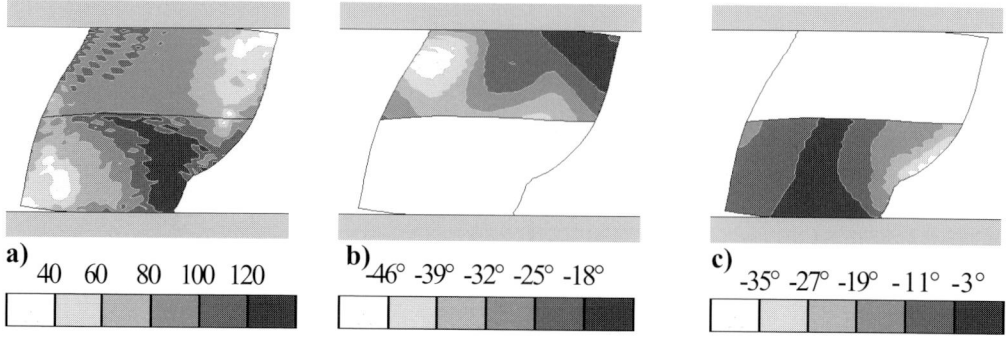

Fig. 11 CPFEM predictions for second case of modified bicrystal (Fig. 9(c)), (a) variations in effective stress (MPa) and variations in lattice rotations about the *y*-direction for, (b) top grain and (c) base grain.

DISCUSSION

The purpose of this work was to investigate the influence of the grain boundary on deformation. The results are discussed firstly in terms of the ability of the CPFEM to correctly predict the deformation behaviour and to assess its ability at dealing with the neighbouring grain interactions. The second half of the discussion attempts to clarify the role of the grain boundary and the influence of neighbouring grains, which naturally includes grain boundary influence.

ASSESSMENT OF THE MODEL

The prediction of lattice rotations in both the single crystal and the bicrystal was in good agreement with the experimental results. The rotations shown by the pole figures of Figs 4(a) and (b) correspond well with the rotations predicted by the model, shown by the plots of Figs 7(c) and (d). Similarly, the predictions of Fig. 7c and d resemble the microstructures shown in Fig. 3. Although it was not measured experimentally, the effective stress is expected to be a maximum for the T crystal where shearing was observed in the specimen (compare Figs 7(a) and 3(a)). The orientation gradients predicted by the model for the B crystal are high (Fig. 7(d)) and these are indicative of deformation banding. Although the locations of the deformation banding of Fig. 7(d) are not identical to those of the experimental specimens (Fig. 3(b)), the angles of lattice rotation that cause the orientation splitting correspond quite well.

In the bicrystal, the lattice rotation predictions are also relatively accurate. For the T grain, the lattice rotations predicted by the model (Fig. 8(b)) are in the same direction as they are in the experiment (Fig. 6(a)). In addition, some initial orientation is retained in both the experimental specimen and in the prediction. This implies good agreement between the two. The orientation gradients predicted by the model are thought to be indicative of the boundaries of deformation bands. The deformation banding shown in the T grain of the bicrystal (Figs 5(a) and (b)) is not predicted by sharp orientation gradients (Fig. 8(b)). The model does not predict large orientation gradients in the B grain, implying, correctly, that diffuse deformation banding is expected. Although the banding in the T grain is not well predicted, there is good agreement with the experiment for the B grain.

To summarise, the model is in relatively good agreement with the experimental results – the single crystal deformation behaviour is predicted accurately and the model can deal with the mechanical interaction of the two grains correctly in terms of the lattice rotations. The deformation banding is not so well predicted. This is because the model is based on slip processes and does not include the movement of individual dislocations during deformation. This factor is thought to be important for the development of even very coarse deformation banding features. It is possible to infer the banding behaviour from the predicted orientation gradients. The variation in effective stress is important in terms of subsequent recrystallisation events, but comparison between the model prediction of this and that of the experimental specimens was not carried out during the course of the present work. To examine this further, investigations of the substructure are required. Measurements of the cell size and misorientation can lead to an estimate of the stored energy which, in turn, can be used to estimate the flow stress in the deformed specimens.

INFLUENCE OF THE MECHANICAL INTERACTIONS OF NEIGHBOURING GRAINS AND THE GRAIN BOUNDARY

The mechanical interaction of grains during deformation, whether it is due to the neighbouring grain orientations or the grain boundary character was seen to be very important in both the experiment and the model. This section begins with a comparison between the experimental single crystals and the experimental bicrystal. The modelling results are then discussed and the grain boundary influence versus the mechanical interaction of grains is evaluated.

Although the initial orientation was the same, little orientation spreading (Fig. 4(a)) and fine diffuse deformation banding were observed in the T crystal, but the reverse was true for the T grain of the bicrystal (coarse, highly misoriented deformation bands, Figs 6(a) and 5(b)). This initial orientation is expected by the Taylor theory to be stable during the present deformation conditions. The Taylor theory is correct for the single crystal of T orientation, but for the T orientation in the bicrystal, although some initial orientation is retained, sharp deformation banding gives rise to large orientation gradients and orientation spreading. It is suggested that this is due to the specimen boundary constraints. Each orientation of the single crystal has two rigid die constraints whereas each orientation of the bicrystal has one rigid die and the neighbouring orientation (or grain boundary) as constraints. Therefore, it is evident that the mechanical interaction of two grains and so the grain boundary, may control the microstructural evolution during plane strain compression. In contrast, for the single crystal of B orientation, the initial orientation splits causing complimentary lattice rotations of ±20°. This orientation splitting provides coarse, well defined deformation bands. The B orientation of the bicrystal however, gives diffuse banding, little orientation spreading and only a small lattice rotation from the original orientation, thus, demonstrating the importance of the mechanical interactions of the orientations of adjoining grains.

Figure 10 describes results of a bicrystal where the orientations are identical to that of the original bicrystal (Fig. 9(b)) but the grain boundary relating the two grains is different. The results described by Figs 8 and 10 are similar, and imply that grain boundary character is a less important parameter on deformation behaviour than the orientations of the neighbouring grains. However, other factors may influence the deformation behaviour of the modified bicrystal. In this case, the specimen boundary conditions are altered – the proportion of unconstrained boundary of each orientation and the grain boundary area has changed.

The second modification to the bicrystal (Fig. 9(c)) had the same grain boundary description as the original bicrystal, but the grain boundary and the orientations of each grain were tilted. The results of this bicrystal were very different to the original bicrystal and the first modified bicrystal in terms of both the specimen geometry and lattice rotations. It is likely that the different initial orientations have different deformation characteristics, i.e., their resistance to slip and their ability to form banded features changes with initial orientation. In this case, these factors appear to greatly affect the overall expected deformation behaviour. However, it must be emphasised that the results are not conclusive – the specimen boundary conditions had altered from the original bicrystal and are thought likely to alter the deformation behaviour as well. The results imply that the grain boundary character alone does not greatly influence deformation and that the mechanical interaction of neighbouring orientations is more important. However, since two neighbouring grains must have different orientations in order to *be* two neighbouring grains, they must always be related by a grain boundary. Perhaps, in this case a simple '2-dimensional' surface description of the grain boundary is not adequate. In dealing with the deformation microstructure, it is suggested that the concept of a grain boundary needs to be generalised to include a description of the two orientations that the grain boundary relates. This would account for the mechanical interactions of the adjoining grains.

CONCLUSION

The lattice rotations predicted by the CPFEM for both the single crystals and the bicrystal are in good agreement with the experimental results but predictions of deformation banding have to be inferred from expected orientation gradients. This is because the model does not take into account the motion of individual dislocations which are likely to be partially responsible for such microstructural features.

Comparisons between single crystals and bicrystals have shown that the deformation behaviour can significantly change depending on the constraints placed on the orientation. The results imply that the influence of the mechanical interactions of the two orientations is more important than the influence of the grain boundary character, although separation of these two factors is difficult. It is suggested that a more generalised description of the grain boundary that can account for the orientations of the two grains would explain the deformation behaviour of the bicrystals in this instance more fully.

ACKNOWLEDGEMENTS

The work is supported by the Engineering and Physical Sciences Research Council and Alcan International (Banbury) UK Ltd. Thanks are also given to P. S. Bate for many valuable discussions.

REFERENCES

1. H. E. Vatne, S. Benum, R. Shahani and E. Nes, 'On the Nucleation of Recrystallised Grains from Grain Boundary Regions', *Proc. 16th Risø Int. Symp.,* Risø National Laboratory, Roskilde, 1995, 573–579.
2. S. V. Harren and R. J. Asaro, 'Nonuniform Deformation of Polycrystals and Aspects of the Validity of the Taylor Model', *J. Mech. Phys. Solids*, 1989, **37**, 191–214.
3. S. J. Lillywhite, M. Aindow and P. S. Bate, 'Grain Boundary Influence on Texture Evolution During Deformation and Recrystallisation', *ICOTOM12*, 1999, **2**, 1619–1624.
4. A. Akef and J. H. Driver, 'Orientation Splitting of Cube Oriented Face Centred Cubic Crystals in Plane Strain Compression', *Mat. Sci. and Eng.,* 1991, **A132**, 245–255.
5. P. Ambrosi and Ch. Schwink, 'Plasticity of FCC Single Crystals Oriented for Multiply Slip', *Proc. 5th Int. Conf. Strength of Metals and Alloys,* Vol. 1, Pergamon Press, 1979, 29–34.
6. G. B. Sarma and P. R. Dawson, 'Effects of Interactions among Crystals on the Inhomogeneous Deformations of Polycrystals', *Acta Met.,* 1996, **44** (5), 1937–1953.
7. N. Yu. Zolotorvsky, Yu. F. Titovets and G. Yu. Dyatlova, 'Lattice Rotations in Single Grains of Large Grained Aluminium Polycrystal During Tension', *Scripta Met.,* 1998, **38** (8), 1263–1268.
8. T. Mizushima, T. Minegishi, H. Saikawa, T. Yamane, H. Araki and S. Saji, 'Contribution of Grain Boundary to Tensile Deformation of Aluminium Bicrystal', *J. Mat. Sci. Letters,* 1993, **12**, 239–240.
9. S. J. Lillywhite, M. Aindow and P. S. Bate, 'Nucleation of Recrystallisation Near Prior Grain Boundaries', *4th Int. Conf. Recrystallisation,* Japan Institute of Metals, 1999, 143–148.
10. P. Dawson, D. Bloyce, S. MacEwen and R. Rogge, 'Intercrystalline Stresses in Metal Polycrystals: Comparing Simulations to Diffraction Experiments' (1999), *ICOTOM12*, 1999, **1**, 505–510.
11. P. Bate, 'Modelling Deformation Microstructure with the Crystal Plasticity Finite Element Method', *Phil. Trans. A,* 1999, **357** (1756), 1589–1601.

Grain Boundary Junctions in Twin-Bearing Materials

V. RANDLE

Department of Materials Engineering, University of Wales Swansea, Swansea SA2 8PP, UK

ABSTRACT

Recently it has been realised that not only is the proportion of 'special' boundaries in a polycrystal important in the control of intergranular degradation, but also the spatial distribution of these elements. Therefore it is important to study the constitution and distribution of grain junctions, that is, where three interfaces join. The focus of attention in this paper is twin-bearing materials and the $\Sigma3^n$ family of boundaries, since subsets of the $\Sigma3$ classification are types known to show reliably special behaviour. After presenting some analysis on the characteristics of coincidence site lattice (CSL) statistics from an extensive survey of cubic materials, the paper reports data concerning analysis of triple junctions in nickel and copper after various thermomechanical treatments to promote enhanced proportions of $\Sigma3$s. It was found that there is no correlation between $\Sigma3$ proportions and 'special' junctions. The role of grain junctions in improving overall material behaviour is discussed.

INTRODUCTION – GRAIN BOUNDARY CATEGORISATION IN POLYCRYSTALS

Various thermomechanical processing regimes can be used to manipulate the proportions of boundary types in metals and alloys. In particular, face-centred cubic materials having a low stacking-fault energy can be produced with overall better response to intergranular degradation. This response is accompanied by a sharp increase in the proportion of annealing twins, generated either during recrystallisation or after strain annealing, and usually accompanied by a small grain size.[1] These interfaces are characterised by having extremely low energy and free volume, and are therefore almost immune to attack. Annealing twins are a subset of '$\Sigma3$' boundaries, where Σ denotes the reciprocal density of coinciding sites in the coincidence site lattice (CSL) nomenclature, which is usually used to categorise grain boundaries in cubic polycrystals.

This paper addresses firstly characteristics of CSL statistics found in the literature, and then focusses on the significance of interactions between $\Sigma3^n$ boundaries at triple junctions, including some recent experimental data. Before examining the geometrical effects which occur where grain boundaries join in polycrystals, it is instructive to consider briefly the statistics of grain boundary types which occur, as categorised by the CSL system, in polycrystals. These data have been extracted from a major review of more than 200 experimental investigations.[2] Figure 1 gives an overview of the materials found which have been the subject of CSL investigations. The first group of materials are face-centered cubic (fcc), followed by a group of body-centered cubics (bcc), and a category of review articles which by definition include several investigations, often on a range of materials. With regard to the fcc groups, the major groups investigated are pure (or nearly pure) materials aluminium, copper and nickel, and fcc steels (denoted 'fcc st' on Fig. 1). Al+, Cu+ and Ni+ signify alloys or compounds of these elements. Ni_3Al has been placed in a separate category because there have been a relatively high number of investigations on this intermetallic. Turning to the bcc group, there have been only about one-quarter of the number of investigations performed as on fcc materials. The 'bcc' group includes iron, tungsten and molybdenum, and the 'bcc st' group includes a range of ferritic steels. A major group in the bcc category is the Fe–Si alloys.

The distribution of materials reflects both basic research and commercial interests, with the more fundamental studies of grain boundary property/CSL behaviour generally carried out on pure materials. Investigations on alloys are usually in response to specific performance-related phenomena of industrial importance, such as boron-associated ductilisation of Ni_3Al[3] or growth of Goss grains in Fe–3%Si.[4] From these statistics it is clear that twin-bearing materials (ie low stacking-fault energy fcc

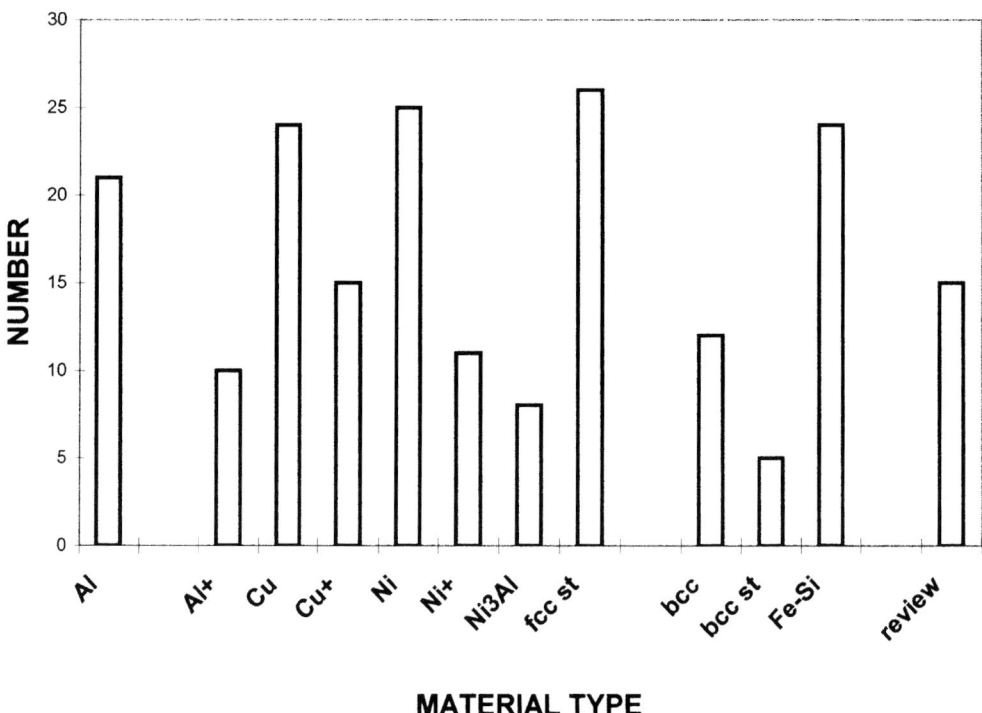

Fig. 1 Frequency distribution of material types analysed during collection of CSL statistics on polycrystals (see text for details).

materials) comprise the largest category of polycrystalline materials of interest for CSL categorisation.

The two factors which have been observed to exert the most influence on the number of CSLs are the presence of a strong texture, resulting in $\Sigma 1$ boundaries, and the amount of twinning, resulting in $\Sigma 3$ and $\Sigma 3''$ boundaries. It is generally true to say that where low stacking-fault energy materials are quoted as having a high proportion of CSLs, this is composed almost exclusively of $\Sigma 3''$ (n = 1 to 3) boundaries.[2] Figure 2 gives more insight into the origins of average CSLs for each material by showing the average proportions of $\Sigma 1$s and $\Sigma 3$s for each material group. It should be noted that not all investigations reported $\Sigma 3$ and $\Sigma 1$ fractions separately, and compilation of the data was therefore restricted to cases where these figures were available. The amount of $\Sigma 3$s in the fcc categories correlates to a first-order approximation with the stacking fault energy. Turning to the $\Sigma 1$ category, proportions of $\Sigma 1$s are most significant for bcc materials, and aluminium. This can be explained by the fact that bcc materials and aluminium often exhibit strong textures, and furthermore annealing twinning is not expected.

The maximum value of Σ to choose as a cut-off when calculating proportions of CSLs has been a long-standing enigma. Essentially, there is no physically correct answer to this question, since it implies a simple (and monotonic) connection between Σ-value and properties which is not the case in reality. We are left, then,

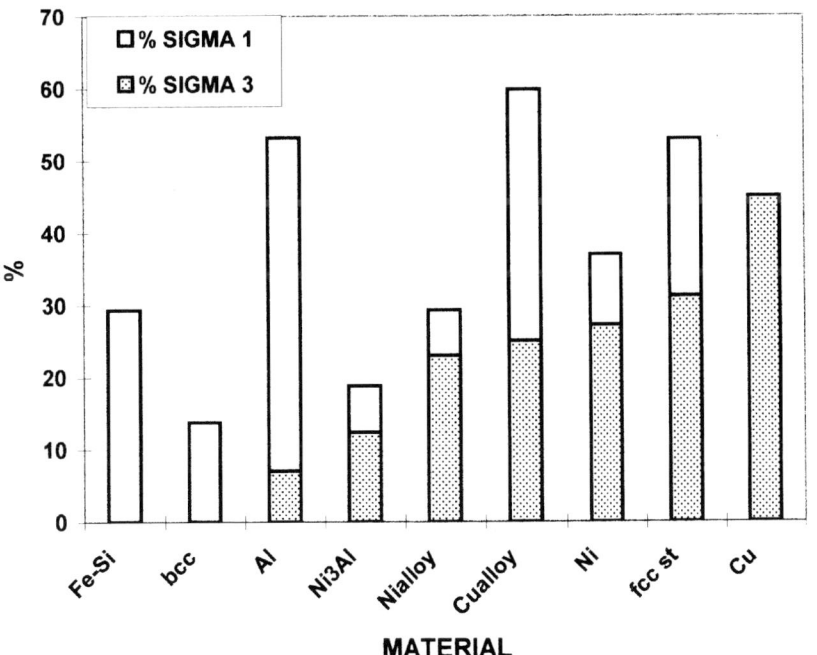

Fig. 2 Proportions of $\Sigma 1$ and $\Sigma 3$ boundaries in polycrystals from various material types.

with empirical choices. Although there is some evidence for special properties associated with 'higher' Σ-values, eg $\Sigma = 31-61$,[5] it is generally assumed that on the whole most special property boundaries will have lower Σ-values than this and moreover most CSL distributions show far fewer boundaries with 'higher' Σ-values than low-Σs. Figure 3, which indicates the choices that have been made for maximum Σ, reflects this trend. There are outstanding maxima at $\Sigma29$ and $\Sigma49$, that is, most workers have selected one of these values as a cut-off. The higher value is more relevant where orientation relationships are being discussed rather than the property of a boundary. This is because there is usually an attempt to correlate properties with low Σ-values, whereas orientation relationships relate more to geometry than properties. Strictly, the different values of maximum Σ selected invalidate comparisons of total CSLs between different investigations. However, since in practice there are usually few CSLs which have higher Σ-values, qualitative comparions are still effective. Where multiple twin interaction are concerned, it is necessary to have a Σ cut-off of at least $\Sigma27$, ie $\Sigma3$.[3]

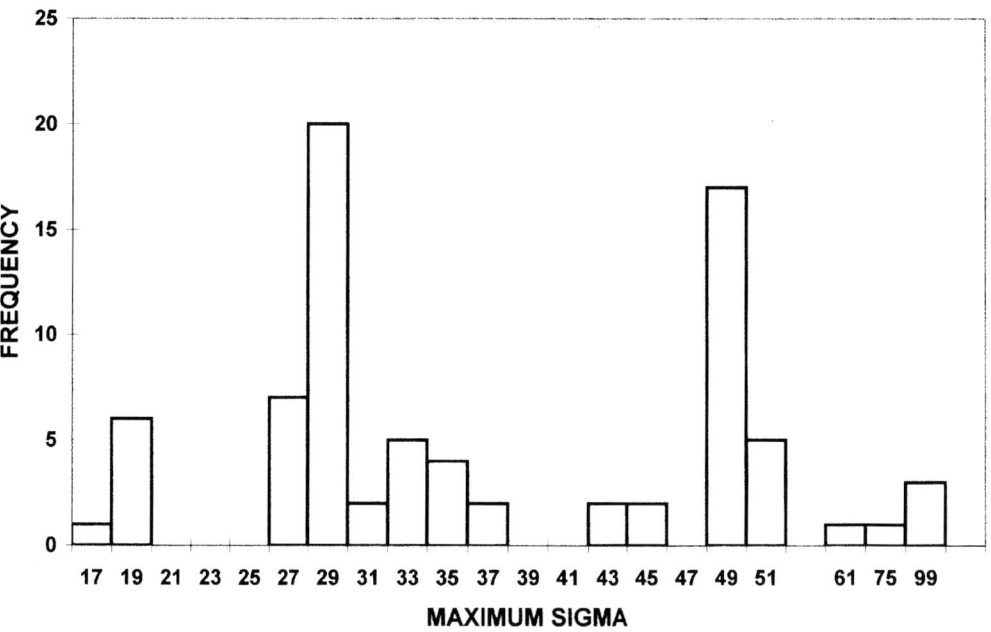

Fig. 3 Distribution of the maximum value of Σ used in studies involving collection of CSL statistics in polycrystals.

TRIPLE JUNCTIONS

Much of the research into controlling grain boundary performance has focussed on proportions of boundary types, particularly CSLs. More recently it has become realised that the *spatial arrangement* of these elements is an additional important factor in material behaviour, that is, where interfacial transport phenomena are involved it is desirable for the connected pathway of boundaries to be interrupted by 'special' boundaries.[6] Twins are the best example of 'special' boundaries, although the twins themselves do not actually form part of the grain boundary network. Their effect must, therefore, work in an indirect way.[7,8] One influence that the twins have is at the site where they join the boundary network, thus forming a *triple junction*, so called because it is almost always three junctions which meet in polycrystals.

There are geometrical rules concerning the relationship between three joining boundaries which, for CSLs, are that they share a common misorientation axis; the sum of two of the misorientation angles gives the third; and the product (or quotient, if applicable) of two of the Σ-values gives the third. As indicated in the previous section, the CSL combination which has the most important practical application is the $\Sigma 3^n$ family. The most common example is the meeting of two $\Sigma 3$ boundaries, where the $\Sigma 3$s could be 'coherent twins', ie on {111}, 'incoherent twins' or $\Sigma 3$-related grain boundaries:

$$70.53°/110 + 70.53°/110 \rightarrow 141.06°/110 \quad \text{ie } \Sigma 3 + \Sigma 3 \rightarrow \Sigma 9 \qquad (1)$$

These geometrical rules for grain junctions have a significant effect on the CSL distribution, particularly when $\Sigma 3$ boundaries are involved. A CSL analysis of grain boundaries in polycrystals can be extended to grain junctions by counting the proportion of CSLs in each triple junction (ie 1,2 or 3 CSLs) and analysing the CSL types.[9] Where deviation from the CSL configuration of a grain boundary at a grain junction is displaced from near-exact matching, the other boundaries at the junction may not fall within a CSL range, and so CSL multiplication is not perpetuated.[10]

Triple junctions having at least two special boundaries (ie $\Sigma 3$s) are beneficial elements of microstructure because interfacial transport phenomena cannot be transmitted at these junctions. They may also be stable against grain growth, and have been called 'secure' junctions.[11] For example it has been demonstrated that the connectivity parameters of low energy boundaries can be improved in nickel by judicious processing, such that proportions of secure junctions have been increased from 37% to 61%.[11] The theoretical limit to the proportion of $\Sigma 3$s in a microstructure is 2/3, ie two at every junction,[12] although this is not achieved in practice. In this theoretical limit the third boundaries at each junction would be $\Sigma 9$ or $\Sigma 1$.

The $\Sigma 3^n$ family gives rise to a whole hierarchy of interactions at grain junctions, both in terms of grain boundary dissociations and perpetuation of CSLs by interactions at junctions. Both of these reactions have been commonly observed in several low-stacking fault energy systems for the $\Sigma 3$–$\Sigma 3$–$\Sigma 9$ interaction given in Eqn (1) [eg

Refs 13 and 14]. The case for $\Sigma 3^n$ where n = 3,4,5 ($\Sigma 27$, $\Sigma 81$ and $\Sigma 243$) is more complex. For example, it has been observed in a Cu-6at%Si alloy that $\Sigma 27a$ boundaries did not dissociate to give $\Sigma 3$ and $\Sigma 9$ boundaries, but rather gave $\Sigma 3$ and $\Sigma 81d$ boundaries. Similarly, $\Sigma 81d$ boundaries always dissociated to $\Sigma 3$ and $\Sigma 243a$ rather than to $\Sigma 3$ and $\Sigma 27$. These results were explained in terms of both boundary symmetry to retain the {111} coherent plane as the twin boundary, and the kinetics of dissociation, rather than simply low energy.[15] However, these reactions may not be duplicated in other materials; for example in silicon $\Sigma 27a$ boundaries gave rise to a $\Sigma 3$ and $\Sigma 9$ rather than a $\Sigma 81$.[13] These discrepancies between materials highlight the purely geometrical nature of the CSL model.

It should also be noted that the $\Sigma 243a$–$\Sigma 3$–$\Sigma 81d$ grain junction configuration, which has been identified in an austenitic steel, could also have been interpreted as a $\Sigma 1$–$\Sigma 3$–$\Sigma 3$ trio since the misorientations for $\Sigma 243a$ (7.4°/110) and $\Sigma 81d$ (60.4°/443) are within the limits of $\Sigma 1$ and $\Sigma 3$ respectively, according to the Brandon criterion.[16] The authors decided on the former designation from examination of the geometry of adjacent grain junctions and morphological features of the boundary.

Clear evidence for the relationship between $\Sigma 3^n$ boundaries, $\Sigma 3$, $\Sigma 9$, $\Sigma 27$ in this case, has been shown in experiments on pure nickel involving some 3000 boundaries. Specimens were strain-annealed cyclically using low temperatures and strains for increasing times. Details are given in Reference 17. Although the proportion of $\Sigma 3^n$ – approximately 30% $\Sigma 3$, 5% $\Sigma 9$ and 2% $\Sigma 27$ – did not vary greatly during the thermomechanical treatments, the proximity to the exact CSL misorientation (expressed here as v/v_m, where v is the measured angular deviation from the exact CSL and v_m is the maximum allowable deviation according to the Brandon criterion, ie 8.7° for a $\Sigma 3$ CSL) changed dramatically. These results are shown on Fig. 4, where all other CSLs except $\Sigma 3^n$ are unlabelled for clarity. Not only are the v/v_m values for $\Sigma 3^n$ lower than the other CSLs, but there is a trend for further evolution towards exact CSL with annealing time. This result was interpreted as grain boundary energy reduction at $\Sigma 3$ boundaries, with concomitant 'geometrically necessary' reductions at $\Sigma 9$ and $\Sigma 27$ because they conjoin $\Sigma 3$s at grain junctions.

TRIPLE JUNCTION ANALYSIS IN COPPER

Following the recognition of the importance of considering not only grain boundary types but also how they join at junctions, some experiments have been carried out recently to investigate specifically proportions of $\Sigma 3^n$ boundaries in a material which twins readily, namely copper. Specimens of pure copper underwent various recrystallisation cycles of cold work and annealing, designed to induce an enhanced population of $\Sigma 3^n$ boundaries while controlling the grain size to below approximately 30 μm. Table 1 gives details of the thermomechanical treatments performed. Proportions of CSLs were then obtained from electron back-scatter diffraction (EBSD) and crystal orientation mapping (COM). Analysis of triple junction components was

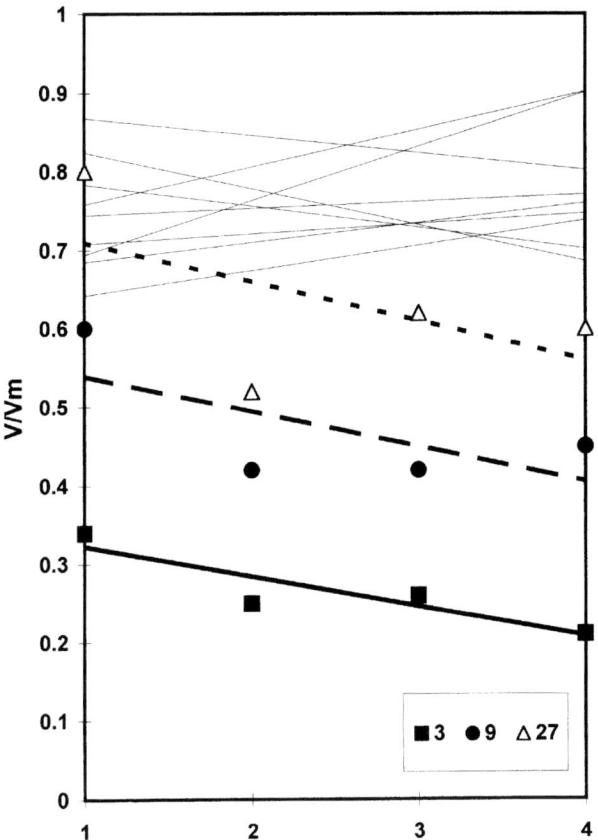

Fig. 4 Average deviation from exact CSL matching, v/v_m, for CSLs with $\Sigma 3$ to $\Sigma 29$ for heat treatments on pure nickel. For clarity only $\Sigma 3^n$ boundaries are represented individually.

Table 1 Details of specimen heat treatments.

Specimen	Heat treatment
A	30% CW, 7 min at 750°C (3 cycles)
B	67% CW, 7 min at 750°C (1 cycle)
C	1.5 h at 450°C (1 cycle)
D	20% CW, 5 min at 750°C (5 cycles)
E	30% CW, 5 min at 750°C + 20 h at 450°C (1 cycle)
F	20% CW, 3 min at 700°C (5 cycles)
G	50% CW, 1 min at 1000°C (1 cycle)

(All specimens given pre-treatment A, 1.5 h at 450°C. CW – cold work).

carried out after extraction of the relevant data from COMs. In total 900 triple junctions were sampled.

Figure 5 shows the statistics of $\Sigma 3^n$ distributions, obtained by a representative sampling line scan, for all the specimens. The proportion of $\Sigma 3$s, which includes annealing twins, varies between 41% and 58% depending on the heat treatment employed. The proportion of $\Sigma 9$s was consistently 6–7% whereas the $\Sigma 27$ proportion was smaller, 2.5% on average. In every case the average relative deviation of the $\Sigma 3$s from the exact CSL configuration was small, 0.1–0.2 v/v_m. The relative deviation for both $\Sigma 9$ and $\Sigma 27$ was larger, which is consistent with experimental data on nickel referred to in the previous section.[17] The statistics for occurrence of $\Sigma 3^n$ boundaries in triple junctions are shown on Fig. 6. These are plotted as proportions of triple junctions having either 1, or 2 or 3, $\Sigma 3^n$ boundaries.

The data have been analysed using the well-known Brandon ($\Sigma^{-1/2}$) criterion. A more restrictive criterion for CSL evaluation, based on sound physical principles, is related to $\Sigma^{-5/6}$.[18] It is instructive to compare analysis of the present data by both criteria. Since most of the $\Sigma 3$s were in fact very close to the exact CSL, calculation of the proportions by either criterion makes little difference. However, for the case of $\Sigma 9$ the proportion is reduced by a third when using the $\Sigma^{-5/6}$ criterion, and by three-

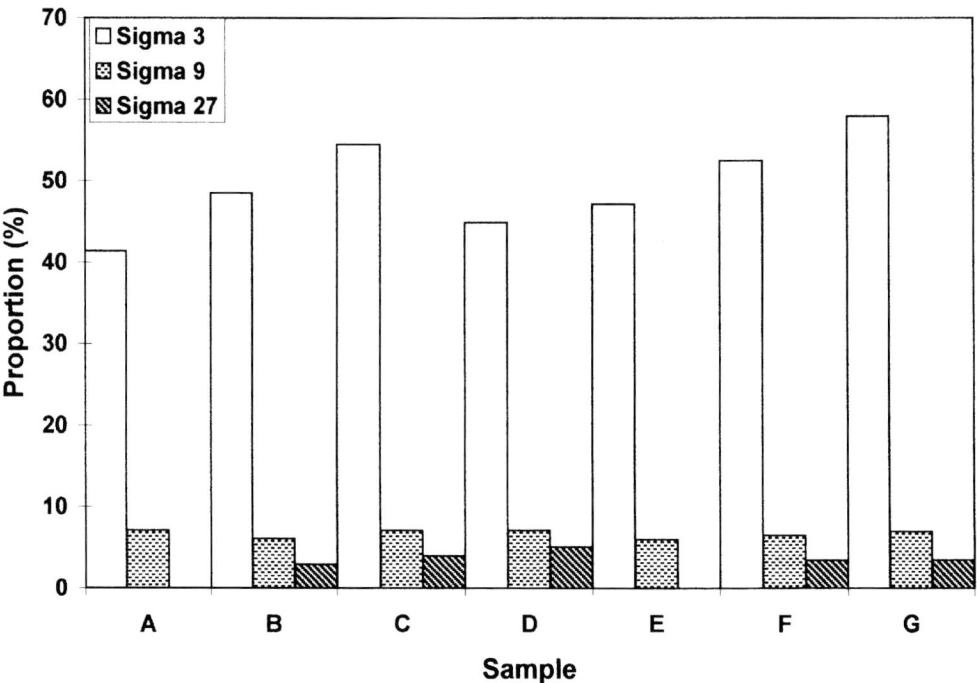

Fig. 5 Proportions of $\Sigma 3^n$ boundaries for each specimen listed in Table 1.

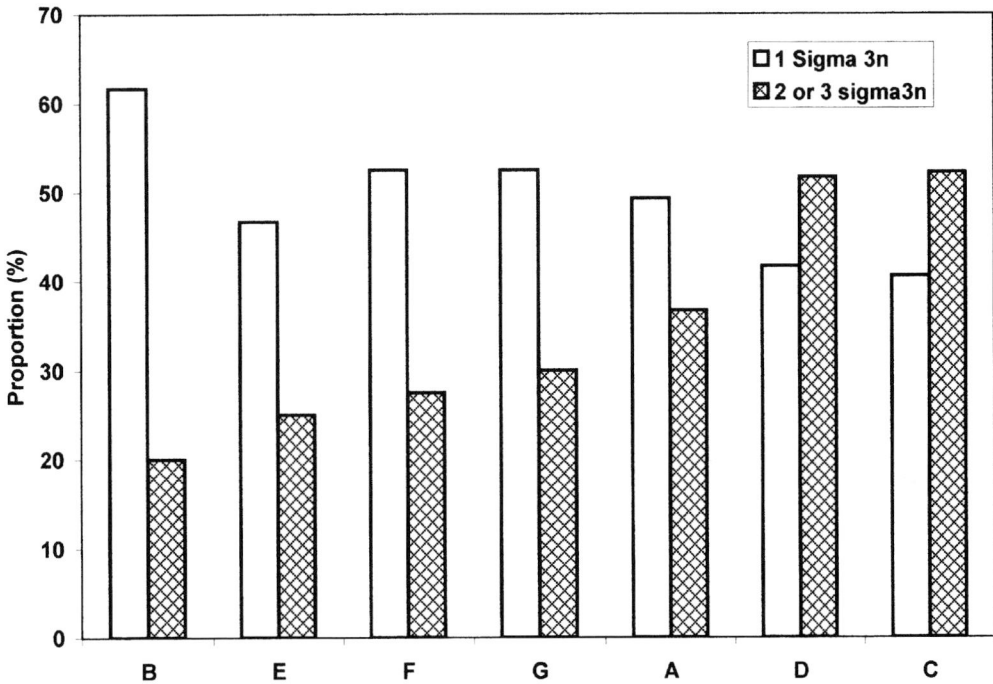

Fig. 6 Proportions of triple junctions containing 1, or 2 or 3, $\Sigma 3^n$ boundaries for each specimen listed in Table 1.

quarters for $\Sigma 27$. Although the $\Sigma^{-5/6}$ criterion is physically the more realistic choice, use of the Brandon criterion allows comparisons with previously reported data.

Turning now to the triple junction analysis, a 'special' triple junction, i.e. one that can arrest intergranular transport, must comprise at least two $\Sigma 3^n$s. The proportion of special junctions thus calculated varies from 20% to 52% as shown on Figure 6. There is no rank correlation between the proportion of $\Sigma 3$s, ie special boundaries, and the proportion of special junctions. Furthermore, there is a much wider variation in the special junction proportion than that for the boundaries. This is quite a striking result since it might have been expected that the proportion of $\Sigma 3$s would produce a concomitant proportion of special junctions. Clearly, it is highlighted that the *spatial distribution* of the boundaries is of prime importance is formulating triple junctions. Necessarily, the grain boundary statistics collected from random linear scans are devoid of any connectivity component. By contrast triple junction data were collected from COMs, and included virtually every triple junction in the map, therefore guaranteeing inclusion of the spatial component. On examining the maps it was noticed that special triple junctions were not distributed at random, although this has not been quantitatively analysed at this time. This clustering has some geometrical origin since neighbouring triple junctions of course include one common boundary.

In collating the statistics for numbers of $\Sigma3^n$s in junctions, for simplification any $\Sigma3^n$s with $n \leq 3$ were included, since the great majority of these was $\Sigma3$, and $\Sigma3$s (plus some $\Sigma9$s) are considered to be special.[19, 20] Hence junctions containing two $\Sigma3^n$s were mostly two $\Sigma3$s. Junctions containing three $\Sigma3^n$s were equally split between $\Sigma3–\Sigma3–\Sigma9$ and $\Sigma3–\Sigma9–\Sigma27$. Of some especial interest are those junctions containing only one $\Sigma3^n$ boundary which was a $\Sigma9$ or $\Sigma27$. Such types comprised 9% of the total junction population. It is generally assumed that $\Sigma9$s and $\Sigma27$ occur only as a consequence of $\Sigma3$ association; this result shows that they also occur for other reasons such as because of suitably-oriented grain impingements, as do some $\Sigma3$s. Alternatively, they could be the result of higher-order $\Sigma3^n$ interactions. Either way, the resulting boundary would be very unlikely to have special characteristics.

ROLE OF TWINS AND GRAIN JUNCTIONS

Although annealing twinning via manipulation of grain boundary crystallography is central to the development of superior intergranular properties in twin-bearing materials, the mechanisms by which this is effected are as yet unclear. Annealing twins on {111} are not, from a properties rather than a crystallographic point of view, part of the grain boundary network – for example they are virtually both immobile and are not a constituent of the intergranular transport network. Therefore, the role which they play in improving intergranular properties such as fracture resistance must be indirect. By contrast other types of $\Sigma3$ can be part of the grain boundary network. They are present as a result of relationships at triple junctions (Eqn (1)), as incoherent portions of twins or from encounters between grains having a $\Sigma3$ orientation relationship. It is unfortunate that most investigations which report $\Sigma3$ fractions do not distinguish between the statistics and distribution of the different categories of $\Sigma3$.

Annealing twinning is thought to occur for one of two reasons: to decrease the overall interfacial energy when the energy of the boundaries between a grain's neighbours and its twin would be less than that of the boundaries between the neighbours and the grain itself,[21] or to reorient grain boundaries so as to facilitate dislocation absorption and mobility during recrystallisation.[22] The former was proposed to account for the presence of twins during grain growth, when reduction in total grain boundary energy is the driving force. However most treatments which increase the proportion of $\Sigma3$s involve recrystallisation or strain-induced boundary migration rather than grain growth and so the energy reduction mechanism of Ref. 21, which is sometimes quoted in this context, is not strictly applicable. Rather, the mobility and dislocation absorption mechanism of reference 22 is more relevant. Hence, the '$\Sigma3$ regeneration model' has been recently proposed which relies on interactions at triple junctions involving a mobile grain boundary and a twin, and provides a mechanism for improvement of the grain boundary network via enhancement of $\Sigma3$s during and immediately following recrystallisation.[8]

The model relies on the condition that an encounter between a $\Sigma 9$ and a $\Sigma 3$ gives a $\Sigma 3$ boundary rather than a $\Sigma 27$. This is supported by experimental evidence since relatively few $\Sigma 27$ boundaries are reported in investigations even when the proportion of $\Sigma 3$s is very high.[2] Furthermore, the generation of an incoherent $\Sigma 3$ rather than a $\Sigma 27$ is generally preferred both on the basis of lower energy and greater mobility. By the same argument if a $\Sigma 9$ does encounter a $\Sigma 27$, a $\Sigma 3$ would again be generated rather than a $\Sigma 243$. In general, then, it can be stated

$$\Sigma 3^n + \Sigma 3^{n+1} \longrightarrow \Sigma 3 \tag{2}$$

which is a method by which *mobile* $\Sigma 3$s – which still have markedly different properties to random boundaries even though they are not the {111} type – enter the microstructure. Once additional $\Sigma 3$s have been generated by this means, it has been shown that further annealing can allow grain boundary planes to approach more closely low volume configurations if grain growth is minimised.[19] The $\Sigma 3$ regeneration model is described in detail elsewhere.[8]

Although the model is successful in rationalising the introduction and proliferation of $\Sigma 3$s into the grain boundary network, it still does not provide the entire reason for the marked improvement in intergranular-related properties observed in a material after enhanced twinning, since probably not sufficient $\Sigma 3$ grain boundaries to break the continuity of random boundaries in the network are introduced in this way. For example, there are only 12% $\Sigma 3$ grain boundaries (not twins) in low-SFE strain-annealed alloy 600 which shows greatly enhanced creep resistance.[23] It has been assumed that when a twinning event occurs, one additional 'special' boundary, which is assumed to mean a lower-Σ CSL (although *all* Σ boundaries are not necessarily special) is added to the microstructure.[1] However, if the boundary was a very high Σ in the first place the twinning reaction makes no essential difference since it is generally agreed that Σ-values above approximately 29 are irrelevant in polycrystals with respect to properties (see above).

It is suggested here that there must be reasons other than those related to Σ-values why the new boundary is 'more special' than the old boundary it replaced as a consequence of the twinning event and the formation of a new triple junction. From a geometrical point of view, it could relate to the fact that a common rotation axis exists at the three conjoined interfaces, and/or that the {111} planes of the twin influence the choice of boundary planes in the adjoining boundaries, especially immediately adjacent to the junction line, constituting a disruption in the transport network. This suggestion is backed up by evidence that boundary portions having symmetrical geometry have been observed adjacent to triple junctions.[24]

The constraint on one grain boundary in a triple junction imposed by an adjoining twin can also be viewed in terms of degrees of freedom, which is illustrated in figure 7. The original grain boundary prior to twinning, labelled GB1, is considered to be unrestrained and therefore none of its degrees of freedom are fixed. On the other hand the new twin has a totally fixed misorientation and boundary plane, ie 70.5°/<110> and {111}. The

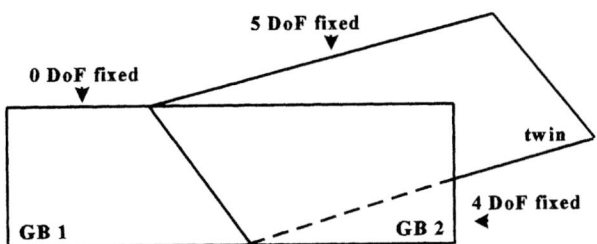

Fig. 7 Schematic diagram showing the effect of annealing twin creation at a general grain boundary (GB1) leading to formation of a triple junction. Since GB 1 is arbitrary in terms of degrees of freedom (DoF), and the twin is entirely fixed because it is a {111} symmetrcial tilt boundary, the newly formed boundary (GB2) has 4 out of 5 fixed DoF. This may lead to GB2 having a better structure to resist intergranular degradation than GB1, ie more 'relative specialness'.

twinning results in a new boundary being formed, GB2, which is constrained to have four out of five degrees of freedom fixed. This geometry is consistent with the observation that triple junction-related $\Sigma 3^n$ boundaries are geometrically constrained in terms of deviations from exact CSL, as shown in Fig. 4.[17] The 'specialness' requirement might be only for one of the non-twin boundaries to be *slightly* improved in terms of properties than the other one to suppress effectively intergranular transport. In other words the critical factor is *relative specialness* rather than *absolute specialness* of conjoined boundaries.

CONCLUSIONS

1. There is not a simple correlation between proportions of $\Sigma 3^n$ boundaries and 'special' triple junctions, where a 'special' triple junction contains at least 2 $\Sigma 3^n$ boundaries.

2. The '$\Sigma 3$ regeneration model' shows how interface interactions at grain boundary junctions in a twin-bearing material generate $\Sigma 3$ boundaries in the grain boundary network. These are mobile types are therefore dominate the next stage of grain boundary migration.

3. A new triple junction formed as the result of twinning imposes restrictions on the newly-formed grain boundary, fixing four out of five of its degrees of freedom, and in this way affects the boundary network.

REFERENCES

1. P. Lin, G. Palumbo, U. Erb and K. T. Aust, *Scripta Metall.*, 1995, **33**, 1387.
2. V. Randle, *The Role of the Coincidence Site Lattice in Grain Boundary Engineering*, Institute of Materials, 1996.

3. T. Watanabe, T. Hirano, T. Ochiai and H. Oikawa, *Proc. IOCTOM10,* H. J. Bunge ed., *Mat. Sci. For.* 1995, **157–162**, 1103

4. R. Shimizu, J. Harase and D. J. Dingley, *Acta Met. Mat.*, 1990, **38**, 973

5. L. S. Shvindlerman and B. B. Straumal, *Acta Metall.*, 1985, **33**, 1735.

6. V. Y. Gertsman, M. Janacek and K. Tangri, *Acta Mater.*, 1996, **44**, 2869.

7. V. Randle, *Proc. IIB'98: Intergranular and Interphase Boundaries, iib'98,* P. Lejcek and V. Paidar eds, *Mat. Sci. For.*, **294–296**, 1998, 51.

8. V. Randle, *Acta Mater,* 47, 1999, 4187

9. P. Fortier, K. T. Aust and W. A. Miller, *Acta Metall. Mater.*, 1995, **43**, 339.

10. V. Randle, *Micros. Microanal. Microstruct.*, 1993, **4**, 349.

11. C. B. Thomson and V. Randle, *J. Mater. Sci.*, 1997, **32**, 1909.

12. G. Palumbo, K. T. Aust, U. Erb, P. J. King, A. M. Brennenstuhl and P. C. Lichtenberger, *Phys. Stat. Sol.*, 1992, **131**, 425.

13. A. Garg, W. A. T. Clark and J. P. Hirth, *Philos. Mag.*, 1989, **A59**, 479.

14. V. Y. Gertsman and K. Tangri, *Philos. Mag.*, 1991, **A64**, 1319.

15. C. T. Forwood and L. M. Clareborough, *Acta Metall.*, 1984, **32**, 757.

16. D. G. Brandon, B. Ralph, S. Ranganathan and M. S. Wald, *Acta Metall.*, 1964, **12**, 813.

17. C.T. Thomson and V. Randle, *Acta Metall.*, 1997, **45**, 4909.

18. G. Palumbo, K. T. Aust, E. M. LeHockey, U. Erb and P. Lin, *Scripta Mater.*, 1998, **38**, 38.

19. V. Randle, P. Davies and B. Hulm, *Philos. Mag.*, 1999, **79A**, 305.

20. V. Randle, *Mater. Sci. Tech.*, 1999, **15**, 246.

21. R. L. Fullman and J. C. Fisher, *J. Appl. Phys.*, 1951, **22**, 1350.

22. G. Gindraux and W. Form, *J. Inst. Metals*, 1973, **101**, 85.

23. G. S. Was, V. Thaveesprungsriporn and D. C. Crawford, *JOM*, February 1998, 44.

24. V. Singh and A. H. King, *Scripta Mater.*, 1996, **34**, 1723.

Grain Boundary Observations as Clues to Atomic Movements

G. W. GREENWOOD*

ABSTRACT

Atomic movements to cause permanent distortion or fracture can occur through sliding processes or through directional diffusion. The former dominate at low temperatures but increase in temperature can alter the form of sliding, changing crystallographic planes of slip and determining the extent to which grain boundaries slide.

Grain boundaries play a major role in processes that involve directional diffusion by providing short circuit paths and through their ability to act as sources and sinks for vacancies. In alloys, such action can give rise to microstructural changes in grain boundary regions and, by allowing displacement between adjacent grains, it requires accommodation by sliding. Such accommodation, however, is distinguishable from that of sliding in bulk and is associated with entirely different distributions of internal stress.

Microstructural observations at grain boundary regions, involving precipitate redistribution, marker line displacement and the recently identified changes in grain boundary profiles, provide clues to the form of atomic movements under stress or other driving forces. The identification of these movements has extensive relevance to the prediction of behaviour at elevated temperatures.

INTRODUCTION

The existence of different mechanisms of deformation, depending upon stress level and temperature, is widely recognised[1] though the precise details of their operation are less well established and information is often lacking to identify their regions of predominance. Problems are further compounded by uncertainties in the variety of formulae proposed[2] to evaluate rates of strain. Some of these result from microstructural instabilities but even when the microstructure is carefully characterised, the ranges of applicability of equations are often ill-defined.

The role of grain boundaries can be crucial in relation to these features and the most distinctive are illustrated schematically in Fig. 1 where the left diagram shows the sliding of a grain boundary in a bi-crystal and the right hand diagram depicts the role of grain boundaries in acting as sources and sinks for vacancies.

The latter example provides the basis of the theory of diffusional creep proposed by Nabarro.[3] Evidence has accumulated to provide firm experimental support,[4] but doubts have been expressed[5] about the range of its validity and the predicted strain rates have not always been established.[6] This form of creep is also expected[7] to produce microstruc-

---

*The author is Emeritus Professor in the Department of Engineering Materials, University of Sheffield, Mappin Street, Sheffield S1 3JD, UK.

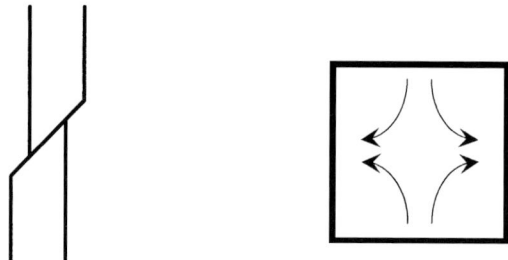

Fig. 1 Schematic illustration of two distinctive features of the behaviour of grain boundaries under low stress at elevated temperatures. The left diagram shows the sliding of a grain boundary in a bi-crystal. The diagram on the right shows the lines of vacancy flux, marked by arrows, when grain boundaries act as vacancy sources and sinks.

tural changes in alloys and discrepancies particularly between the occurrence of precipitate free zones and the expected form of mechanical behaviour have been pointed out.[8]

The resolution of these discrepancies is essential for it is now recognised that quite different internal stress patterns can be created[9] that lead to widely different predictions. Only recently has it been noted that observations of additional features in grain boundary regions can throw valuable light on atomic movements that allow clear distinctions to be made between deformation through bulk sliding and through processes involving directional diffusion.[10] Such observations provide the basis of the approach described here.

GRAIN BOUNDARY PROFILES

It is long established that chemical or thermal etching produces grooves in the surface of a material along lines where it is intersected by grain boundaries.[11, 12] The dihedral angle of these grooves has often been measured since it leads directly to a determination of the ratio between grain boundary and surface free energies.

Internally, since all grain boundaries in polycrystalline materials cannot meet each other at 120°, they are unstable and this results in grain growth when temperatures are high enough for significant atomic movement. A stable state can be maintained, however, at all temperatures where there are only grain boundary/surface intersections. This is most simply achieved in the creation of 'bamboo structures' in thin wires. After suitable annealing, grain boundaries lie predominantly perpendicular to the wire axis.

If a tensile stress is applied along the wire such grain boundaries are unable to slide since there is no shear stress acting upon them. At stresses above a critical level, slip bands are seen on the wire surface revealing the orientation of the slip plane when concurrent measurements are made of grain orientation. With increasing temperature, slip bands generally become more widely spaced.

Wires with such 'bamboo structures' under stresses too low to induce dislocation movememt were employed[13] to provide the early experimental support for the prediction by Nabarro of creep purely by a diffusional process with the deduction that grain boundaries act as vacancy sources. With the simple cylindrical grain geometry, the vacancy fluxes can be precisely evaluated and show remarkably close agreement between theory and experiment.

The application of the Nabarro equation to more complicated grain geometries and particularly to complex alloys, however, has often not shown a comparable level of agreement[14] nd has given rise to extensive debate. In this connection it is important to invoke microstructural observations to provide clues to the deformation mechanisms that are taking place. Some new observations will be described that throw light on this problem.

GRAIN BOUNDARIES IN MATERIALS UNDER STRESS

Considering first the 'bamboo structures' where close numerical agreement between theory and experiment is widely accepted, it may be noted that the thermal etching at the grain boundaries reduces slightly the grain boundary area. The vacancy emission from such an area is associated with a counterflow of atoms plating upon it. As illustrated in Fig. 2, this should result in a widening of the groove where the boundary intersects the surface. Microstructural observations and groove measurements clearly reveal this effect. Moreover, this feature may also be deduced from early work[15] on 'bamboo structures' (that was not commented upon at the time) simply from surface reflectivity. Figure 2 shows the substantially widened groove through the creation of a reflection zone within it. In situations such as this, there is close agreement with the Nabarro equation and correlation with atom deposition causing the effective extension confined to the grain boundary region.

Such illustrations provide striking evidence for the importance of making unambiguous distinctions between deformation arising diffusional processes and those that result from the consequences of sliding. In the former, it is noted that the entire deformation results from deposition and depletion of atoms in the grain boundary regions. This is of major significance in causing a localisation of strain that may be destructive of any applied surface coatings or of adjacent components in a composite material.

Before proceeding to examine this phenomenon in materials with more complex grain geometry, it is appropriate to consider, in a simple situation, how this mechanism may lead to grain shear displacement.

In wire specimens under axial stress with grains completely occupying the wire cross section, a shear stress acts on grain boundaries that are not perpendicular or parallel to the wire axis. A general observation is that of grain boundary sliding when an axial stress is applied. A sharp delineation is then apparent to mark the relative grain displacement as shown in the left hand diagram in Fig. 1. Such sliding, however, requires a shear stress above a certain threshold and the question of the form of behaviour then arises at stresses below this value. At these low levels for wire speci-

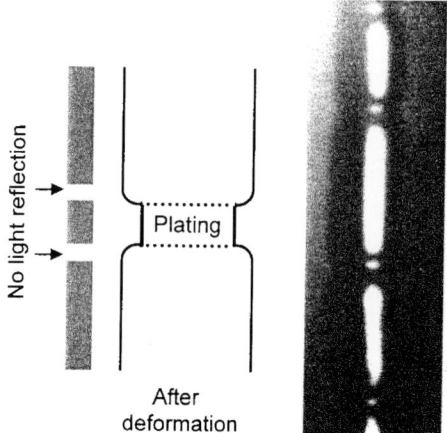

Fig. 2 The plating of atoms in the grain boundary regions of wires with bamboo structures after deformation under a small axial tensile stress can be observed through its effect on the reflectivity of the wire surface. Here the grain boundaries are acting as vacancy sources.

mens surface tension forces are significant. When they are sufficient to balance the applied stress then this represents a zero creep condition. At still axial lower stresses the surface tension forces become dominant and specimens tend to shrink axially and increase in radius to decrease their total surface energy.

It is instructive to observe, from the early work of Jones[15] the wire profile under such conditions in the region of a non-perpendicular grain boundary. This is illustrated in Fig. 3 where a displacement between grains on either side of the boundary is noted, that can be compared to that which would be caused by sliding under a compressive axial stress. The surface profile in Fig. 3, however, is different from that which would occur from grain boundary shear. There is no sharp surface step characteristic of such shear. In Fig. 3 it has occurred through vacancy diffusion to the boundary and absorption through the grain boundary acting here as a vacancy sink. The above observations are a further indication of two distinctive forms of grain displacement across a boundary as originally identified by Lifshitz.[16]

Another manifestation of these differences is apparent in the high temperature deformation of polycrystals. When displacement across grain boundaries takes place to accommodate diffusional processes, the number of grains intersected in the cross sectional area perpendicular to the direction of strain remains constant. When displacement occurs through a slipping process at the interfaces, the number of grains in the cross section progressively diminishes. Marker lines on the surface provide further information on these features and reference will be made to these later, in consideration of deformation modes in polycrystalline material.

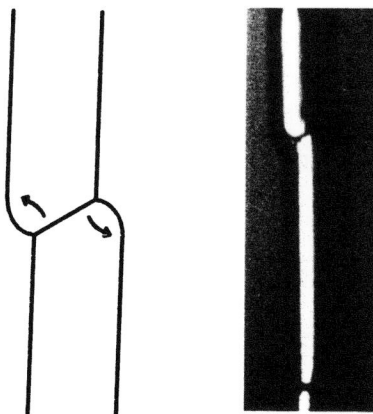

Fig. 3 The effect of an oblique grain boundary across a wire that is shrinking axially through the influence of surface tension. The arrows show the directions of atom flow from the grain boundary taking the shortest path towards the wire surface. The counter flow of vacancies reaches the grain boundary that acts here as a vacancy sink. This causes a shear displacement between the adjacent crystals that is caused entirely by diffusional flux in the manner identified by Lifshitz. Note the surface profile is curved and contrasts with the sharp delineation apparent in the left diagram in Fig. 1 where displacement occurs through sliding at the interface.

IDENTIFICATION OF DEFORMATION MODES IN POLYCRYSTALS

In the search for appropriate strengthening mechanisms, it is important to determine the mechanisms by which deformation takes place. Deformation mechanism maps[1] such as that illustrated schematically in Fig. 4 are available and used for this purpose but in many cases data are insufficient to locate where the boundaries lie that delineate areas of predominance of individual mechanisms.

The problem of mapping these areas precisely through mechanical testing is compounded by the various proposals of stress, strain rate and temperature relationships that continue to be put forward[2] so that unique identification is often inhibited. This is partly due to the simultaneous operation of different processes. Nevertheless, it is important to distinguish these, particularly across the heavily marked boundary in Fig.4 that separates the regions where dislocation glide plays a role from those where purely directional diffusive flow is occurring. Much of the importance of this distinction lies in the totally different patterns of stress and strain distributions that occur within each grain.[17] It will be apparent later that these differences have direct relevance to fracture mechanisms at elevated temperatures, to joining processes, to anisotropic behaviour and to radiation damage.

Because of the problems in deriving uniquely applicable mechanical relationships, it becomes especially important to assess mechanistic changes through microstruc-

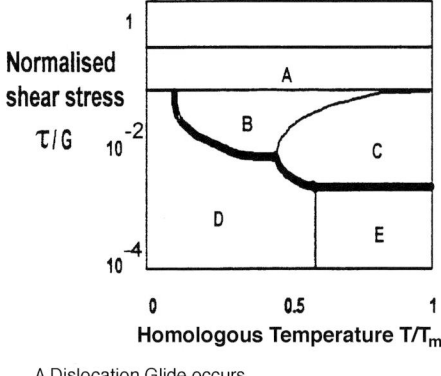

A Dislocation Glide occurs

B Dislocation Glide and Climb by Core Diffusion

C Dislocation Glide and Climb by Lattice Diffusion

D Creep by Grain Boundary Diffusion

E Creep by Lattice Diffusion

Fig. 4 The deformation mechanism map proposed by Ashby illustrates regions in which specific deformation mechanisms are expected to predominate. The heavily marked line separates regions B and C, where lattice dislocation movement occurs, from regions D and E where deformation is expected by diffusional processes.

tural observations. Here, the observation of surface profiles at grain boundaries come to our assistance.

Copper is known[4] to be a material for which the Nabarro formula for diffusional creep applies. It is seen, in Fig. 5, through scanning electron microscopy that the grain boundary grooves nearly perpendicular to the direction of strain are considerably broadened, in the manner expected when diffusional creep conditions are established. The atomic force microscope now provides a convenient and quantitative way of following the change in surface profiles and this has begun to provide more detailed information.[10]

It has long been considered[4] that, in alloys with dispersed precipitates, the creation of precipitate free zones adjacent to grain boundaries lying nearly perpendicular to the direction of strain provides microstructural support for the diffusional creep. Nevertheless, this evidence has been challenged,[18] since in some of these alloys, the Nabarro creep formula was not followed. The proposals of alternative mechanisms,, however, have proved difficult to sustain, especially in the interpretation of the concurrent effect of precipitate pile up on grain boundaries that lie nearly parallel to the strain direction. Nevertheless, such concerns require further examination..

A critical feature is the experimental evidence linking the occurrence of precipitate free zones with the changes in grain boundary profiles in alloys where the occurrence of diffusional creep has previously been in some doubt. Such evidence has

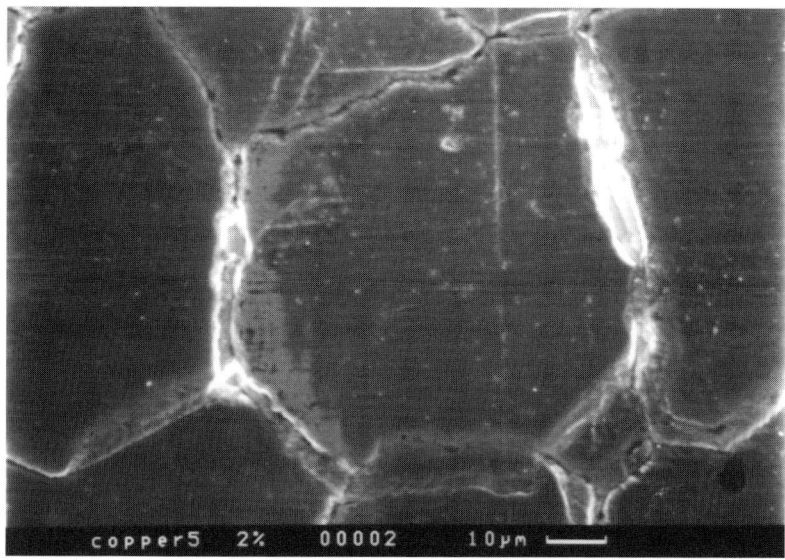

Fig. 5 A scanning electron micrograph of copper after 2% creep strain under a stess of 3.5 MPa at 500°C. The stress axis is horizontal and the widened profiles of grain boundaries nearly perpendicular to the stress are noted, indicative of atom plating on these boundaries through diffusional creep.

been obtained in recent work[10] showing clearly a connection between grain boundary profile changes and precipitate free zone formation, thus providing a firm basis for the identification of diffusional creep. The studies, moreover, have also resulted in a better understanding why the Nabarro equation is not always directly applicable. It results from not taking fully into account the effect of primary creep, microstructural instability in the alloys examined, changes that diffusional creep induces, inhibition of vacancy source and sink action and concurrent dislocation creep operation.

In this work, detailed study was made of the displacement of marker lines, shown schematically in Fig. 6, in the vicinity of grain boundaries. This confirms that diffusional creep processes are accompanied by grain boundary sliding but, as indicated previously, such sliding is an essentially accommodating process. As envisaged by Lifshitz,[16] this type of sliding may be considered to be controlled by diffusional creep and not occur independently of it, without need to modify the Nabarro formula. All the microstructural observations reveal that, when purely diffusional creep is taking place, no deformation occurs within the grains: deformation occurs only through atom plating or removal at grain boundaries, dependent upon their orientation.

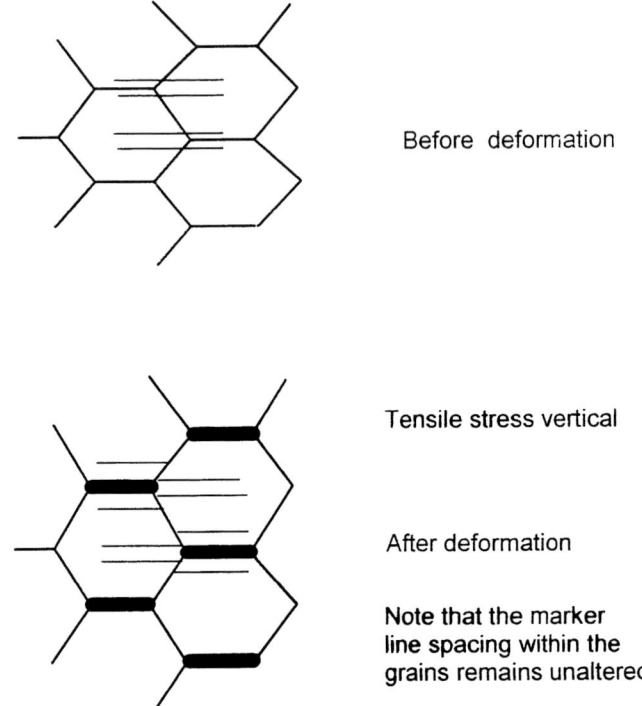

Fig. 6 A schematic illustration of the displacement of marker lines across grain boundaries that results from diffusional creep. Note that the grain displacement arises by diffusional processes in the form proposed by Lifshitz. All the deformation results from atom plating on grain boundaries nearly perpendicular to the tensile stress and atom removal at longitudinal boundaries transferred through the counterflow of vacancies.

INTERNAL STRESS REDISTRIBUTION IN DIFFUSIONAL CREEP

In the more familiar deformation modes at low temperatures involving dislocation motion, grain boundaries can act as barriers to cause impediment. This results in dislocation pile up and in a concentration of stress in the grain boundary regions. Grain boundary sliding is not necessary to maintain grain contact but at higher temperatures grain boundaries are able to slide with the result that the number of grains in the cross section perpendicular to elongation is progressively reduced. In diffusional creep the situation is entirely different. When grain boundaries are operating at high temperatures as vacancy sources and sinks to allow diffusional creep to occur, a stress pattern is developed[9] whereby the maximum shear stresses exist near the grain centres. Grain boundary sliding is then entirely controlled by the diffusional processes. This feature highlights the fundamental distinction between these two deformation mechanisms and illustrates the significance of the heavily marked line on the schematic deformation mechanism map in Fig. 4.

Such considerations assist in understanding the form of the transition between these mechanisms that occurs with increase or decrease in applied stress. Starting with a low stress where diffusional creep operates exclusively, the flux lines of vacancy flow can be readily determined, together with the stress patterns throughout the grains. As the applied stress is increased, a level will be reached at which the internal shear stress within the grain interior is sufficient to cause dislocation motion. The region for this will first be small, but will steadily grow as the applied stress is progressively increased. It will eventually occupy the entire grain, so that there is no region in which diffusional creep occurs. It is evident from this that there will be a range of stress over which dislocation and diffusional creep mechanisms are concurrently operative in different regions of each grain.

THE ROLE OF DIFFUSIONAL PROCESSES IN FRACTURE AT ELEVATED TEMPERATURES

Where grain boundaries are able to act as vacancy sources, they can assist in the growing of cavities when the stress is sufficient for the cavities to become vacancy sinks. Here the predominant vacancy flow may be through the grain boundaries on which the cavities are situated. Such a mechanism was first analysed by Hull and Rimmer[19] and provided clear quantitative predictions. As with the studies of diffusional creep processes previously discussed, such predictions are not always in accordance with experimental results.

Several reasons have been proposed for such discrepancies. Thermodynamic requirements demand pre-existence or creation of cavity nuclei The activation of these depends upon the applied stress and may also be influenced by strain such that their number increases during deformation.[20] Segregation of elements to grain boundaries can affect behaviour. Grain boundaries may not be able to act as vacancy sources and these aspectst must be resolved before the applicability of theoretical proposals is established.

Here microstructural observations at grain boundaries can provide clues to the mechanism of cavity growth. Some valuable observations have been made of precipitate free zones adjacent to the boundaries on which cavities are situated, indicative of a diffusional process for cavity growth. The determination of the widening of the profiles at such boundaries can add further support to this interpretation but work in this area is at an early stage.

As with the case of diffusional and dislocation creep, the contribution of vacancy fluxes to cavity growth may be simultaneously assisted by dislocation glide and climb processes, each operating in their respective regions[21] with the vacancy fluxes confined to grain boundary areas increasingly close to the cavities as the stress is increased.

THE FINAL STAGES OF SINTERING

In reversal of the conditions of cavity growth, in sintering where decrease of internal surface area provides a driving force or when hydrostatic compression is imposed as in HIPping, vacancy fluxes can arise in a reversed direction to cause shrinkage and finally complete elimination of residual pores.[22] Here, the grain boundaries are required to act as vacancy sinks.

The microstructural evidence for such grain boundary operation has long been available in the observation of pore distribution in relation to the boundaries. The pores situated on grain boundaries are subject to preferential rates of shrinkage. When they disappear entirely the grain boundaries are no longer pinned by their presence and so they are more free to migrate. Their migration continues until they become attached to a further set of pores that were initially in the grain interiors. So the shrinkage process continues with no pores in the regions over which the grain boundaries have swept. Eventually, all regions are covered by the migrating boundaries are all the pores disappear.

VACANCY SOURCE DETECTION BY FOLLOWING α-PARTICLE BOMBARDMENT

A further example of the influence of grain boundaries on atomic movements can be observed if α-particles are accelerated and injected into a target material in which helium is formed from electron capture. From such bombardment of copper it was first shown[23] that the helium atoms can acquire vacancies to form gas bubbles in regions adjacent to the grain boundaries from which vacancies can emerge. This provides a further clear indication of the ability of grain boundaries to act as vacancy sources with vacancies fluxes causing bubble growth. An additional significant observation was of the inability of coherent twins to act in a similar way. This has led to recent closer examination of the influence of interfacial structure on vacany source and sink action with the observation that some grain boundaries close to an exact coincident site lattice orientation are inactive.[24]

There remains continuing scope for experiments of this kind in a wide range of materials, for the technique reveals definitive evidence of grain boundary action in vacancy provision. From the other examples discussed, it is apparent that such action is central to the understanding of a wide range of behaviour.

CONCLUSIONS

The observations of changes in surface profiles at grain boundaries dependent upon their orientation with respect to the applied stress, the displacement of marker lines, the absence of precipitates in regions adjacent to grain boundaries nearly perpendicular to the direction of elongation and precipitate pile up on longitudinal boundaries all reinforce the evidence for identification of the operation of diffusional creep mechanisms. It is clear that such mechanisms create creep strain by atom deposition and depletion exclusively in grain boundary regions, leaving undeformed the material within grains. This illustrates an important distinction between deformation resulting from atomic movements of this kind from those of where the strain is accumulated from dislocation motion throughout the entire volume.

ACKNOWLEDGEMENTS

The author is grateful to Prof. H. Jones and Dr K. R. McNee for useful discussions.

REFERENCES

1. M. F. Ashby, *Acta Metall.*, 1972, **20**, 887.
2. J. Cadek, *Creep in Metallic Materials*, Academia, 1988.
3. F. R. N. Nabarro, *Rep. on Conf. on Strength of Solids*, The Phys. Soc., 1948, 75.
4. B. Burton, *Diffusional Creep in Polycrystalline Materials*, Trans Tech Publications, 1977.
5. O. A. Ruano, J. Wadsworth, J. Wolfenstine and O. D. Sherby, *Scr. Metall. Mater.*, 1993, **29**, 515.
6. O. A. Ruano, J. Wadsworth, J. Wolfenstine and O. D. Sherby, *Mater. Sci. Eng.*, 1993, **A165**, 133.
7. R. L. Squires, R. T. Weiner and M. Phillips, *J. Nucl. Mater.*, 1963, **8**, 77.
8. J. Wolfenstine, O. A. Ruano, J. Wadsworth and O. D. Sherby, *Scripta. Metall.*, 1994, **30**, 383.
9. G. W. Greenwood, *Res Mechanica*, 1985, **14**, 61.
10. K. R. McNee, H. Jones and G. W. Greenwood, in *Creep Behaviour of Advanced Materials for the 21st Century*, Rajiv S. Mishra, Amiya K. Mukerjee and K. Linga Murty eds, TMS, 1999, 481.
11. W. W. Mullins, *J. Appl. Phys.*, 1957, **28**, 333.
12. P. Tritscher, *Proc. R. Soc. Lond.*, 1999, **A455**, 1957.
13. A. P. Greenough, *Philos. Mag.*, 1952, **43**, 1075.
14. J. E. Harris, R. B. Jones, G. W. Greenwood and M. J. Ward, *J. Austral. Inst. Met.*, 1969, **14**, 154.
15. H. Jones, Ph.D. Thesis, University of Manchester, UK, 1965.
16. I. M. Lifshitz, *Soviet Physics*, 1963, **17**, 909.
17. G. W. Greenwood, *Philos. Mag.*, 1985, **A51**, 537.
18. O. A. Ruano, J. Wadsworth and O. D. Sherby, *Acta Metall.*, 1988, **36**, 1117.
19. D. Hull and D. E. Rimmer, *Philos. Mag.*, 1959, **4**, 673.
20. G. W. Greenwood, *Scripta Met.*, 1970, **4**, 171.
21. W. Beere and M. V. Speight, *Met. Sci.*, 1978, **18**, 172.
22. H. V. Atkinson and B. A. Rickinson, *Hot Isostatic Processing*, Adam Hilger, 1991.
23. R. S. Barnes, G. B. Redding, and A. H. Cottrell, *Philos. Mag.*, 1958, **3**, 97.
24. J. B. Bilde-Sorenson and P. A. Thorsen, *Creep Behaviour of Advanced Materials for the 21st Century*, Rajiv S. Mishra, Amiya K. Mukerjee and K. Linga Murty eds, TMS, 1999, 441.

Nickel–Boron Cosegregation at Grain Boundaries in Boron-doped γ/γ' Nickel–Aluminium Superalloys

A. H. W. NGAN AND Y. L. CHIU

Department of Mechanical Engineering, The University of Hong Kong, Pokfulam Road, Hong Kong, P.R. China

ABSTRACT

Boron-doping can toughen nickel–aluminium γ/γ' superalloys in a manner that is analogous to the known boron effects in nickel-rich Ni_3Al. The toughening is a result of the formation of a well-defined nickel-rich grain boundary zone at grain boundaries, which also improves slip transmission across grains. Fundamental differences with the Ni_3Al situation are highlighted, and some general conditions for cosegregation-induced toughening of intermetallic materials are discussed.

INTRODUCTION

The notion of cosegregation of two or more elements at defect surfaces in multi-component alloy systems has been receiving much attention in the literature because cosegregation may give rise to important effects in properties. The most well-known example is temper embrittlement of low alloy steels. In such steels, alloying elements such as Ni, Cr, Mn etc may cosegregate with embrittling elements like Sb, P, As or Sn at grain boundaries, causing the toughness of the material to decrease. The cosegregation isothermal kinetics are found to be driving-force-limited at high temperatures and diffusion-limited at low temperatures, so that the waiting time for embrittlement to appear would vary with the annealing temperature according to a C-shape curve. The diminishing driving force at high temperatures is due to configurational entropy effects which favour desegregation at high temperatures.

In low alloy steels, the consequence of grain boundary cosegregation is to *embrittle*, but in another equally celebrated example, the effect is to *toughen*. This concerns the ductilisation of Ni_3Al by boron doping – a phenomenon that once revitalised intermetallics research in the 1980s and one in which Professor Ray Smallman has established a long-term interest.

BORON-INDUCED TOUGHENING IN NI₃AL

Several pieces of key evidence have now been established to support the theory that the boron ductilisation effect in Ni_3Al is caused by boron–nickel cosegregation at grain boundaries. First, only a tiny amount of boron, about 0.02 wt-% depending on the grain size, is found sufficient to cause tremendous improvement in ductility. Further ductility gain on increasing boron content is very marginal and in fact, the ductility suffers a slight drop at higher boron levels exceeding, say 0.1 wt-%. The small quantity of boron required rules out any toughening mechanisms based on general slip, since the average boron concentration would have been negligibly small if the boron were to distribute uniformly. The ductilisation effect must therefore be owing to some segregation behaviour, and this is indeed confirmed directly by Auger electron spectroscopy investigations, which showed that boron tends to segregate to grain boundaries in Ni_3Al rather than to free surfaces. Secondly, for the ductilisation to work, the Ni_3Al must be nickel-rich. Liu *et al.* have found that the ductility of boron doped Ni_3Al remains at the undoped low level when the nickel content is smaller than stoichiometry, but as stoichiometry is exceeded, the ductility increases steadily.[1] This suggests that free nickel rather than free aluminium must be present for boron to do its job. More direct evidence on boron–nickel cosegregation is also available but the situation is controversial. It was reported in one study that direct TEM imaging had revealed a disordered phase about 20 nm wide along grain boundaries in ball-milled Ni_3Al alloys.[2] No such disordered phase, however, was observed in conventionally case Ni_3Al alloys,[3] but this may be because the cosegregated layer is too thin to be observable. The monolayer-scale segregation of boron at grain boundaries in Ni_3Al has been confirmed by atom probe spectroscopy in the field-ion microscope.[4]

While it is quite certain that boron does segregate to grain boundaries, and it is likely that nickel may cosegregate with it, the exact role of boron and nickel with regard to ductilisation remains controversial. In one earlier mechanism supported by more recent atomistic simulations,[5–7] boron is believed to possess the right chemistry to decrease the grain boundary energy relative to the free surface energy, thus improving the grain boundary cohesive strength. Another mechanism is based on the observation that the Hall–Petch slope decreases upon boron doping.[8] This reflects that the effectiveness of grain boundaries in hindering intergranular slip is reduced by boron doping, and the reason for this is conjectured to be owing to the formation of a disordered cosegregated layer at the grain boundaries. A third group of mechanisms, which is also supported by compelling evidences, focuses on the ability of boron in suppressing environmental embrittlement in moist environments.[9, 10] The role of boron is believed to either suppress the dissociation of H_2O into hydrogen which is the embrittling reagent, or inhibit the diffusion of hydrogen along grain boundaries.

Since the above ductilisation mechanisms are all supported by hard experimental facts or simulations, and they are not mutually inclusive or exclusive, nor are they controllable by independent factors, it remains a very difficult, if not impossible task to attempt to resolve the discrepancies among them. The established evidences on boron–nickel cosegregation in Ni_3Al nevertheless pose an interesting question that has never been addressed in the literature before. As stated above, the richer the Ni_3Al is in nickel, the stronger the ductilisation effect of boron. This leads to the question of what will happen to the cosegregation effects if the nickel content exceeds the single phase γ regime and enters the two-phase γ/γ′ regime.

BORON-NICKEL COSEGREGATION IN γ/γ′

To answer the above question, the authors have recently performed a series of experiments to investigate the effects of boron doping on γ/γ′ superalloys. The full account of the investigation will be published elsewhere and here only the main findings will be presented. The alloy system studied contained only nickel and aluminium as the metallic species with composition $Ni_{85}Al_{15}$. One such ingot was doped with ~0.5 at.-% of boron, while another was kept boron free as a control. It was discovered that after prolonged heat treatment at 1100°C, the doped alloy exhibited a grain boundary structure distinctly different from the undoped control. This is illustrated in Fig. 1, which shows that in the boron-doped alloy, the grain boundaries are decorated by a zone much richer in nickel content. Along such grain boundaries are scattered large γ′ particles, which are conjectured to be the source of nickel, or sink of aluminium, for the formation of the nickel-rich zone. In the undoped control, the grain boundaries remain sharp without any clear sign of nickel segregation.

In a rather indirect experiment, the strong affinity between nickel and boron in the γ/γ′ system was clearly demonstrated. This involved heat-treating a piece of undoped γ/γ′ alloy while in contact with some boron powder in a vacuum furnace at 1100°C. The result is illustrated in Fig. 2, which shows that boron after diffusing into the alloy formed a new phase significantly richer in nickel than the alloy composition. At the outskirts of the nickel-rich phase lied a nickel-depleted region with much reduced γ′ volume fraction as compared with the surrounding matrix. The morphology here

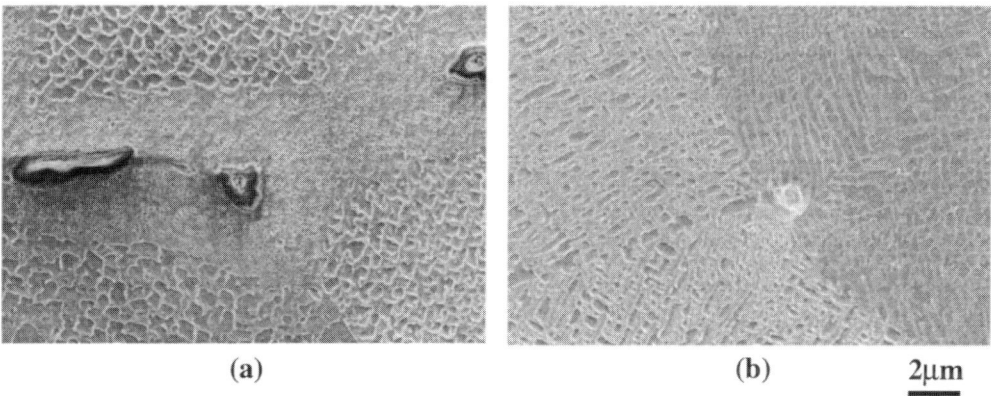

(a) (b) 2μm

Fig. 1 Grain boundary microstructures of (a) boron-doped and (b) undoped $Ni_{85}Al_{15}$ γ/γ′ superalloys after heat-treatment at 1100°C for 48 hours.

suggests the extraction of nickel from what eventually ended up as the nickel-depleted zone towards the surface region. Since in this experiment, it is definitely known there was more boron at the surface region, the observation clearly indicates a much stronger affinity between boron and nickel relative to that between boron and aluminium in the γ/γ' system.

BORON-INDUCED TOUGHENING IN γ/γ'

The boron-doped alloy above was found to be significantly tougher than the undoped control. Table 1 shows the various mechanical properties of the two alloys. The toughness elevation by boron-doping is consistent with fractography observations shown in Fig. 3, from which it is obvious that the doped sample fractured transgranularly while the undoped one fractured intergranularly. Interesting to note is the significant reduction in work-hardening rate upon boron doping, and this is also consistent with slip trace observation. Figure 4 shows the deformed states of the free surfaces of the two alloys. Severe accumulation of slip at grain boundaries leading eventually to intergranular cracks can be seen to occur on the undoped surface. In the doped sample, the slip traces are much more continuous across grains, indicating that intergranular slip transmission is much easier.

The formation of a well-defined cosegregated zone at grain boundaries is expected to retard the grain growth rate. This is indeed found to be the case in boron-doped γ/γ'. Figure 5 shows the recrystallisation and grain growth kinetics of the doped alloy and the control at 1200°C following cold-rolling to 80% reduction. The grain growth rate of the boron-doped alloy is found to be significantly slower than the boron-free alloy, indicating that the presence of boron will retard the grain growth kinetics of γ/γ'.

In such a grain growth experiment is implied an interesting point, namely the interrelation between grain growth and segregation. Since the overall doping level of

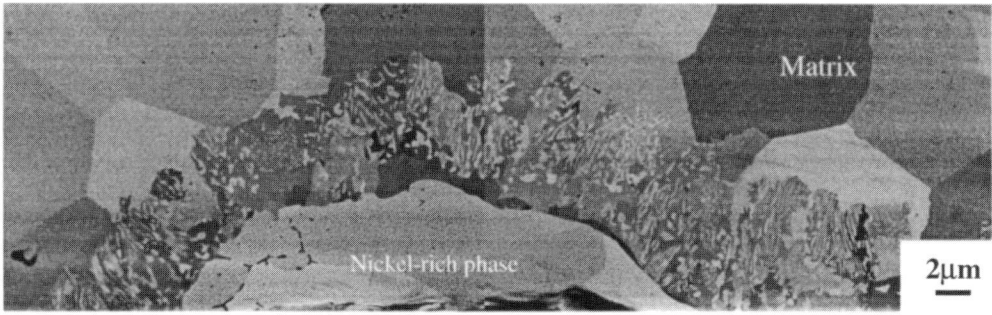

Fig. 2 A nickel-rich phase formed at the surface in contact with boron powder after heat-treatment in vacuum at 1100°C for 48 hours.

Table 1 Mechanical properties of boron-free and boron-doped γ/γ′ alloy.

	Ni$_{85}$Al$_{15}$(B)	Ni$_{85}$Al$_{15}$
Charpy impact energy absorption	10×10^5 N m^{-1}	1.8×10^5 N m^{-1}
0.2% proof stress	400 MPa	380 MPa
Work hardening rate at 1% strain	1949 MPa	5374 MPa
Hardness	212 Hv	233 Hv

boron is held fixed, the equilibrium concentration of boron at grain boundaries will increase with the grain size. Thus at the early stage of grain growth, the grain boundary boron concentration is very low, so that no well-defined cosegregated zone should be expected to form. To investigate this the authors have carefully noted the grain boundary microstructure development as the grain grew in the experiment. As expected, it was found that no well-defined grain boundary zones of the type shown in Fig. 1a for short to intermediate annealing times in the boron-doped alloy. The zone, however, appeared only after prolonged treatment for ~165 hours at 1200°C. The occurrence of the grain boundary zone after prolonged anneal corresponded well to a sudden increase in the ductility and a sudden drop in the hardness of the doped alloy as shown in Fig. 6. The grain boundary microstructural observation and the property jump in Fig. 6 suggests a sudden speed up of the cosegregation kinetics at a later stage of cosegregation. The interesting point is that such a sudden speed up of kinetics happened when the grain size had more or less been stabilised (cf Figs. 5 and 6), so that it cannot be explained solely by an increase in grain size. Such as sudden speed up in cosegregation kinetics is indeed in line with the calculations by Tyson,[11] who has shown that a sudden rise in kinetics is quite common during the later stage of co-segregation in a ternary system. In fact, the jump in mechanical properties in Fig. 6 is analogous to the phenomenon of temper embrittlement in low alloy steels mentioned in the introduction. The kinetics of temper embrittlement in low alloy steels are found to exhibit a time-delay effect during isothermal heat treatment and

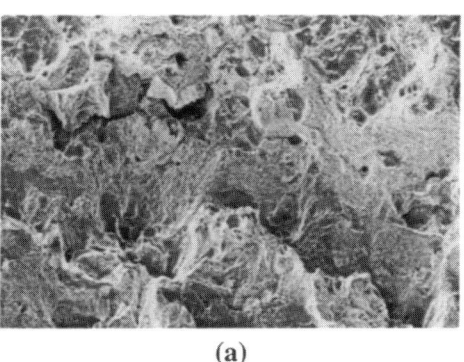

(a) (b) 2μm

Fig. 3 Fractography of (a) boron-doped and (b) boron-free γ/γ′.

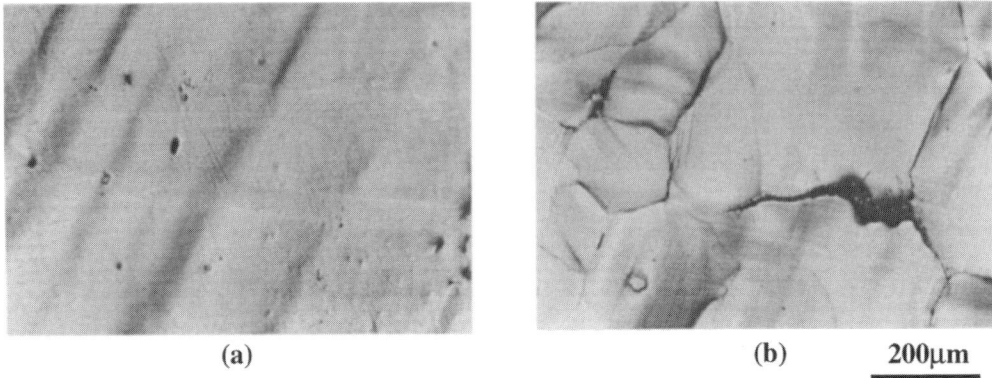

(a) (b) 200µm

Fig. 4 Free surface deformation marks in (a) boron-doped and (b) boron-free γ/γ'. Overall deformation was ~7% compression for both samples.

this can be characterised by a C-shape TTT curve. Current work is in progress to investigate the TTT behaviour in boron-induced toughening in γ/γ'. The results in Fig. 6 nevertheless indicate that, for a fixed overall impurity level, cosegregation induced toughening works when the grain size is large enough. This leads to the significant point that, if beneficial toughening impurities can be found, coarse-grained and hence more creep resistant alloys can be conveniently toughened this way.

COMPARISON OF BORON-EFFECTS IN NI₃AL AND γ/γ'

The observed effects of the nickel-rich grain boundary zone on toughening and ductilisation of boron-doped γ/γ' alloys are very similar to those reported for the grain boundary disordered phase in Ni_3Al(B) alloys. The amelioration of intergranular slip transfer as observed in Fig. 4 and the lowering of work-hardening rate upon doping in γ/γ' suggest that the nickel-rich zone facilitates dislocation transmission across grain boundaries in a similar way as that suggested for boron-doped Ni_3Al.[12, 13] There are, however, fundamental differences between the boron effects in Ni_3Al and in γ/γ'. First, Choudhury *et al.*[14, 15] observed that in Ni_3Al, boron desegregated from grain boundaries at high temperatures, say 1050°C. In our investigation, the retarded grain growth behaviour in Fig. 5 and the direct observation of the nickel-rich grain boundary zone upon prolonged heating at 1200°C indicate boron segregation at grain boundaries at 1200°C rather than desegregation. It should be noted that 1200°C is already $0.9T_m$ of the alloy (T_m = absolute melting temperature), and hence the present results suggest the likelihood of no desegregation before melting in boron-doped γ/γ'. The second difference with Ni_3Al concerns the width of the nickel-rich grain boundary zone in Fig. 1a. In the present experiment, such a zone in γ/γ' is found to be several micrometres wide, but that reported for Ni_3Al is only about 20 nm wide

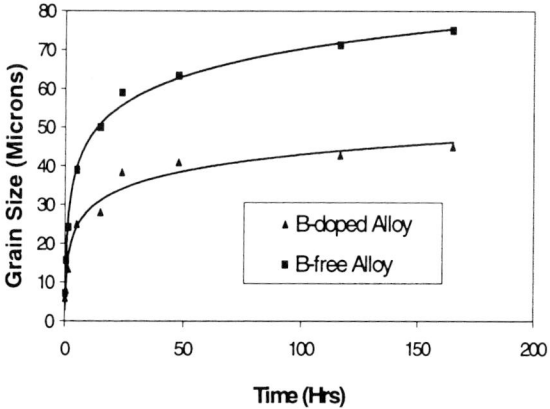

Fig. 5 Grain size *v*. annealing time of the boron-doped and boron-free γ/γ′ at 1200°C after cold-rolling by 80%.

in ball-milled Ni_3Al,[2] or atom probe field-ion microscopy has confirmed that boron segregation in Ni_3Al is monolayer wide.[4] In other words, the cosegregated zone in our γ/γ′ is at least two orders of magnitude wider than the widest disordered zone ever reported in Ni_3Al.

A more thorough understanding of the mechanisms controlling the zone width and its composition is desirable because the zone directly controls the grain boundary cohesive strength and hence the toughness of the material. In the literature, segregation or cosegregation is often discussed by reference to two contexts, ie equilibrium and non-equilibrium segregation. Equilibrium segregation, also called reversible

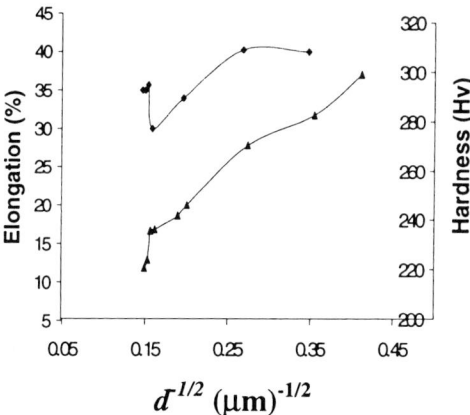

Fig. 6 Tensile elongation (upper curve) and hardness (lower curve) *v*. grain size of γ/γ′ doped with 0.5 at.-% boron.

segregation, exhibits two characteristics. The first is that the segregation can be made reverse, ie to desegregate, by heating at a high temperature. This is due simply to entropy effects. Secondly, the driving force for equilibrium segregation is a reduction in lattice misfit or bond energies by locating the impurity at the grain boundary instead of in the bulk. Hence, the equilibrium segregation zone must be one to at most a few monolayers thick. It is therefore clear that the situation in boron-doped Ni_3Al exhibits both of these characteristics, and hence boron segregation in Ni_3Al may be classified as the reversible equilibrium type. On the other hand, observations in boron-doped γ/γ' indicate neither of these characteristics. The observed segregation at $0.9T_m$ indicates no sign of reversibility, and the persisted appearance of the micrometre-wide grain boundary zone after prolonged heat treatment indicates that boron segregation in this material system does not equilibrate into a monolayer at the grain boundary.

As reviewed by Hondros *et al.*,[16] micrometre-wide segregated zones can be produced in certain material systems by quenching, and this is usually referred to as the non-equilibrium type of segregation in the literature. Quenching produces excess vacancies in the bulk which then diffuse towards the grain boundaries where they are annihilated. If the impurity atoms form complexes with the vacancies, and if the complexes diffuse faster than the vacancies and impurity atoms, the drop of the vacancy concentration in the grain boundary vicinity will attract complexes in the bulk to diffuse towards the grain boundary, resulting in net impurity enrichment there.[17] The enrichment is non-equilibrium in the sense that it depends critically on the cooling rate. If the cooling rate is too slow, the vacancies and the impurity atoms will homogenise everywhere, causing no segregation. If the cooling rate is too fast, the impurity-vacancy complexes have not time to diffuse to the grain boundaries. The enrichment should also disappear upon re-heating. In the grain growth experiments on γ/γ', the recrystallisation heat treatment was conducted at $0.9T_m$, and the nickel enrichment was observed only after prolonged heating, ie after ~165 hours. It is thus inconceivable that the nickel enrichment, which results from heating at such a high temperature and for such a long time, would be non-equilibrium in nature. Also, since all samples were water-quenched after the heat treatments, the chance that the enrichment zone was formed during cooling is very slim. The present results therefore indicate that the boron–nickel cosegregation in γ/γ' is not the usual non-equilibrium type as discussed in the literature.

The strong boron–nickel coupling in the Ni–Al–B system is best interpreted within a cosegregation theory for a ternary system. Such a theory has been proposed by Guttmann.[18] A basic assumption of the Guttmann theory is that the segregated layer is a two-dimensional phase for which the usual thermodynamic variable *pressure* for a bulk solid has become meaningless and has to be replaced by *surface tension*. Notably, this theory breaks down when the cosegregated layer is thicker than what would be meaningful for the concept of *surface tension*, ie at most a few layers of atoms. The observed micrometre-sized thickness of the grain boundary zone in boron-doped γ/γ' in the present work is obviously not compatible with the concept of a two-dimensional phase; in other words, an interpretation is needed that is beyond the two-dimensional phase assumption. Here, an earlier isotherm developed by Seah and Hondros[19] may seem useful, although it has to be admitted that such an isotherm has not been applied or tested to the extreme regime of micrometre-sized segregated zone. The isotherm was developed by making analogy with the BET theory for multilayer gas adsorption, and it states that

$$\frac{X_{bo}}{X_b} \frac{X_c}{X_{co} - X_c} = \frac{1}{K} + \frac{K-1}{K} \frac{X_c}{X_{co}} \tag{1}$$

where X_{bo} is the quantity of segregation sites available per unit area of one monolayer, X_b is the quantity of impurity atoms segregated in unit area of grain boundary surface, X_c and X_{co} are the concentration and solubility limit of the impurity in bulk respective. $K = \exp(E/RT)$, where E is the heat of segregation. This isotherm can predict finite segregation thickness, i.e. $X_b > X_{bo}$, but it takes no account of cosegregation in a ternary system. Let us assume for simplicity its validity in the present case, perhaps after introducing the interactions between element species in E as was done by Guttmann in his two-dimensional phase theory. From equation (1), the number of segregated layers is given by

$$\frac{X_b}{X_{bo}} = \frac{p}{1-p} \frac{1}{\frac{1-p}{K} + p} < \frac{1}{1-p} \text{ where } p = \frac{X_c}{X_{c_o}} \leqslant 1 \tag{2}$$

The number of segregated layers turns out to be rather insensitive to the expected variance in K (about 1 to 20) at any constant p. The segregated zone remains a few monolayers thick for unsaturated impurity levels $p < 1$, but a micrometre-wide segregation zone can be produced by a p value very close to 1, or a bulk impurity level very close to or already at the solubility limit. In other words, in terms of Seah and Hondros' isotherm, the presently observed nickel-rich zone is best viewed as a region of saturated solid solution, or even precipitation of a new phase, owing to a saturated level of boron. This implies that the 0.5 at.-% doping level has already reached the solubility limit in the bulk. The solubility of boron in γ/γ' is not very well-known, but as an indication, the solubility limit of B in Ni_3Al is estimated to be at least 0.93 at.-% in Ni–24 at.-%Al at room temperature,[1] or up to 1.12 at.-% in stoichiometric Ni_3Al,[20] while that in Ni is only 0.3 at.-% at 1093°C.[21] Assuming a 50:50 volume fraction ration of γ and γ', the solubility limit of B in our γ/γ' alloy is likely to be about 0.6 at.-%, which, taking into account the crude estimates of the solubility in the pure phases, is very close to the 0.5 at.-% doping level in the experiment. The boron doping level may thus correspond to saturation, and hence the micrometre-wide nickel-rich zone can be reconciled with equation (2). In the case of $Ni_3Al(B)$ for which the boron solubility is around 1 at.-%, a comparable doping level, say 0.5 at.-%, would yield a p value significantly smaller than unity, say 0.5. At such an unsaturated impurity level, the thickness of the segregated zone calculated from equation (2) is of the order of a few monolayers. The small observed size of the cosegregated zone in boron-doped Ni_3Al may thus be explained this way.

The phenomenological Seah and Hondros isotherm unfortunately does not provide enough physics about the saturation, and for this purpose, we need to look at the phase equilibria of the problem. Figure 7 shows the 1000°C-section of the Ni–Al–B ternary phase diagram. It can be seen that as the boron content exceeds the γ–γ' two-phase field, the composition will enter the γ–γ'–τ three-phase field. The τ phase has a cubic unit cell which is almost exactly three times that of the γ or γ' fcc unit cells (see Table 2), and hence it may well be coherent with γ and γ'. τ has the composition $Ni_{20}Al_3B_6$, ie richer in nickel than γ'. With this information, it may therefore be conjectured that the nickel rich zone seen in, for example, Fig. 1a was a region consisting of a distribution of fully coherent clusters of the τ phase in a γ/γ' matrix. This implies that the 0.5 at.-% boron content in the experiment either lies close to or has already passed the dividing line between the γ–γ' two-phase and the γ–γ'–τ three-phase fields. Homogeneous nucleation of τ in the bulk could not occur as yet because

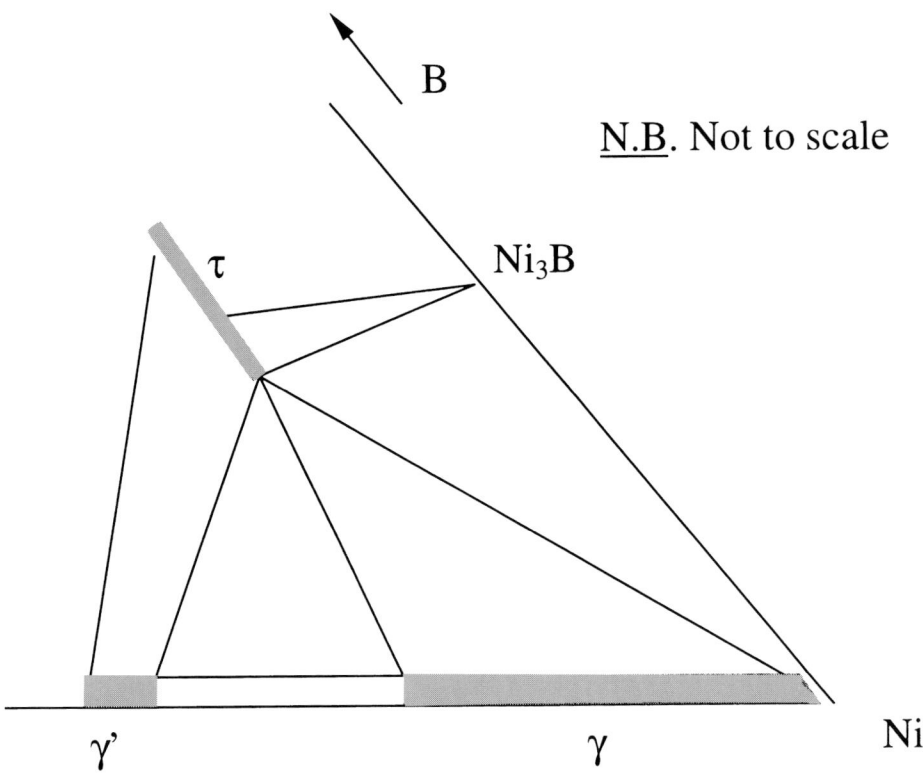

Fig. 7 Ni-rich section of the Ni–Al–B phase diagram at 1000°C (adapted from Ref. 22).

Table 2 Crystallographic data

Phase	Structure	Lattice constant
γ	cF4	0.352 nm
γ'	cP4	0.357 nm
τ	cF116	1.048 nm

of the limited driving force, but heterogeneous nucleation at grain boundaries could be enhanced for a number of possible reasons. First, the lattice strain as a result of nucleation may be relieved partially at grain boundaries. Secondly, the number of Ni and Al atoms per unit volume in the τ unit cell is only about 20% of that in γ or γ'. Nucleation of τ phase may therefore be favoured at a region with more vacancies. At a high temperature such as $0.9T_m$, owing to preferential thermal agitation, the formation energy of vacancies in the grain boundary vicinity, especially the first few monolayers, may be significantly lower than that in the bulk. Hence a higher equilibrium vacancy concentration and easier nucleation of the τ phase in the grain boundary vicinity may be expected. Thirdly, boron segregation and perhaps nickel cosegregation of the reversible type can be expected to have already occurred at the first few monolayers of the grain boundary. As the nucleation of τ requires clustering of boron and nickel atoms, the already formed cosegregated zone would provide ideal nucleation sites. Equilibrium cosegregation of boron and nickel at the first few monolayers may therefore be a prerequisite of the formation of the τ phase. The kinetics of the nickel-rich zone formation may even be controlled by the kinetics of the monolayer equilibrium segregation.

GENERAL CONDITIONS FOR COSEGREGATION-INDUCED TOUGHENING

Having seen that the nickel rich zone in boron-doped γ/γ' can toughen the material significantly, let us come to the question of whether other intermetallic base systems can be similarly toughened by cosegregation of a coherent phase. In the $Ni_3Al(B)$ system, there is a vast amount of first principles calculations devoted to predicting the segregation of boron and its cosegregation with nickel (eg Refs. 5–7), but the tendency for boron and nickel to cosegregate in boron-doped Ni_3Al or γ/γ' can be understood in a much simpler way by comparing the binary phase diagrams for Ni–B and Al–B. In the Ni–B system, nickel and boron have a strong tendency to form compounds, line phases like NiB, Ni_4B_3, Ni_2B and Ni_3B all exist, and their melting temperatures are all above 1000°C, indicating their high stability. In the Al–B system, however, there is no line phase of the form Al_nB with $n \geq 1$. The two phases AlB_2 and AlB_{12} do exist but these require much higher content of boron, which may not be available in microalloying. It may therefore be conjectured in a hand-waving

125

manner that in the Ni–Al–B system, Ni–B bonds are more stable than Al–B ones, and hence boron will tend to cosegregate with nickel rather than with aluminium.

Summing up the experience in the Ni–Al–B system, several conditions have to be met for a general binary system A–B to be toughened by impurity X though cosegregation effects.

(i) There should be a large difference in compound formation tendency or general solubility between A–X and B–X. Since X is available in a minor quantity in microalloying, the compounds that are relevant should be rich in metallic content, eg Ni_3B is a useful indicator for Ni–B cosegregation in boron-doped Ni_3Al but AlB_{12} is less relevant as an indicator for Al–B cosegregation.

(ii) Suppose (i) above is met, say A–X has much stronger tendency to form solid solution or compounds than B–X, then the A–X phases should have comparable melting points with the bulk phase A–B. Otherwise, X may fail to draw out excess A from the bulk phase to form the cosegregated phase.

(iii) The element to be cosegregated with the impurity X, eg A, should form a grain boundary phase with X which can improve the grain boundary cohesion of the bulk phase. The A–B alloy to be toughened should also be richer in A relative to stoichiometry.

This view of cosegregation is similar to but is broader than an earlier one by Briant,[23] who viewed segregation or cosegregation as being a result of the tendency for an impurity species to form compounds with structural units that are only available at the grain boundary layer owing to the special CSL structure there. Briant's view is only pertinent to monolayer segregation/cosegregation. For multilayer cosegregation, the suggestion here is that the tendency to form a large number of A_nX phases but no B_nX phases with $n \geq 1$ would serve as an indicator for strong A–X cosegregation in the A–B–X system to form an extended coherent phase at the grain boundary vicinity.

An analysis of the toughening potential of a number of $L1_2$ systems by microalloying in terms of the relative phase stability is presented in Table 3, which shows the melting or decomposition temperatures of the binary phases A_3B, A–X and B–X. It can be seen that those material systems that can be ductilised by microalloying in general exhibit very good ability to form compounds or extended solutions between the dopant and the major metallic element. As an example, Ni and Pd are completely miscible into an fcc solution coherent with the Ni_3Al and the melting temperature of Ni–Pd is 1237°C, which is comparable to the melting temperature of 1395°C of Ni_3Al. On the other hand, Pd only forms the Al_4Pd and $Al_{21}Pd_8$ phases with melting temperatures 580°C and 790°C respectively, which are much lower than that for Ni_3Al. Thus, Ni and Pd have good potential to cosegregate at Ni_3Al grain boundaries to form a coherent phase there. In fact, Pd-doping has been found to be able to toughen Ni_3Al.[28] As a second example, it has been shown that carbon cannot toughen Ni_3Al. In the Ni–C phase diagram, Ni and C are immiscible, indicating that C has low affinity for Ni. Although C forms the very stable Al_4C_3 phase with Al, extracting Al out

Table 3 Toughening of $L1_2$ compounds by microalloying

Bulk phase A_3B	Dopant X	A_nX (T_m)	B_nX (T_m)	Comments	Experimental Observation[34,35]
Ni_3Al (1395°C)	B	Ni_3B (1156°C; Ni_2B (1125°C); Ni_4B_3 (1031°C); NiB (1035°C)	No Al_nB compound with $n \geq 1$	Good potential for Ni–B cosegregation	Ductility improved[24]
	C	Immiscible; no compound	Al_4C_3 (2173°C)	Likelihood of Al–C cosegregation but condition (iii) may not be met	Ductility not improved[25,26]
	Be	CsCl-type β phase between 25 and 51.6 at.-% Ni (1605°C); fcc Ni solid solution with max. 15 at.-% Be (1150°C)	Immiscible; no compound	Good potential for Ni–Be cosegregation	Ductility improved[25,27]
	Pd	Complete miscibility up to 1237°C	Al_4Pd (580°C); $Al_{21}Pd_8$ (790°C)	Al–Pd compounds less stable. Good potential for Ni–Pd cosegregation	Ductility improved[28]
	Fe	Complete miscibility up to 1440°C; Ni_3Fe (517°C)	Al_3Fe (1160°C); Al_5Fe (1169°C); Al_2Fe (1160°C); AlFe (1310°C)	Ni–Fe affinity seems to be stronger than Al–Fe, hence good potential for Ni–Fe cosegregation	Ductility improved[29]
	Mn	Complete miscibility up to 1020°C; a few lower melting compounds	$Al_{12}Mn$ (507°C); Al_5Mn (706°C); Al_4Mn (918°C); $Al_{31}Mn_4$ (989°C); Al_8Mn_5 (1160°C)	Ni–Mn affinity seems to be stronger than Al–Mn, hence good potential for Ni–Mn cosegregation	Ductility improved[25,29,30]
	Zr	Ni_5Zr (1300°C); Ni_7Zr_2 (1440°C); Ni_3Zr (920°C); $Ni_{21}Zr_8$ (1160°C); $Ni_{10}Zr_7$ (1150°C); NiZr (1280°C)	Al_3Zr (1580°C); Al_2Zr (1660°C); Al_3Zr_2 (1590°C); AlZr (1276°C)	Al–Zr have more line phases than Ni–Zr, hence good potential for Ni–Zr cosegregation	Ductility improved[24]
	S	Ni_3S_2 (560°C); Ni_7S_8 (560°C); NiS (1000°C)	No Al_nB compound with $n \geq 1$	Ni–S compounds less stable except NiS, but this requires much S	Ductility not improved[30]
Ni_3Ga (1220°C)	B	Ni_3B (1156°C); Ni_2B (1125°C); Ni_4B_3 (1031°C); NiB (1035°C)	Immiscible; no compound	Good potential for Ni–B cosegregation	Ductility improved[30-32]
Ni_3Si (1035°C)	B	Ni_3B (1156°C); Ni_2B (1125°C); Ni_4B_3 (1031°C); NiB (1035°C)	No Si_nB compound with $n \geq 1$	Good potential for Ni–B cosegregation	Ductility improved[30,33]
Co_3Ti (1190°C)	B	Co_3B (1126°C); Co_2B (1280°C); CoB (1480°C)	TiB (2200°C)	TiB very stable but requires much B. Good potential for Co–B cosegregation but whether condition (iii) can be met is not sure	Ductility improved[30]

from the bulk in C-doped Ni_3Al may not help strengthen the grain boundaries of the latter. Other examples in Table 3 similarly satisfy the above suggested conditions to a reasonable extent.

The argument here is indeed very simplistic, and by no means this can replace first principles calculations of the full A–B–X problem. Nevertheless, the criteria above may serve as a rough guide for the search of suitable systems for future experiments or computations.

CONCLUSIONS

In this paper, evidence has been presented that nickel–aluminium γ/γ' alloys can be toughened by boron doping, and that the toughening is caused by the formation of a nickel–boron cosegregated gain boundary zone. The results may be viewed as a logical extension to known effects of boron in nickel-rich Ni_3Al, but there are fundamental differences between the two situations in terms of the reversibility and phase equilibria of the cosegregation. The micrometre-wide cosegregated zone γ/γ' seems to be reconcilable with the existing understanding of grain boundary segregation as a saturated zone owing to excess boron with respect to the solubility in the $\gamma–\gamma'$ field. The authors have also suggested some generic, phenomenological conditions for cosegregation-induced toughening to work. Finally, the authors share with Professor Smallman the view that there is still a great deal to progress by doing further research in this field.

ACKNOWLEDGMENTS

AHWN would like to thank Professor Ray Smallman for introducing intermetallics as a PhD topic ten years ago, and for his continuous encouragement throughout the years. The interest in intermetallics was subsequently spread to YLC, whom Ray has also taught during this frequent visits to The University of Hong Kong in the last few years. Helpful comments on the work from Professor Ian Jones are gratefully acknowledged, and the authors also thank Professor Brian Duggan for suggesting to Ray to include this work in the workshop. This research was carried out under financial support from the Hong Kong Research Grants Council (Project no. HKU 7078/98E) and a HKU CRCG Grant (#337/064/0065).

REFERENCES

1. C. T. Liu, C. L. White and J. A. Horton: *Acta Metall.*, 1985, **33**, 213.
2. I. Baker and E. M. Schulson: *Scr. Metall.*, 1989, **23**, 1883.
3. J. W. Cohron, E. P. George, L. Heatherly, C. T. Liu and R. H. Zee: *Scr. Mater.*, 1998m, **38**, 847.
4. M. K. Miller: *Int. Met. Rev.*, 1987, **32**, 221.

5. S. P. Chen, A. F. Voter, R. C. Albers, A. M. Boring and P. J. Hay: *Scr. Metall.*, 1989, **23**, 217.
6. O. Ita and H. Tamaki: *Acta Metall. Mater.*, 1995, **43**, 2731.
7. F. H. Wang, C. Y. Wang and J. L. Yang: *J. Phys.: Condens. Matter*, 1996, **8**, 5527.
8. T. P. Weihs, V. Zinoviev, D. V. Viens and E. M. Schulson: *Acta Metall.*, 1987, **35**, 1109.
9. E. P. George, C. T. Liu and D. P. Pope: *Acta Mater.*, 1996, **44**, 1757.
10. J. W. Cohron, E. P. George, L. Heatherly, C. T. Liu and R. H. Zee: *Acta Mater.*, 1997, **45**, 2801.
11. W. R. Tyson: *Acta Met.*, 1978, **26**, 1471.
12. H. J. Frost: *Acta Metall.*, 1987, **35**, 519.
13. A. H. King and M. H. Yoo: *Scr. Metall.*, 1987, **21**, 1115.
14. A. Choudhury, C. L. White and C. R. Brooks: *Scr. Metall.*, 1986, **20**, 1061.
15. A. Choudhury, C. L. White and C. R. Brooks: *Acta Metall. Mater.*, 1992, **40**, 57.
16. E. D. Hondros, M. P. Seah, S. Hofmann and P. Lejcek: *Physical Metallurgy*, R. W. Cahn and P. Haasen, ed., 4th edn, Elsevier Science BV, Amsterdam 1996, Chapter 13.
17. T. M. Williams, A. M. Stoneham and D. R. Harries: *Met. Sci.*, 1976, **1**, 14.
18. M. Guttmann: *Surf. Sci.*, 1975, **53**, 213.
19. M. P. Seah and E. D. Hondros: *Proc. R. Soc. (London) A.*, 1973, **335**, 191.
20. N. S. Stoloff: *Int. Mater. Rev.*, 1989, **34**, 157.
21. J. D. Schobel and H. H. Stadelmaier: *Z. Metallkd*, 1965, **56** (12), 856.
22. G. Petzow and G. Effenbery (ed.): *Ternary alloys: a comprehensive compendium of evaluated constitutional data and phase diagrams*, VCH Verlagsgesellschaft, Weinheim, 1988, 201.
23. C. L. Briant: *Metall. Trans*, 1990, **21A**, 2339.
24. K. Aoki and O. Izumi: *Nippon Kinzoku Gakkaishi*, 1979, **43**, 1190.
25. N. Masahashi, T. Takasugi and O. Izumi: *Acta Metall.*, 1988, **36**, 1823.
26. S. C. Huang, C. L. Briant, A. I. Taub, K. M. Chang and E. L. Hall: *J. Mater. Res.*, 1986, **1**, 60.
27. T. Takasugi and O. Izumi: *Acta Metall.*, 1986, **34**, 607.
28. A. Chiba, S. Hanada and S. Watanabe: *Acta Metall. Mater.*, 1991, **39**, 1799.
29. T. Takasugi and O. Izumi: *Acta Metall.*, 1985, **33**, 1259.
30. T. Takasugi, H. Masahashi and O. Izumi: *Acta Metall.*, 1987, **35**, 381.
31. A. I. Taub, K. M. Chang and S. C. Huang: *Proc. ASM Conf. on Rapidly Solidified Material*, American Society for Metals, Metals Park, OH, 1986, 297.
32. Y. Xu and E. M. Schulson: in 'High Temperature Ordered Intermetallic Alloys IV', *MRS Symp. Proc.*, 1993, **288**, 635.
33. E. M. Schulson, L. J. Briggs and I. Baker: *Acta Metall. Mater.*, 1990, **38**, 207.
34. E. M. Schulson: *Physical Metallurgy and Processing of Intermetallic Compounds*, N. S. Stoloff and V. K. Sikka, ed., Chapman and Hall, London, 1996, 72.
35. C. L. Briant: *Intermetallic Compounds: Vol. 1 – Principles*, J. W. Westbrook and R. L. Fleischer, ed., Wiley, New York, NY, 1995, 899.

Grain Size and Shape Effects during Creep of Oxide-Dispersion-Strengthened Alloys

B. WILSHIRE

Department of Materials Engineering, University of Wales, Swansea SA2 8PP, UK

ABSTRACT

While dispersions of fine particles must impede dislocation movement, the volume fractions of particles in oxide-dispersion-strengthened (ODS) alloys are usually very low, typically about 0.5%. Moreover, the results of prestrain studies on polycrystalline nickel parallel many features of the creep characteristics often displayed by ODS alloys. These observations suggest that the dispersoid-stabilised dislocation substructures retained after thermo-mechanical processing, rather than the dispersoids themselves, control the creep properties of ODS materials. The complex behaviour patterns exhibited as a consequence of the pre-existing dislocation substructures can then be explained in terms of the varying contributions of the grain interiors and the grain boundary zones to the overall creep rate. Predictions based on these concepts are shown to account for the seemingly-anomalous stress/creep rate relationships found for the ODS ferritic steel, Incoloy MA956, produced with different grain sizes and grain aspect ratios.

INTRODUCTION

At temperatures above about half of the absolute melting point, T_m, oxide-dispersion-strengthened (ODS) alloys usually display creep strengths which are markedly superior to those of the equivalent dispersion-free matrix materials. In addition, one of the most striking features of dispersoid strengthening is that not only the grain size but also the grain shape appears to exert a major influence on the observed creep behaviour. Specifically, the creep resistance has been shown to increase rapidly with increasing grain aspect ratio (ie the grain length divided by the grain width), as illustrated by the results shown in Fig. 1 for ODS nickel alloys.[1]

Further striking features are revealed when power-law relationships are used to describe the creep properties of ODS alloys with large grain aspect ratios (GAR) ie when the variations of the minimum or secondary creep rate ($\dot{\epsilon}_m$) with stress, σ, and temperature, T, are represented by expressions of the form

$$\dot{\epsilon}_m = A\sigma^n \exp{-Q_c/RT} \tag{1}$$

where A is a constant, n is the grain size exponent and Q_c is the activation energy for creep (in units of J mol^{-1} when the gas constant, $R = 8.314$ J K^{-1} mol^{-1}).

131

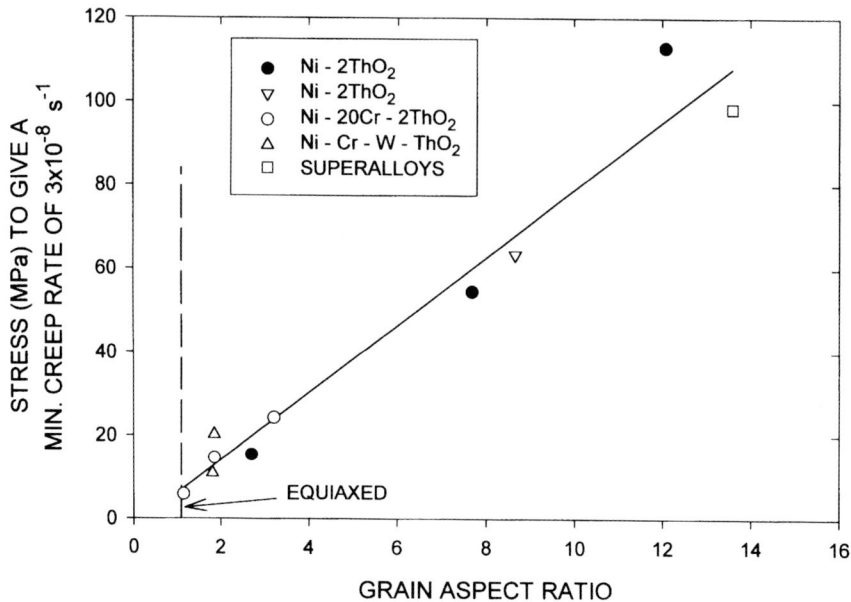

Fig. 1 The relationship between creep strength and grain aspect ratio for dispersion-strengthened nickel alloys.[1]

With pure metals and most single-phase alloys, $n \cong 4$ or so at high stresses, decreasing to $n \cong 1$ at low stresses. Similarly, $Q_c \cong Q_L$ (where Q_L is the activation energy for lattice diffusion) at temperatures above about $0.6\ T_m$, whereas Q_c decreases to around $0.5Q_L$ when diffusion along preferred paths such as dislocations and/or grain boundaries become important as the creep temperature is decreased towards $\sim 0.4\ T_m$. In contrast, large values of n and Q_c are often recorded for ODS alloys having elongated grain structures. In fact, n values up to 50 or more and Q_c values many times those for matrix diffusion have been found repeatedly for high GAR polycrystals.[2]

An early attempt to rationalise the anomalously large n and Q_c values obtained for ODS alloys suggested[3] that creep occurs not under the full applied stress (σ) but under a reduced stress ($\sigma - \sigma_o$). With the σ_o approach, the stress and temperature dependences of the creep rate can be written as

$$\dot{\epsilon}_m = A^* \, (\sigma - \sigma_o)^p \exp{-Q_c^*/RT} \qquad (2)$$

where $A^* \neq A$ and $p \cong 4$, with Q_c^* representing the creep activation energy derived at constant ($\sigma - \sigma_o$) rather than at constant σ, as in the determination of Q_c (equation 1). In relation to equations 1 and 2, $n \cong p$ and $Q_c \cong Q_c^*$ when $\sigma_o \cong 0$ or when $\sigma_o \propto \sigma$, whereas $n > p$ and $Q_c > Q_c^*$ when σ_o is large and decreases with increasing temperature.[4]

This idea has been modified and extended considerably over recent decades, with σ_o now almost universally referred to as a 'threshold stress'. However, σ_o cannot be measured or predicted reliably, so attention has been focused on dislocation theories of dispersion strengthening, often involving threshold stress concepts.[5] Unfortunately, despite their increasing complexity, these models have yet to provide a realistic description of the unusual combination of creep properties displayed by various ODS alloys. In particular, the influence of grain size and grain aspect ratio on the creep properties of oxide-dispersion-strengthened (ODS) alloys has proved especially difficult to interpret. As a result, a comprehensive review by Arzt[5] concluded

(a) that the 'creep behaviour of polycrystalline dispersion-strengthened materials is insufficiently understood' and
(b) that 'the rôle of grain boundaries during creep of dispersion-strengthened materials is not clear-cut'.

For these reasons, the creep behaviour patterns typically observed for ODS alloys are now re-analysed in relation to the recent idea[6] that the principal rôle of dispersoid particles is to stabilise the retained dislocation substructures present after thermomechanical processing. In this context, particular emphasis is placed on clarification of the effects of variations in grain size and shape on the stress/creep rate relationships recorded for the ODS ferritic steel, Incoloy MA956.

CREEP CHARACTERISTICS OF ODS ALLOYS

To interpret the apparently-complex creep properties of dispersion-strengthened materials, several features typical of the behaviour of ODS alloys can be compared with equivalent characteristics found for other metals and alloys. Three distinctive features merit consideration.

(i) There is ample evidence to prove that the creep resistance of ODS alloys increases with increasing grain aspect ratio (Fig. 1). However, this relationship between grain shape and creep strength does not seem compatible with the behaviour of nickel-base superalloys. These materials can be produced with equiaxed or elongated grain structures by conventional casting (CC) and by directional solidification (DS) respectively. Inspection of the early stages of creep curves recorded for the same γ/γ' superalloy manufactured by these two routes then shows that the rates of creep strain accumulation are very similar, ie comparable $\dot{\epsilon}_m$ values are obtained under the same stress-temperature conditions for CC and DS materials with equivalent γ' distributions.[7] Hence, the improved performance of DS over CC aeroengine turbine blades is not a consequence of improved creep resistance. Instead, longer creep lives are a result of the increased difficulty of crack development with elongated grain structures, postponing creep fracture to higher creep strains. Thus, while the creep lives of γ/γ′ superalloys are affected, the creep resistance is independent of the grain aspect ratio. On this basis, it appears that the dependence of the creep strength of ODS alloys on their grain aspect ratio (Fig. 1) is a consequence not of grain shape *per se* but of some indirect relationship between strength and some microstructural feature(s) related to the grain aspect ratio.

(ii) Although dispersoid particles must obviously be capable of impeding the movement of dislocations, the volume fractions of dispersoids in ODS alloys are usually very low (typically <1%). For this reason, it has been proposed that the dispersoid-stabilised dislocation substructures retained after thermomechanical processing, rather than the dispersoid particles themselves, determine the creep properties of ODS alloys.[6] With single-phase materials, pre-existing dislocation substructures can be eliminated by recrystallisation at relatively low creep temperatures. However, the presence of even relatively small volume fractions of alumina particles result[8] in the recrystallisation temperatures of cold worked Cu–Al$_2$O$_3$ alloys being much higher than those for pure wrought copper (Fig. 2). Thus, dispersions of fine oxide particles inhibit recrystallisation, raising the temperature at which dislocation strengthening continues to be effective.

(iii) The influence of pre-existing dislocation substructures, uncomplicated by the presence of particle dispersions, can be clarified[6] by considering the effects of varying prestrain treatments[9] on the creep properties of polycrystalline nickel (containing 0.1 at % gold to avoid the extensive grain growth which occurred[10] during creep of high-purity nickel under equivalent test conditions). The testpieces were annealed for 60

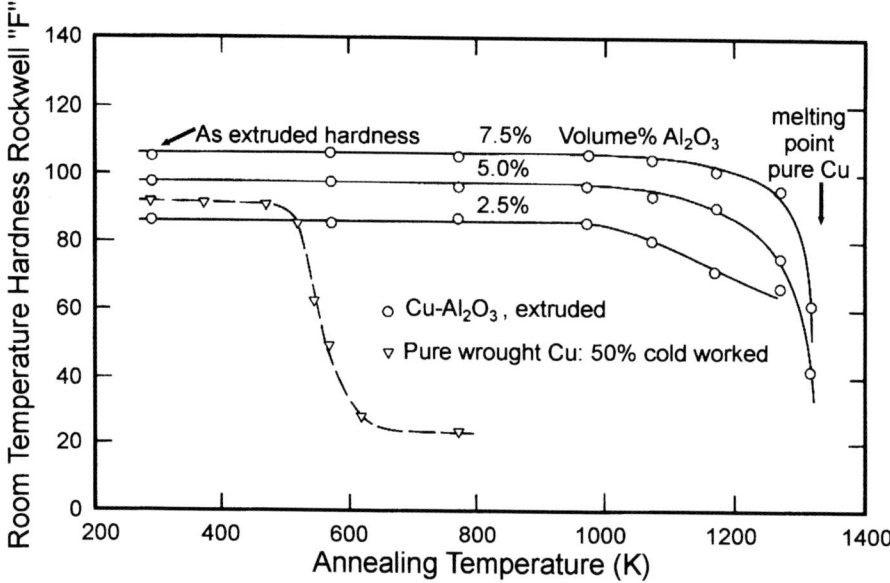

Fig. 2 Softening of strained copper and copper–alumina alloys.[8]

ks at 1273 K, resulting in a mean grain diameter of ~150 μm. The annealed samples were then prestrained by 0 to 25% in tension at room temperature. After prestraining by a fixed amount, each testpiece was heated to the creep temperature of 773 K and held at this temperature for 2.7 ks before loading. This procedure allowed dislocation recovery and rearrangement to occur without causing recrystallisation. Thus, the dislocation density just prior to loading increased with increasing prestrain. Moreover, the grain aspect ratio increased from 1.0 with zero prestrain to ~1.4 with 25% prestrain. The stress/creep rate plots in Fig. 3 then suggest that, as with ODS alloys having high grain aspect ratios (Fig. 1), the creep strength increases with increasing GAR (Fig. 4). Even so, detailed analysis of the data produced in this experiment[9] demonstrates that the dislocation substructure introduced by prestraining, rather than the resulting GAR value, causes the improvement in creep resistance (Figs 3 and 4).

PRESTRAIN EFFECTS DURING CREEP

Inspection of the results in Fig. 3 shows that significant improvements in creep resistance were achieved only with prestrains of 10% or more but, even then, the creep rates for the prestrained samples approached those of the non-prestrained material at

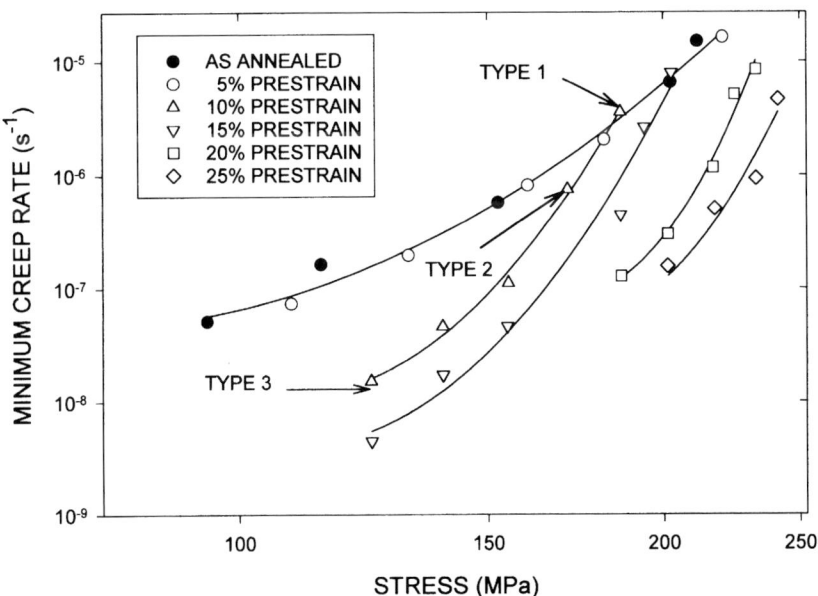

Fig. 3 The stress dependence of the minimum creep rates at 773 K for a polycrystalline nickel–0.1 at. % gold alloy prestrained 0 to 25% at room temperature.

135

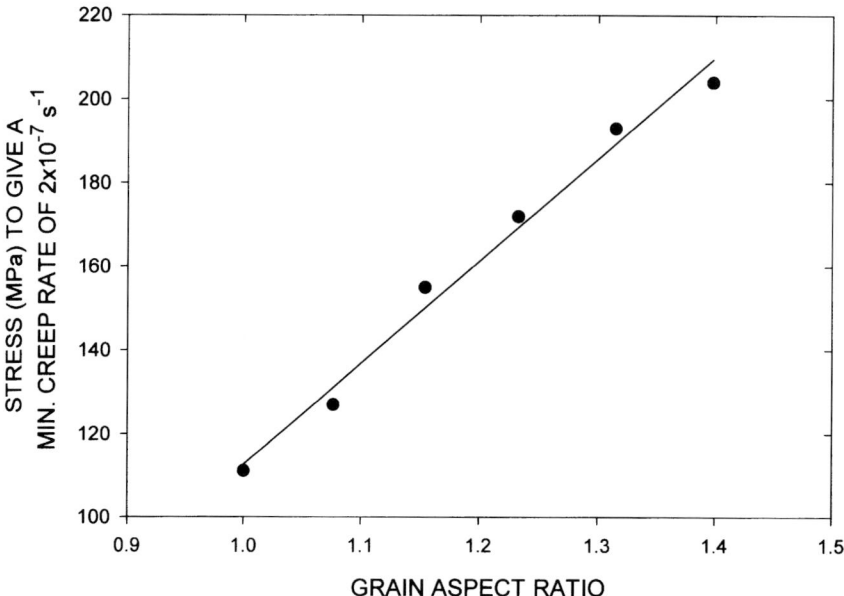

Fig. 4 The relationship between the stress to give a creep rate of $2 \times 10^{-7}\,\mathrm{s}^{-1}$ at 773 K and the grain aspect ratio of the nickel–0.1 at % gold alloy prestrained by different amounts at room temperature.

the highest stresses studied. These trends can be explained by reference to the stress dependences of the initial strains on loading (ϵ_o) for each prestrain level (Fig. 5).

The minimum creep rates for the prestrained samples equalled those for the non-prestrained specimens (Fig. 3) when application of the creep stresses resulted in initial strains on loading having plastic as well as elastic components ie when $\epsilon_o > 0$ for the prestrained samples (Fig. 5). When this condition was met, the creep curves obtained for the prestrained and non-prestrained samples were virtually identical, ie although the ϵ_o values were reduced by prestraining (Fig. 5), the primary creep curves and the minimum creep rates (Fig. 3) were similar at the same creep stress. In contrast, when the prestrains eliminated the plastic components of the initial strains on loading, giving near-zero ϵ_o values (Fig. 5), very low primary strains were displayed and the minimum creep rates became progressively lower than those for the non-prestrained nickel as the prestrain level was increased (Fig. 3). Three conclusions can be drawn from this prestrain study.[6]

(i) When the initial specimen extension had a plastic component, virtually identical creep curves were observed because comparable dislocation structures were present immediately after loading at the same stress for prestrained and non-prestrained samples. However, when prestraining eliminated the initial plastic strain on loading, the dislocation densities at the start of the tests were higher for the prestrained test-

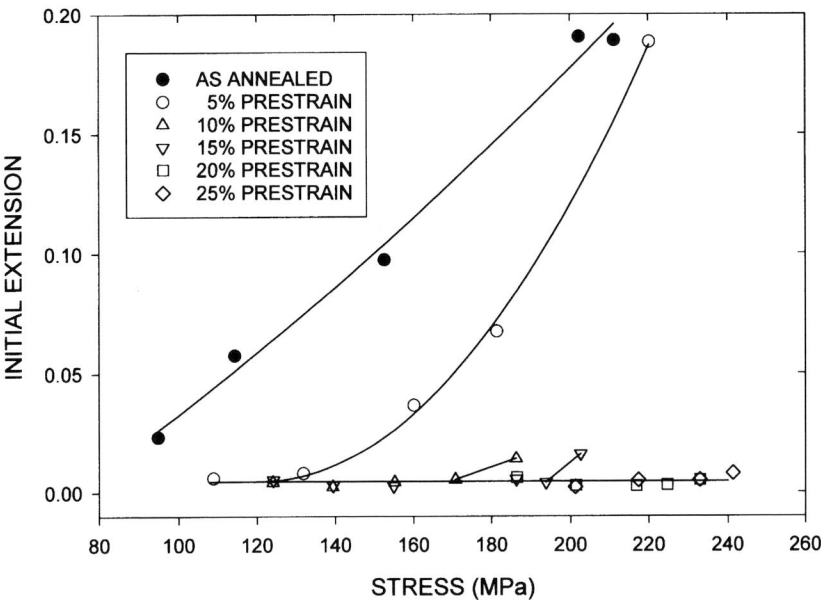

Fig. 5 The variation of the initial strains on loading at 773 K as a function of the creep stress for samples of the nickel–0.1 at % gold alloy prestrained 0 to 25% at room temperature.

pieces than for the non-prestrained specimens, reducing the primary strains and improving the creep resistance (Fig. 3).

(ii) The creep characteristics of the prestrained nickel were explained[9] by distinguishing between the relative contributions to the overall creep rate made

(a) by the grain interiors and
(b) by the grain boundaries zones, with zone deformation including grain boundary sliding and associated deformation in regions of the grains adjacent to boundaries.

When the prestrains eliminated the plastic components of ϵ_o (Fig. 5), high pre-existing dislocation densities increased creep resistance (Fig. 3) by restricting deformation of the grain interiors. However, detailed measurements taken from fine grids deposited on the surfaces of testpieces indicated that the zone deformation is comparatively unaffected by prestrain, even when the prestrain improved creep resistance.[11] This effect appears to be attributable to dislocation recovery processes being more rapid within the grain boundary zones. Thus, with zone deformation being relatively unaffected, the creep rate at a given stress decreases with higher prestrain levels, primarily because of a decrease in the grain contribution to the overall creep rate.

(iii) The idea that zone deformation is less affected than grain deformation, even when prestraining improves creep resistance, is supported by the creep rupture data obtained for the testpieces prestrained 0 to 25%. In all cases, fracture occurred by the nucleation and growth of intergranular cavities and cracks. Although no general agreement has been reached on the mechanisms governing intergranular creep failure, quite obviously, it is a grain boundary phenomenon. It is then interesting to note that, while prestraining could result in the creep rate being decreased by well over an order of magnitude (Fig. 3), only very limited improvements in creep rupture life were found (Fig. 6). In fact, the product of the minimum creep rate ($\dot{\epsilon}_m$) and the time to fracture (t_f) was $\dot{\epsilon}_m.t_f \cong 0.10$ when prestraining did not increase the creep resistance, compared with $\dot{\epsilon}_m.t_f \cong 0.01$ when an increase in creep strength was achieved (Fig. 3). Improvements in creep resistance can then be linked to decreases in the grain contribution to the overall creep rate, while the comparatively small changes in creep rupture strength (Fig. 6) confirm that boundary zone behaviour is much less influenced by prestraining.

CLASSIFICATION OF PRESTRAIN EFFECTS

By distinguishing between the grain and the boundary zone contributions, a straightforward explanation can be provided for the variations in the stress-dependence of the creep rates displayed by the prestrained samples (Fig. 3).

(i) A prestrain of 5% had no significant effect on creep resistance. Hence, $n \cong 4$ for samples prestrained 0 and 5%, but with n increasing gradually at the highest stresses

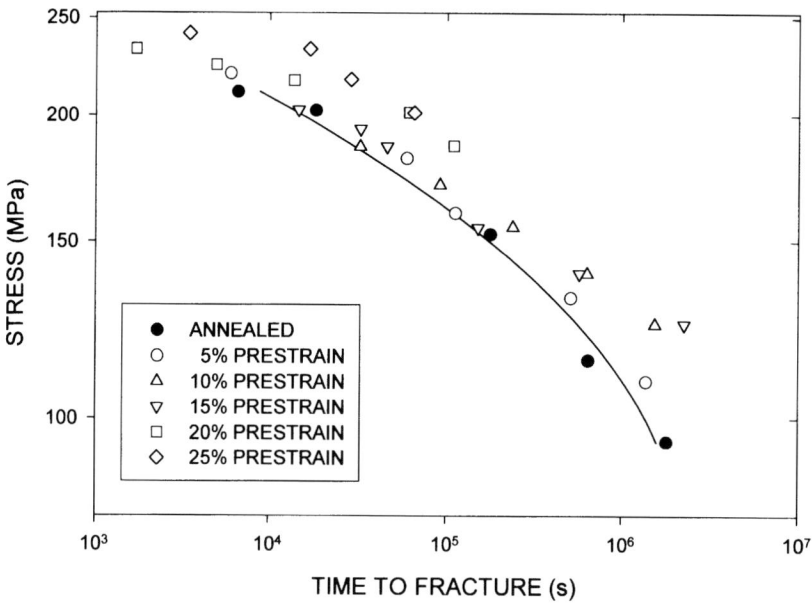

Fig. 6 The dependence of the time to fracture on stress at 773 K for specimens of the nickel–0.1 at. % gold alloy prestrained 0 to 25% at room temperature.

studied (Fig. 3). This deviation from $n \cong 4$ in tests of very short duration appears to be associated with 'power-law breakdown', which occurs when equation 1 does not properly represent the stress/creep rate curves at high stresses.[2] Even so, over the entire stress ranges covered for samples prestrained 0 and 5%, both the grains and the boundary zones contribute to the overall creep rate. This condition can be designated as Type 1 behaviour.

(ii) With a prestrain of 10%, the observed behaviour pattern depends on the stress ranges considered, designated as Types 1, 2 and 3 in Fig. 3. At high stresses, when $\epsilon_o > 0$ in Fig. 5, the measured creep rate lies on the stress/creep rate curve for the samples prestrained 0 and 5%. Thus, the grain and boundary zone contributions are unaffected by prestrain when Type 1 behaviour is displayed at high stresses. However, when prestraining results in an improvement in creep resistance (when $\epsilon_o \cong 0$ in Fig. 5), the curving $\log \sigma / \log \dot{\epsilon}_m$ plots are characterised by regions where $n \cong 20$ at intermediate stresses, but with n decreasing towards $n \cong 4$ at low stresses (Fig. 3). When $n \cong 20$, the contribution of the grains to the overall creep rate decreases rapidly with decreasing stress. This can be termed Type 2 behaviour. Type 3 behaviour then describes the region where the grain contribution is minimal. Even so, since zone deformation is relatively unaffected by the prestrain, comparable n values are observed for the prestrained and non-prestrained samples at low stresses (Fig. 3).

(iii) A stress/creep rate plot similar in shape to that for 10% prestrain was also found for a prestrain of 15%, albeit with the regions showing Type 1, 2 and 3 behaviour displaced to higher stresses. In contrast, with prestrains of 20 and 25%, considerable improvements in creep strength were achieved at all stresses studied (Fig. 3). Hence, even at the highest stresses, $\epsilon_o \cong 0$ for the prestrained specimens (Fig. 5), so that Type 1 behaviour was never exhibited. Yet, as with samples prestrained 10 and 15%, testpieces prestrained 20 and 25 do show Type 2 behaviour (with $n \cong 20$ at intermediate stresses) and a transition to Type 3 behaviour (with n decreasing towards 4 at low stresses).

THE RÔLE OF GRAIN SIZE AND SHAPE

Although the grain aspect ratio at the start of the tests increased with increasing prestrain (Fig. 4), it was the dislocation substructure imposed by the prestrain not the resulting GAR which governed the creep properties shown in Figs 3 to 6 inclusive. In a similar manner, the complex creep behaviour displayed by ODS alloys can then be explained in terms of the contributions of the grains and the boundary zones, as influenced by the dislocation substructures retained after thermomechanical processing.[6] Simultaneously, this concept clarifies the dependence of the creep strength on the grain size and shape of the ODS alloys (Fig. 1). With this approach, three distinct categories of creep behaviour can be defined for ODS alloys, as illustrated by reference to the schematic stress/creep rate relationships shown in Fig. 7.[6]

(i) Category A properties are found when the thermomechanical processing

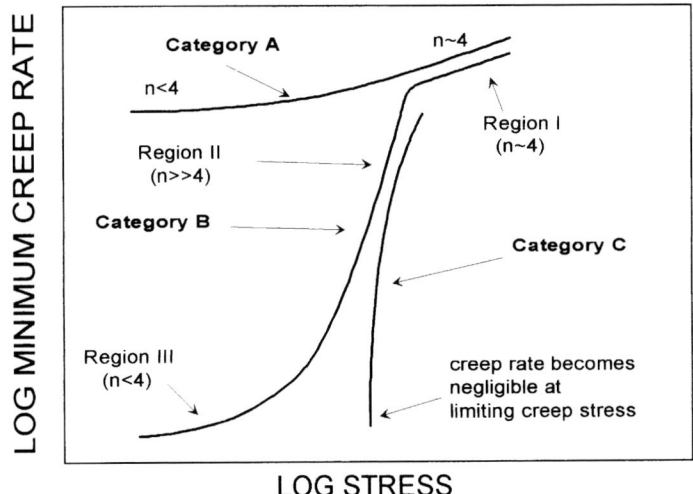

Fig. 7 Schematic representation of the stress dependence of the minimum creep rate for oxide-dispersion-strengthened alloys produced with differing grain sizes and grain aspect ratios. Category A characteristics are found for ODS alloys with fine equiaxed grain structures and low retained dislocation densities. Category B behaviour is often observed with fine-grain ODS alloys with large grain aspect ratios and high retained dislocation densities. Category C curves are expected for ODS single crystals or coarse-grain ODS alloys with large grain aspect ratios and high retained dislocation densities.[6]

operations result in complete recrystallisation, producing equiaxed grain structures (GAR = 1) with low retained dislocation densities. Without a dislocation substructure limiting grain deformation, both the grains and the grain boundary zones contribute to the overall creep rate. Consequently, these ODS alloys behave like their equivalent dispersion-free matrix materials, ie the creep strengths are not impressive and high values of n and Q_c are not exhibited. Instead, with Category A behaviour, the observed stress exponents decrease from $n \cong 4$ at high stresses towards $n \cong 1$ at low stresses, just as with pure metals. Clearly then, when complete recrystallisation during the processing treatment eliminates the dislocation substructure developed during the working operations, the oxide particle dispersion has only a very limited ability to impede dislocation movement and to improve creep resistance. Thus, Category A behaviour emphasises that the particle-stabilised dislocation substructure not the particle dispersion determines the creep properties of ODS alloys.

(ii) Category B behaviour is found for high GAR polycrystals with relatively fine grain sizes, ie for thermomechanical processing conditions resulting in high retained dislocation densities, seemingly, with the dislocation density increasing with increasing grain aspect ratio. In this way, as with the prestrained nickel (Fig. 4), the creep strength would increase with initial dislocation density, which is scaled by the grain aspect ratio of the ODS alloys (Fig. 1).

Category B materials often show sigmoidal stress/creep rate curves,[5] illustrated as Region I, II and III behaviour in Fig. 7. Interestingly, the transition from Regions I to II for Category B alloys is very similar to the Type 1 to 2 transition in the stress/creep rate plot for nickel prestrained 10 or 15%, ie with the prestrained nickel, the stress exponents change from $n \cong 6$ at high stresses to $n \cong 20$ at intermediate stresses (Fig. 3), mirroring the change from $n \cong 4$ in Region I to $n >> 4$ in Region II with fine-grain high-GAR ODS alloys (Fig. 7). This comparison suggests that, if the creep stresses are high enough to cause a plastic strain on loading, high GAR polycrystals behave in Region I just like their Category A equivalents. Only when the retained dislocation substructures eliminate the plastic components of the loading strains are high creep strengths and high n values recorded in Region II for high GAR polycrystals.

Category B materials also show a transition from Region II to III behaviour with decreasing applied stress (Fig. 6). This transition has been attributed to dissolution, coarsening or deformation of the dispersoids as the test duration increases,[5] but such processes must be considered most unlikely given the strength and stability of fine oxide particles. However, the parallel between the Type 2 to 3 transition with nickel prestrained 10 to 25% (Fig. 3) and the Region II to III transition with high GAR polycrystals (Fig. 7) again supports the idea that the pre-existing dislocation substructure determines the creep properties of ODS alloys. Thus, in Region II, the contribution of the grain interiors to the overall creep rate decreases rapidly with decreasing stress (giving high n values). In Region III, creep is then a result only of grain boundary zone deformation. Since the zone deformation is largely unaffected by the elimination of the grain contribution , the range of stress exponents found in Region III are comparable with those displayed by substructure-free material (Category A), even though the creep rates are much lower because of the absence of a significant contribution from the grain interiors (Fig. 7).

(iii) Category C properties are observed for ODS single crystals and high GAR polycrystals with very coarse grain sizes.[6] With these materials, numerous studies have indicated that the creep rate decreases rapidly with decreasing applied stress, so that $n >> 4$, as in Region II with Category B materials (Fig. 7). Moreover, n is often considered to increase with decreasing stress, such that creep does not occur below some limiting value. These observations have usually been taken as evidence for 'threshold stress' ideas. Yet, even this type of behaviour would be expected by distinguishing between the contributions made to the overall creep rate by the grains and the grain boundary zones. With single crystals and coarse grain polycrystals having a dislocation substructure retained from thermomechanical processing, large n values would be anticipated as the 'grain contribution' decreases rapidly with decreasing stress. Since grain boundaries are absent or rare, the 'zone contribution' would then be zero or negligible. Hence, with no clear transition from Region II to III as with Category B materials, Category C alloys should exhibit extended Region II behaviour, with creep rates seemingly decreasing towards zero with decreasing stress (Fig. 7).

MICROSTRUCTURES OF INCOLOY MA956

Many sets of tensile creep data are available for ODS alloys, allowing checks to be made on the validity of the schematic stress/creep rate plots in Fig. 7. In particular, numerous investigations have been carried out on the ODS ferritic steel, Incoloy MA956, produced by mechanical alloying. The nominal composition (wt. %) of this alloy is Fe–20Cr–4.5Al–0.5Ti–0.5Y_2O_3, resulting in microstructures with yttria particle dispersions in the ferritic Fe–Cr–Al–Ti matrix. In relation to the present analysis, tensile creep properties have been determined for MA956 having three different grain structures.[12,13] The detailed compositions of these materials are given in Table 1.

(i) Material A was obtained in the form of rolled bar, described as being in the extruded, rolled and recrystallised condition.[12] The processing operations resulted in a fine equiaxed grain structure, with a mean grain diameter of ~12 μm.

(ii) Material B was produced in the form of tubes, with outside diameters of 25.4 mm and wall thicknesses of 1.95 mm.[13] The average transverse grain diameter was measured as 60 μm, with a longitudinal grain length of 1.4 mm (giving GAR ≈ 23).

(iii) Material C had a coarse elongated grain structure, with a mean grain diameter of ~15 mm and a grain aspect ratio of 3.2.[12]

Although the grain size and shape varied markedly for these MA956 samples, the distribution of Y_2O_3 particles appeared to be similar. Thus, the average particle size was ~10 μm, but with particles ranging from ~5 to 500 nm.[12,13] Moreover, for Materials A, B and C, the dislocation densities in the as-received condition were reported as low but, in all cases, the dislocations appeared to be associated with the dispersoid particles.

CREEP OF INCOLOY MA956

The stress/creep relationships observed for the MA956 samples manufactured with different grain sizes and shapes, designated as Materials A, B and C in Table 1,[12,13] are shown in Fig. 8. Clearly, these results are directly in line with the predicted behaviour patterns presented in Fig. 7.

(i) The fully-recrystallised Material A behaves in a manner similar to that widely reported for pure metals and many single-phase alloys, namely, $n \cong 4$, with the results in Fig. 8 possibly indicating that n is decreasing with decreasing stress (exactly equivalent to Category A behaviour in Fig. 7). Since the Y_2O_3 particle dispersions in

Table 1 Compositions of Incoloy MA956 (wt %)

	Fe	Cr	Al	Ti	C	S	Y_2O_3	Ref.
Material A	75.3	19.42	4.34	0.29	0.02	0.003	0.48	12
Material B	74.7	19.70	4.70	0.38	–	–	0.50	13
Material C	73.1	20.69	5.09	0.32	0.02	0.017	0.76	12

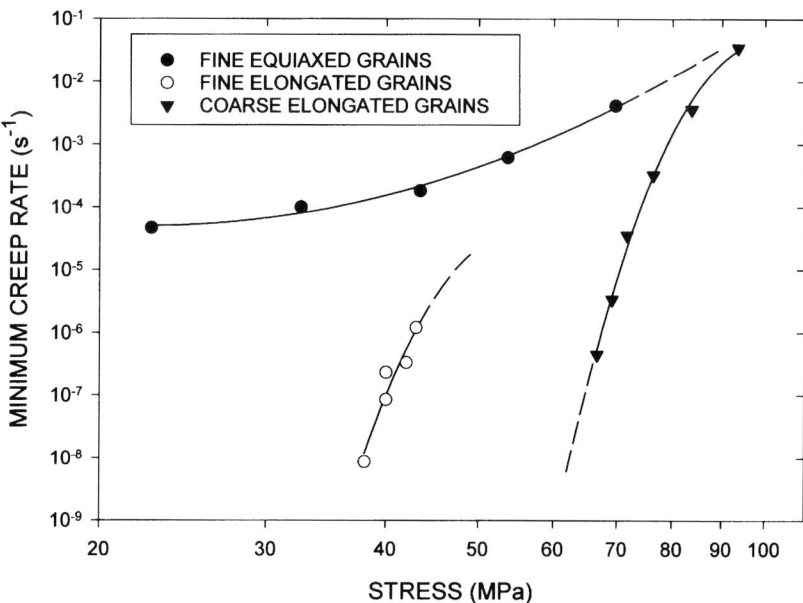

Fig. 8 The stress dependence of the minimum creep rate at 1373 K for Incoloy MA956 produced with fine equiaxed grains, fine elongated grains and coarse elongated grains.[12, 13]

Material A are essentially similar to these in Materials B and C, the low n values and low creep strength of the equiaxed Material A must mean that the particles themselves have little effect on dislocation movement. This observation therefore reinforces the view that the dispersion-stabilised dislocation substructures present after thermomechanical processing, rather than dispersoid pinning of moving dislocations, are responsible for the high creep strengths and high n values exhibited by Materials B and C (Fig. 8).

(ii) Over the stress range studied, the results in Fig. 8 show that $\dot{n} \cong 35$ for Material B, produced with a fine elongated grain structure (equivalent to Region II behaviour for Category B alloys in Fig. 7). Interestingly, extrapolation of the data obtained for Material B indicates that the stress/creep rate plots for Materials A and B would intersect at high stresses (mirroring to the Region I to II transition for Category B alloys in Fig. 7). Although the data for Material B in Fig. 8 shows no evidence for the Region II to III transition predicted in Fig. 7, ample information exists for other ODS alloys[5] to suggest that even this trend would occur in tests conducted at lower applied stresses for MA956 manufactured with fine elongated grain structures.

(iii) With Material C, having a coarse elongated grain structure, the creep rate recorded at the highest stress is similar to that expected for Material A in Fig. 8, ie at stresses sufficiently high to give a plastic component of the initial strain on loading, Categories C as well as Category B materials display the Region I to II transition in creep rate depicted in Fig. 7. Furthermore, as indicated in Fig. 7 for Category C alloys, the n values recorded for Material C in Fig. 8 seem to increase with decreasing stress, with $n \cong 60$ at the lowest stresses studied. While this behaviour is often taken as support for threshold stress concepts,[5] there is no doubt that creep does occur at stresses well below the limiting value apparently indicated by short-term test results. Moreover, all theories predict that no significant minimum threshold stress exists.[5] For these reasons, it must be considered likely that Category C as well as Category B materials will show a transition from Region II to Region III behaviour at sufficiently low stresses, although the extent of Region II will be greater for coarse-grain high-GAR polycrystals (Fig. 7 and 8).

CONCLUDING COMMENTS

Recent dislocation models for creep of ODS alloys can be categorised into two groups,[5] depending on whether (a) surmounting the dispersoids by climb (climb control) or (b) the breakaway of dislocations from dispersoids (detachment control) is defined as the rate controlling event. Yet, despite their complexity, the available models do not properly describe the creep properties of dispersoid-strengthened materials. This outcome is hardly surprising because most theoretical approaches have tried to model dislocation behaviour within crystals, whereas the vast majority of experimental programmes have considered polycrystalline samples. Indeed, the present analysis suggests that a valid description of the creep characteristics of ODS alloys must incorporate two key features.

(i) Although a dislocation moving through a crystal would be impeded if it encountered a dispersoid particle, the volume fractions of dispersoids typically present are very low. For this reason, fully recrystallised ODS alloys with fine equiaxed grain structures have poor creep strengths, also displaying the low n values found for dispersion-free materials. In contrast, ODS alloys with high grain aspect ratios (GARs) exhibit high creep strengths, as well as very large n values over intermediate stress ranges. Moreover, the creep properties commonly observed for high-GAR polycrystals mirror those reported for polycrystalline nickel prestrained 0 to 25% at room temperature, suggesting that the pre-existing dislocation substructure not the GAR value determines the creep resistance, i.e. dispersoid-stabilised dislocation substructures retained after thermomechanical processing govern dislocation movement during creep of ODS alloys. These observations also suggest that the grain aspect ratio provides some measure of the retained dislocation density, which would then account for the dependence of creep strength on GAR commonly noted for ODS alloys (Fig. 1).

(ii) As with the prestrained nickel, the retained dislocations substructures in ODS alloys influence the contributions of the grain interiors and the grain boundary zones to the overall creep rate. In this way, as exemplified by data available for the ODS ferritic steel, Incoloy MA956, a straight-forward interpretation is provided for the seemingly-anomalous stress/creep rate relationships reported for dispersoid-strengthened alloys produced with different grain sizes and grain aspect ratios.

REFERENCES

1. B. A. Wilcox and A. H. Clauer, *The Superalloys,* J. Wiley, New York, 1972.
2. O. D. Sherby and P. M. Burke, *Prog. Mater. Sci.,* 1968, **13**, 325.
3. K. R. Williams and B. Wilshire, *Metal Sci.,* 1973, **7**, 176.
4. J. D. Parker and B. Wilshire. *Metal Sci.,* 1975, **9**, 248.
5. E. Arzt, *Res Mech.,* 1991, **31**, 399.
6. B. Wilshire, *Proc. 5th Inter. Conf. on Creep and Fracture of Engineering Materials and Structures,* J. C. Earthman and F. A. Mohamed eds, TMS, 1997, 19.
7. R. W. Evans and B. Wilshire, *Creep of Metals and Alloys,* The Institute of Metals, London, 1985.
8. D. McLean, *Mechanical Properties of Metals,* J. Wiley, New York, 1962.
9. P. W. Davies, J. D. Richards and B. Wilshire, *J. Inst. Metals,* 1961–62, **90**, 431.
10. J. P. Dennison and B. Wilshire, *J. Inst. Metals,* 1962–63, **91**, 343.
11. J. D. Parker and B. Wilshire, *Mater. Sci. Eng.,* 1977, **29**, 219.
12. R. Petkovic-Luton, D. S. Srolovitz and M. J. Luton, in *Frontiers of High Temperature Materials II,* J. S. Benjamin and R. C. Benn eds, INCO Alloys Inter., 1983, 73.
13. J. C. Healy, M. Rees, J. D. Parker and R. C. Hurst, *Proc. 5th Inter. Conf. on Creep and Fracture of Engineering Materials and Structures,* J. C. Earthman and F. A. Mohamed eds, TMS, 1997, 719.

Grain Boundaries in High Performance Magnets: Reasons for Poor or Excellent Properties

JOSEF FIDLER, THOMAS SCHREFL and DIETER SUESS

Institute of Applied and Technical Physics, Vienna University of Technology, Wiedner Hauptstr. 8–10, A–1040 Austria.

ABSTRACT

Advanced $Nd_2Fe_{14}B$-based permanent magnets exhibit a complex, multiphase microstructure and show the highest values of coercivity and energy density products, obtained so far. The hysteresis properties are governed by a combination of the intrinsic properties of the material, such as saturation polarisation, exchange and magneto-crystalline anisotropy. The other important factors are the microstructural parameters, such as grain size, the orientation of the easy axes of the grains and the distribution of phases. The grain size of the magnets and the alignment of the grains strongly depend on the processing parameters. The formation and distribution of the phases is determined by the composition of the magnets and the annealing treatment. The intergranular structure between the grains plays a significant role in determining the magnetic properties. The coercivity is determined by the long range dipolar interaction and short range exchange coupling between neighbouring grains. The doping of elements changes the phase relation and favours the formation of new phases. Additional intergranular phases decrease the remanence and interrupt the magnetic interactions between the grains, thereby improving the coercivity of large grained sintered magnets. Non magnetic phases, which replace the Nd-rich intergranular phase, considerably improve the corrosion resistance and are of great technological interest. Exchange interactions between neighbouring soft and hard grains lead to remanence enhancement of isotropically oriented grains in nanocrystalline composite magnets. Micromagnetic finite element simulations show that the magnetic properties of the disturbed intergranular region strongly deteriorate the coercive field of the magnet. Insufficient temperature stability and poor corrosion resistance are the main factors limiting applications of $Nd_2Fe_{14}B$-based magnets.

INTRODUCTION

Hard magnetic materials are divided into the group of conventional metallic and oxide magnets and the group of modern magnets based on intermetallic compounds of rare earth elements with cobalt and/or iron. The importance of newly developed permanent magnetic materials in many electro-, magnetomechanical and electronic applications can be attributed to the drastic improvement of the magnetic energy density product and coercive field of the new hard magnetic materials. The rare earth intermetallic phases $SmCo_5$, Sm_2Co_{17}[1] and $Nd_2Fe_{14}B$[2, 3] are the basis for these high

147

performance magnets. Rare earth-Co magnets exhibit the highest coercive fields and (Nd,Dy)-(Fe,Co)-B:(M1,M2) magnets show the highest value of energy density product obtained so far. High performance $SmCo_5/Sm_2Co_{17}$- and $Nd_2Fe_{14}B$-based permanent magnets exhibit a complex, multiphase microstructure. The grain size of the magnets and the alignment of the grains strongly depend on the processing parameters. The formation and distribution of the phases is determined by the composition of the magnets and the annealing treatment. High performance $Nd_2Fe_{14}B$-based permanent magnets are produced with different composition and various processing techniques,[4,5] which influence the complex, multiphase microstructure of the magnets, such as grain size, the orientation of the easy axes of the grains and the distribution of phases:

- grain sizes in the range between 10 nm and 500 nm are obtained by melt-spinning, mechanical alloying and the HDDR (hydrogenation–disproportionation–desorption–recombination) process
- sintered and hot worked magnets exhibit grain sizes above 1 μm

For the better understanding and the further development of high performance permanent magnets and for the search for new phases a detailed understanding of the microstructure and grain boundaries is necessary. This also includes the exact knowledge of the crystallography, phase diagram and phase relation of the magnet system. The hysteresis properties are governed by a combination of the intrinsic properties of the material, such as saturation polarisation, magnetic exchange and magnetocrystalline anisotropy and the influence of the microstructure on the magnetisation reversal process. The role of intergranular structure between the grains plays a significant role determining the magnetic properties, especially if the grain diameter is on the nanometre scale.[6]

EXPERIMENTAL PROCEDURE

Highest energy density Nd-Fe-B permanent magnets (>400 kJ m^{-3}) exhibiting a low oxygen content have been prepared by the sintering route according to the following experimental procedure.[7,8] The $(Nd,Pr)_{12.6-15.1}Fe_{72.8-80.5}B_{5.8-6.1}$ alloys were strip cast in an argon atmosphere from Nd-Fe, Fe-B, Fe and (Cu,Al) pre-alloys. The crushed flakes exhibit a homogeneous microstructure of fine dispersed Nd-rich phase within a $Nd_2Fe_{14}B$ matrix phase.[9] Using the hydrogen decrepitation (HD) process coarse powders of less than 500 μm were prepared. After jet milling under nitrogen-gas atmosphere fine powders of about 2.2–3.3 μm in mean FSSS particle size (Fisher Sub Sieve Size) were obtained. The oxygen content of the milling gas was in the order of 10 ppm O_2. The distribution of the size of fine powders was measured with a scanning electron microscope.

The fine powders were pre-pressed and pre-aligned with a pulsed field in the glove box under argon atmosphere to a rectangular compact. The encapsulated magnets were pressed in a transverse magnetic field press (combination of RIP and TDP) under a magnetic field of 1200 kA m^{-1}. This new pressing technique showed an improved alignment of the hard grains and therefore a high value of remanence. These compacts were sintered at 1233–1373 K for 10.8 ks in vacuum and were subsequently annealed at about 1073 K for 900 s and 783–843 K for 3.6 ks in order to obtain optimum density and coercive field.

Magnetic properties were measured by a B–H hysteresis tracer after being fully magnetised in a pulsed field up to 7 T. After each step of the preparation route the oxygen content of the alloy, power and the magnet, respectively was measured with a LECO oxygen analyser. The microstructure (grain size, phases and degree of alignment of grains) and the magnetic domain structure of the magnets was investigated by Kerr microscopy, electron probe microanalysis, scanning and transmission electron microscopy.

INTERGRANULAR PHASES IN DOPED AND SUBSTITUTED ND-FE-B MAGNETS

Substituent and dopant elements influence the microstructure, coercivity and corrosion resistance of advanced (Nd,S1)–(Fe,S2)–B:(M1,M2) magnets. The multicomponent composition of the magnets leads to the formation of non-magnetic and soft magnetic phases. Generally, two types of substituent elements, which replace the rare earth element or the transition element sites in the hard magnetic phase, and two types of dopant elements are distinguished.[6] Selected substituent elements replace the Nd-atoms (S1 = Dy,Tb) and the Fe-atoms (S2 = Co,Ni,Cr), respectively, in the hard magnetic ϕ-phase and considerably change intrinsic properties, such as the spontaneous polarisation, the Curie temperature and the magnetocrystalline anisotropy. The main difference between substituent and dopant elements is the solubility range within the $Nd_2Fe_{14}B$ phase. Our previous, systematic TEM studies performed on sintered, melt-spun, mechanically alloyed and hot worked magnets have shown that two different types of dopants can be distinguished independently of the processing route. Depending on the type, the dopant elements form additional intergranular rare-earth-containing or boride phases:[7, 8]

- *Type 1 dopants* (M1 = Al, Cu, Zn, Ga, Ge, Sn)
  form binary M1-Nd or ternary M1-Fe-Nd phases
- *Type 2 dopants* (M2 = Ti, Zr; V, Mo; Nb, W)
  form binary M2-B or ternary M2-Fe-B phases

The doping changes the microstructure of Nd-Fe-B sintered magnets in the following way: if there exists a solubility at the high temperature (1100°C), the dopant element is partly dissolved in the hard magnetic $Nd_2Fe_{14}B$ phase. This is the case for most of the M1-dopant elements (Al and Ga). The dopant element partly replaces the Fe-atoms and therefore also changes the spontaneous polarisation, Curie temperature and anisotropy field. If the solubility at sintering temperature is low, precipitation within the 2:14:1-phase occurs. This is mainly the case with M2-dopants.

TEM investigations have been carried out in order to identify the various phases in doped and substituted $Nd_2Fe_{14}B$ based magnets. The main effect of the addition of dopant elements is the formation of new intergranular phases. In summary the following phases were identified:[10]

- *Primary hard magnetic phase*: $Nd_2Fe_{14}B$
- *Secondary phases:*
  Nd-rich liquid sintering phase
  $Nd_{1+\varepsilon}Fe_4B_4$
  $M1-Nd(CuNd, GaNd_3, Ga_3Nd_5)$
  $M1-Fe-Nd(M1_{1+x}Fe_{13-x}Nd_6)$
  $M2-B(TiB_2, ZrB_2)$
  $M2-Fe-B(V_{2-x}Fe_{1+x}B_2, NbFeB)$
  $\alpha-Fe$
  Oxide-phases $(Nd_2O_3,..)$
- *Additional* phases in Dy- and Co-substituted and M1- and M2-doped *magnets*:
  $Nd(Co,Fe)_4B$
  $Nd(Co,Fe)_2$
  Nd_3Co
  Dy-containing phases

Intergranular phases change the coupling behaviour between the hard magnetic grains. Non-magnetic phases eliminate the direct exchange interaction and also reduce the long-range magnetostatic coupling between the hard magnetic grains; both effects lead to an increase of the coercive field. The Kerr optical micrograph of Fig.1 shows the magnetic domain contrast inside the grains parallel to the alignment direction of the magnet. The formation of intermetallic, soft magnetic Nd-(Fe,S2) phases, such as the Laves type $Nd(Fe,S2)_2$-phase, deteriorate the coercivity of the magnets. If dopant elements M1 or M2 are added to Nd–Fe–B, generally the coercivity is increased and the corrosion resistance is improved. This is the case, if the Nd-rich intergranular phase is replaced by other phases, such as $AlNd_6Fe_{13}$ and Nd_3Co, especially in large-grained magnets. Figure 2 is a TEM micrograph showing the additional intergranular $Al_1Fe_{13}Nd_6$ and Nd_3Co phases in a Nd-(Fe,Co)-B:(Al,Mo) magnet. On the other hand, the decrease of the volume fraction of the hard magnetic phase within the magnet decreases the remanence.

The processing route of the magnet strongly influences the grain size and grain size distribution. The coercive field in sintered magnets strongly depends on the sintering parameters, such as temperature and time. Nanocrystalline and submicron magnets are obtained by the melt-spinning route, or by mechanically alloying, or by the HDDR process.[11-13] Hot pressing and die upsetting of Nd-Fe-B ribbon materials reveals a densely packed, anisotropic magnetic material. Platelet-shaped grains with diameters less than 1 micrometre are observed by TEM-investigations. The degree of

Fig. 1 Kerr optical micrograph showing the magnetic domain contrast inside the grains parallel to the alignment direction of the optimised Nd–Fe–B magnet. The average grain size is in the order of 2–5 micrometres.

orientation of the platelets, which are stacked transverse to the press direction with the easy c-axis perpendicular to the face of each grain, determines the remanence and coercive field of the magnet. These magnets are fully dense, and energy density products as large as 360 kJm^{-3} have been attained.[14] The degree of alignment, size and shape of the grains and the intergranular regions within the ribbons control the macroscopic magnetic properties. Die-upsetting modifies the spheroidal grains[15] after hot-pressing to platelets as shown in the TEM micrographs of Fig. 3. Misaligned grains, which are clearly visible, deteriorate the remanence. The c-axis for each grain runs perpendicular to the straight elongated edge. Nd-rich phase is found among the platelet shaped grains as a fine layer between the straight edges or as pockets at the end of the platelets or between the misaligned and aligned grains. On the other hand the magnets with a lower remanence show a microstructure with more equiaxed grains. In most of the melt-spun magnets regions with abnormally grown, large grains were found. Some of these grains were fully developed, platelet shaped grains.

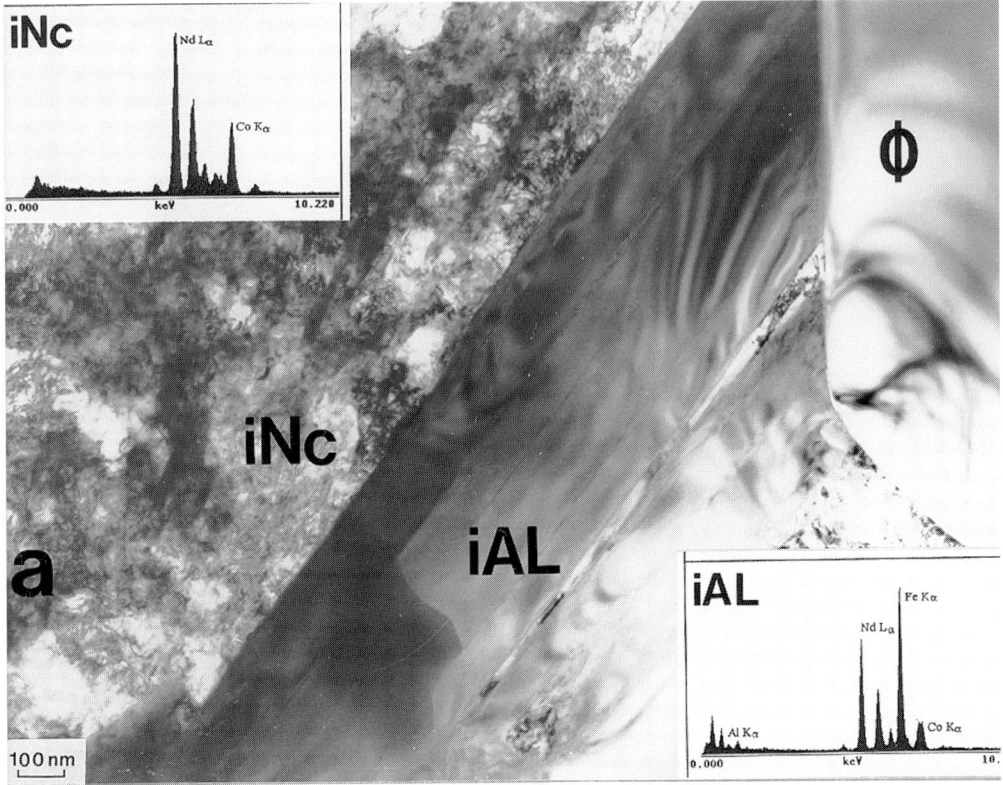

Fig. 2 TEM micrograph showing the additional intergranular $Al_1Fe_{13}Nd_6$ (iAL) and Nd_3Co (iNc) phases besides the $Nd_2(Fe,Co)_{14}B$ (Φ) phase in a Nd-(Fe,Co)-B.(Al,Mo) magnet.

HIGHEST ENERGY DENSITY Nd-Fe-B MAGNETS

Nd–Fe–B sintered magnets possessing outstanding magnetic properties have developed into a major permanent magnet material in the past 15 years since their invention. The drastic increase of the energy density product of newly developed $Nd_2Fe_{14}B$ based magnets enabled the invention of many new applications of permanent magnets. Applications of highest energy density magnets (> 400 kJ m^{-3}) are expanding; voice coil motors for hard disk drives, magnetic resonance imaging, small sized motors, electrical devices and so on. These magnets are produced by a conventional powder metallurgical process which is essentially based on alloy-making, coarse-milling, pulverising, pressing in a magnetic field, sintering, heat-treatment and surface coating. In this process, it is very important to keep the processing atmosphere either in a vacuum or in an inert gas because rare earth elements easily oxidise. The theoretical maximum value of the energy density product is in the case of a perfectly square demagnetisation curve given by:

$$(B \cdot H) \, {}^{theor.}_{max} = \frac{1}{4 \cdot \mu_0} \cdot J_r^2 = \frac{1}{4 \cdot \mu_0} \cdot \left(J_s \cdot \frac{\rho}{\rho_0} \cdot V_{hm} \cdot \cos\varphi \right)^2 \qquad (1)$$

$$\text{if } | _j H_c | \geqslant \frac{1}{2 \cdot \mu_0} \cdot J_r \qquad (2)$$

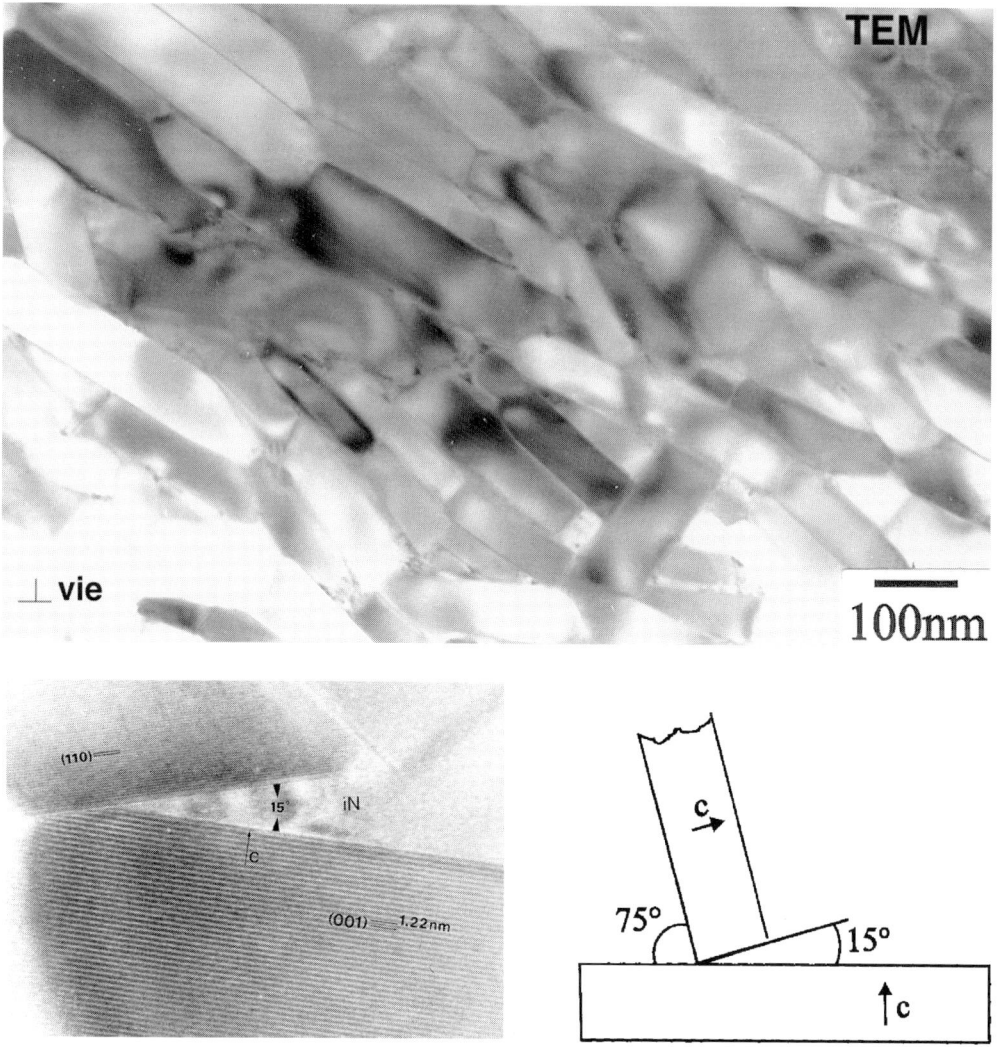

Fig. 3 TEM micrographs of a hot pressed and die-upset $Nd_{14}Fe_{72}Co_7B_6Ga1$ magnet with $J_r = 1.32$ T, $_jH_c = 1240$ kA m^{-1} (perpendicular to c-axis view).

153

where J_s is the saturation magnetisation of the hard magnetic phase (1.61 T), V_{hm} and φ are the volume fraction and the degree of alignment of the hard magnetic grains, respectively. In order to enhance J_r and therefore the energy density product, it is necessary to avoid pores and to densify the magnets up to the theoretical value, increase the volume fraction V_{hm} and achieve a high degree of alignment φ. In anisotropic sintered magnets, the orientation of each grain from the easy direction of the magnet is usually described by the relation:

$$(B \cdot H)_{max}^{theor.} = \frac{1}{4 \cdot \mu_0} \cdot J_r^2 = \frac{1}{4 \cdot \mu_0} \cdot \left(J_s \cdot \frac{\rho}{\rho_0} \cdot V_{hm} \cdot \cos\varphi \right)^2 \tag{1}$$

$$\text{if } |_{J}H_c| \geq \frac{1}{2 \cdot \mu_0} \cdot J_r \tag{2}$$

The theoretical value of the maximum energy product of $Nd_2Fe_{14}B$ based magnets is calculated to be 512 kJ m^{-3} assuming 100% perfect alignment and 100% volume fraction of the hard phase. The origin of this magnetic property lies in the $Nd_2Fe_{14}B$ ternary tetragonal compound as a main phase. In addition, according to the ternary Nd–Fe–B phase diagram this magnet contains also a certain amount of $Nd_{1.1}Fe_4B_4$-phase and a Nd-rich phase. which is essential for sintering with liquid phase. In order to densify the magnets up to the theoretical density, it is very important to control the composition of magnets thus generating a sufficient amount of liquid phase at sintering.

Several authors have reported obtaining Nd–Fe–B based magnets with a 400 kJ m^{-3} energy density product by

- keeping the oxygen content low[16]
- using the powder mixing technique[17]
- increasing the magnetising field and reducing the pressure during compaction[18]

A new technology – Rubber isostatic pressing (RIP) – has been developed by Sagawa et al.[19,20] to improve the orientation of the particles in the green compact to obtain sintered magnets with perfect orientation. Rubber isostatic pressing (RIP) is one of the key technologies to approach the theoretical limit of the magnets based on $Nd_2Fe_{14}B$. In RIP, the magnet powder is subjected to such a strong pulsed field just before the compaction that the powder in the rubber mould is thoroughly oriented. Then the powder is compacted isostatically, while the orientation is completely held. In the conventional die pressing which uses no rubber moulds, the pressure applied to the powder is uniaxial. The uniaxial pressure tends to disturb the orientation of the particles during the pressing. To prevent this orientation disturbance, the powder has to be subjected to a strong magnetic field throughout the pressing

The high orientation of the magnet produced by RIP is attributed to:

- Application of a strong pulsed field which dissolves the agglomeration of the magnet powder particles and then impulsively orients the particles
- Isostatic pressing which holds the orientation tight during the pressing

Magnets of the composition $Nd_{15.1-x}Fe_{78+x}B_6Cu_{0.03}Al_{0.7}$ [x = 0–2.5] were prepared by the powdermetallurgical sintering route.[21] The energy density product > 400 kJ m^{-3} and the coercive field of 800 kA m^{-1} were obtained after a combination of rubber isostatic and transverse die pressing methods under optimised sintering conditions. The misalignment of the hard magnetic grains with a diameter of ~2–5 μm was calculated in the best case to be of the order < 14°. The high oxygen content of the magnets was gradually decreased from values of 4000–6000 ppm to a value < 1000 ppm. This high oxygen content was one limiting factor to decrease the Nd-content in order to improve the volume fraction of the hard magnetic phase.

The demagnetisation curves of two optimised magnets with a low and a high oxygen content and a composition of $Nd_{13.5}Fe_{bal}B_{5.95}Cu_{0.03}Al_{0.7}$ are shown in Fig. 4. The magnets produced were sintered between 960°C and 1100°C. The sintering temperature was varied to get optimum density (7.5–7.6 g cm^{-3}) and (B.H)$_{max}$. The density of the samples and the remanence increased with increasing sintering

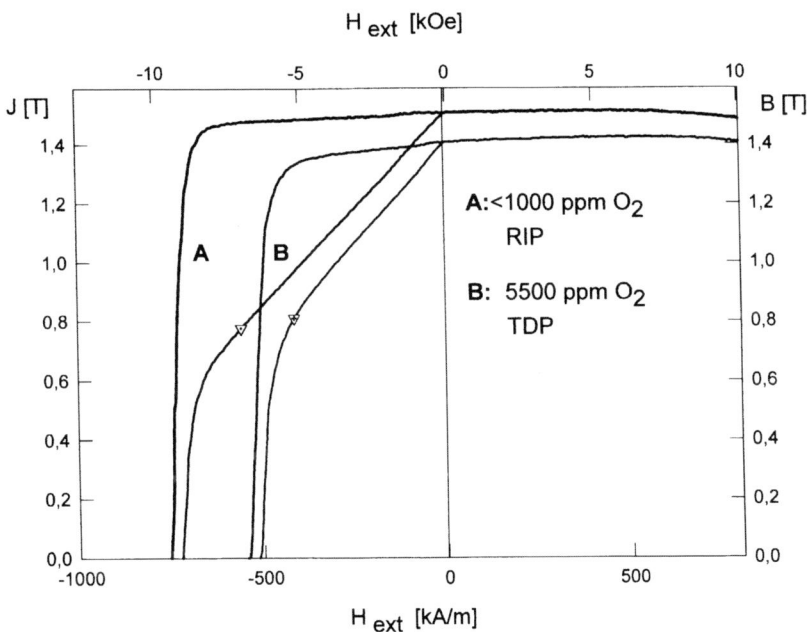

Fig. 4 Comparison of demagnetisation curves of $Nd_{13.5}Fe_{bal}B_{5.95}Cu_{0.03}Al_{0.7}$ sintered magnets manufactured by rubber isostatic die pressing (RIP) and transverse die pressing in air (TDP). (A) J_r=1.505 T, (B.H)$_{max}$= 432 kJ m^{-3} (B) $J_r \nabla$1.406 T, (B.H)$_{max}$= 333 kJ m^{-3}

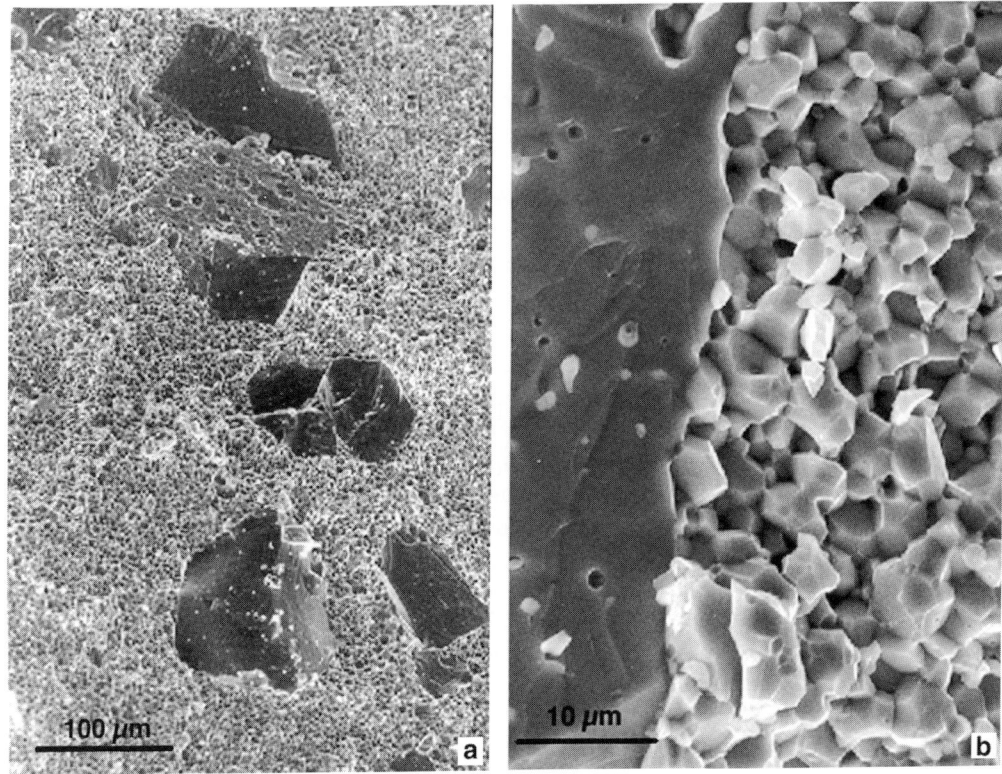

Fig. 5 Scanning electron micrograph of a fractured surface showing the abnormally grown grains in a low oxygen (< 1000 ppm) $Nd_{13.7}Fe_{bal}B_{5.95}Cu_{0.03}Al_{0.7}$ magnet sintered at 1000°C for 3 hours.

temperature on keeping the sintering time constant (3 hours), while the squareness of the demagnetisation curve only partly increased and drastically decreased as abnormal grain growth of the 2:14:1 grains occurred.[22] Abnormal grain growth (AGG) of the 2:14:1 grains occurred preferentially in magnets with low oxygen content (Fig. 5), thus the squareness of the demagnetisation curve drastically decreased. The oxygen content strongly affects the AGG and the magnets with higher oxygen content have higher critical temperatures at which the AGG occurs. On the other hand, isotropic magnets tend to have lower critical temperatures than anisotropic magnets by 10–20°C. A single or a two step (800°C plus quenching) annealing treatment at 500–600°C for 60 min increased the coercive field by about 200 kA m^{-1} or about 20%. Besides the density and coercive field the squareness of the demagnetising curve also has to be optimised in order to get the highest values of energy density products.

The influence of oxygen on the hard magnetic properties is more complex. Kim *et al.*[23] reported that a controlled doping with oxygen improved grain alignment and resulted in an increase in remanence, coercivity and loop squareness. Endoh and Shindo [18] reported an energy density product of 410 kJ m^{-3} of a low oxygen (< 2000 ppm) $Nd_{12.7}Fe_{81.3}B_6$ sintered magnet. One possibility to improve the alignment factor is to optimise the alignment field and/or pressure during transverse pressing. Even the sintering process influences the degree of alignment of the grains.[24] On the other hand several authors[20, 25] found that the highest alignment is obtained by isostatic die pressing (φ = 11–14°), followed by transverse field die pressing (φ = 18–20°)and axial field die pressing (φ = 25–27°).

MICROMAGNETIC SIMULATION OF MAGNETIC DOMAIN NUCLEATION AT GRAIN BOUNDARIES

The coercive field is determined by the long range dipolar interaction and short range exchange coupling between neighbouring hard magnetic grains. The doping of elements changes the phase relation and favours the formation of new phases. Additional secondary non-magnetic intergranular phases decrease the remanence and interrupt the magnetic interactions between the grains, thereby improving the coercivity of large grained sintered magnets. For magnets with higher Nd-concentrations grain size, misorientation and distribution of grains control coercive field. Numerical micromagnetics help to understand the correlation between the intrinsic magnetic properties, the microstructure and the magnetic interaction that determines coercivity and remanence. The dipolar interactions considerably reduce the coercive field of ideally oriented particles with respect to $_JH_c$ of an isolated particle. Exchange coupling between misaligned grains drastically reduces the coercive field. The higher the Nd-content of the magnet and therefore the volume fraction of the Nd-rich intergranular phase is, the more reduced is the contribution of the exchange and also dipolar coupling between the grains.

The numerical simulation starts from the total magnetic Gibbs free energy E_t which is the sum of the sum of the exchange energy, the magneto-crystalline anisotropy energy, the magnetostatic energy, and the Zeeman energy of the magnetic polarisation in an external field H_{ext}:[26]

$$E_t = \int \left[A \sum_{i=1}^{3} (\nabla \beta_i)^2 - K_l (\mathbf{J} \cdot \mathbf{u})^2 - \tfrac{1}{2} \mathbf{J} \cdot \mathbf{H}_d - \mathbf{J} \cdot \mathbf{H}_{ext} \right] dV \qquad (3)$$

where A is the exchange constant, H_d is the demagnetising field, K_1 is the anisotropy constant and u denotes the unit vector parallel to the c-axis. When the direction cosines of the magnetisation β_i are approximated by piecewise linear functions on the finite element mesh, the energy functional (3) reduces to an energy function with the nodal values of the direction cosines as unknowns. Its minimisation with respect to the direction cosines of the magnetisation at the nodal points, subject to the constraint $\beta_1^2 + \beta_2^2 + \beta_3^2 = 1$, provides an equilibrium distribution of the magnetisation. To satisfy the constraint, the magnetisation can be represented by polar coordinates.

The resulting algebraic minimisation problem is solved using a quasi-Newton conjugate gradient technique.[27] The demagnetising field $\boldsymbol{H}_d$ follows from the magnetic scalar potential which is calculated using a hybrid finite element / boundary technique.[28]

Micromagnetic 3D finite element calculations were used to simulate the influence of Nd-rich phases located at grain boundary junctions, reduced anisotropy near the grain boundaries and the degree of alignment on the nucleation of reversed domains. The numerical results show that all three effects are correlated and contribute in a complex way to the measured coercivity. The finite element models used for the simulations are based on TEM investigations. Figure 6a shows a TEM image of a grain boundary junction between three grains and a small amount of a Nd-rich intergranular phase. The 3D drawing of Fig. 6b shows schematically the Nd-rich phase along grain boundary edges and the direct contact between the neighbouring 2:14:1-grains in the middles of the grain faces. Figure 7 shows the 3D model of the finite element model and the generated mesh near the junction of four neighbouring grains. For the calculations, the misalignment of the grains was varied from 8° to 16°.

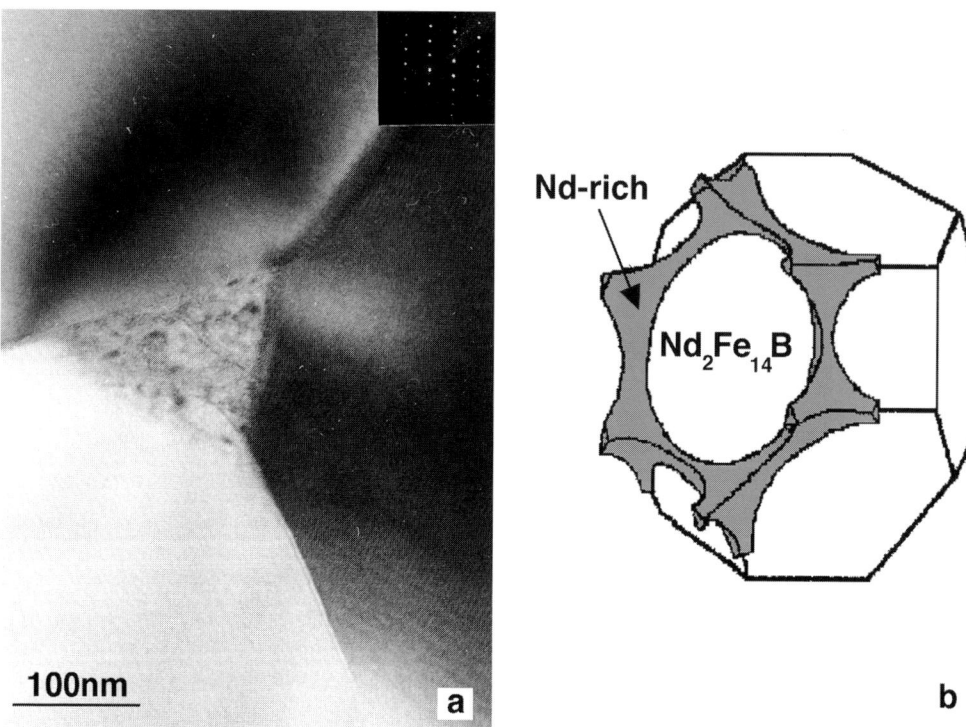

Fig. 6 TEM image of a grain boundary junction of a $Nd_{13.7}Fe_{bal}B_{5.95}Cu_{0.05}Al_{0.7}$ sintered magnet with $(B.H)_{max} = 432$ kJ m^{-3} (a) c-axis of grains is perpendicular to the image plane. A sketch of the 3D grain surrounded by a Nd-rich phase at the grain edges is shown in (b).

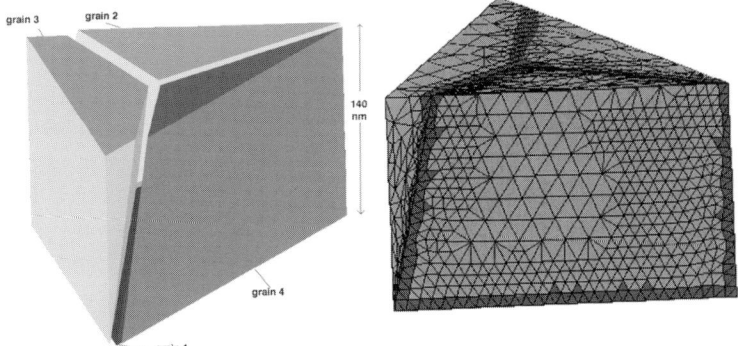

Fig. 7 3D model of the finite element model and the generated mesh near the junction of four neighbouring grains. For the calculations, the misalignment of the grains was varied from 8° to 16°.

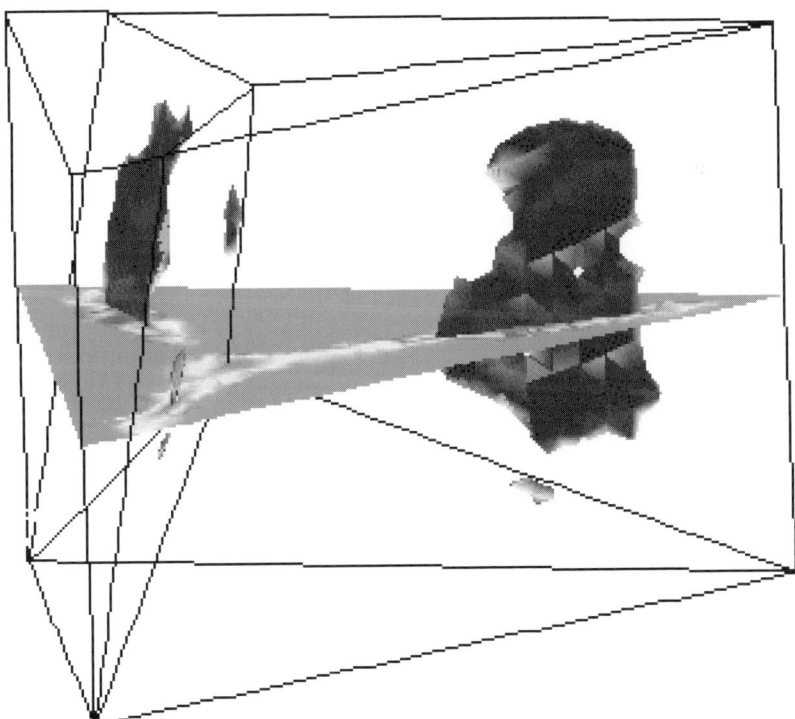

Fig. 8 Finite element simulations allow the identification of the regions within the microstructure where reversed domains are nucleated at H_{ext} = 960 kA m^{-1}. The nucleation of reversed domains starts in the intergranular region assuming K_1 = 0.

159

The difference in the intrinsic magnetic properties between the bulk of the grains and the grain boundary region is taken into account. For a perfect microstructure the numerical results agree well with the Stoner–Wohlfarth theory. The most misoriented grain, which has the largest angle between the c-axis and the alignment direction, determines the coercive field. The coercive field decreases with increasing misalignment. A reduction of the magneto-crystalline anisotropy near the grain boundaries leads to a linear decrease of the coercive field. The coercive field decreases from 3200 kA m^{-1} to 900 kA m^{-1} as the anisotropy constant in a 6 nm thick region near the grain boundaries is reduced from its bulk value to zero. The reduction in the magneto-crystalline anisotropy reverses the dependence of the coercive field on the degree of alignment. The coercive field increases by about 80 kA m^{-1} as the misalignment angle is changed from 8° to 16°. This effect has to be attributed to a higher demagnetising field in the well aligned sample which initiates the nucleation of reversed domains into the defect region. The coercive field of Nd–Fe–B sintered magnets increases with increasing Nd content.[29] The finite element simulations confirm that non-magnetic Nd-rich phases at grain boundary junctions significantly increase the coercive field. The coercive field increases by about 15% as a non-magnetic Nd-rich phase near the grain boundary junctions is taken into account. The simulations show that the presence of the Nd-rich phase significantly changes the exchange and the magnetostatic interactions. As a consequence the nucleation of reversed domains is suppressed. The simulations allow the identification of the regions within the microstructure where reversed domains are nucleated (Fig. 8). The comparison with

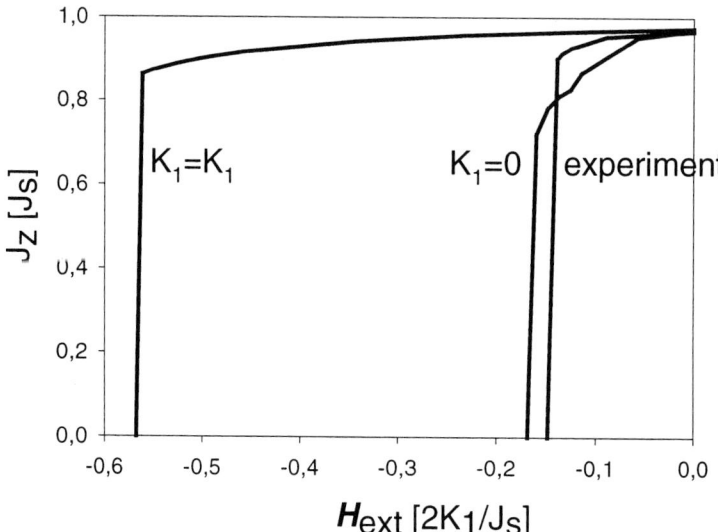

Fig. 9 Effect of the magnetocrystalline anisotropy of the intergranular region between the Nd$_2$Fe$_{14}$B grains on the demagnetisation curve and coercive field compared with experimental data.

experimental data provides a detailed understanding of magnetisation reversal in high energy density permanent magnets. The demagnetisation curves of Fig. 9 compare perfect grain boundaries, distorted grain boundaries and experimental data. The influence of the magnetocrystalline anisotropy of the intergranular region on the coercive field is clearly shown. The simulations explain experimental data[23] which show a decrease of the coercive field with increasing misalignment for high-coercive, Dy-containing Nd–Fe–B magnets, whereas a slight increase of coercivity with misalignment is observed in Dy-free Nd–Fe–B magnets. This effect may be attributed to a 'surface hardening' due to the addition of Dy or Tb. Similar numerical results were obtained by Bachman *et al.,*[30] who used a finite element method combined with an atomic exchange interaction model based on the Heisenberg model for localised interacting magnetic moments and found a decrease of the intrinsic coercivity with an increase in the degree of the grain misalignment assuming a perfect grain boundary. In order to obtain an accurate result for the magnetisation near the grain boundary, Bachman et al. combined an atomistic calculation with a continuous finite element model. Thus it was possible to predict the arrangement of the magnetic moments within the grain boundary and the resulting coercive field precisely.

Soft magnetic secondary phases, such as α-Fe, destroy coercivity in large grained magnets. Numerical, micromagnetic calculations have shown that the critical diameter for the curling process of the polarisation vector inside the α-Fe phase, embedded between hard magnetic grains, is about 90 nm. This is the reason why the formation of large soft magnetic grains considerably reduces the coercive field.[31]

The coercive field of high performance Nd–Fe–B based magnets is determined by the high uniaxial magnetocrystalline anisotropy as well as the magnetostatic and exchange interactions between neighbouring hard magnetic grains. The dipolar interactions between misaligned grains are more pronounced in large-grained magnets, whereas exchange coupling reduces the coercive field in small-grained magnets.

CONCLUSIONS

Microstructural parameters, such as grain size, shape of grains, crystal structure, crystal defects, orientation and composition of various phases, were investigated by means of analytical and high resolution electron microscopy. The complex microstructure considerably influences the magnetic reversal process. Special attention was laid on determining the role of intergranular phases in the coercive field of the magnets. The Nd-rich intergranular phase is necessary for the liquid phase sintering process, but deteriorates the corrosion stability and reduces the remanence. Detailed TEM analysis revealed various additional binary and ternary phases in the intergranular region between hard magnetic grains in doped and substituted Nd–Fe–B magnets.

The investigation of high remanence and low Nd-content sintered magnets reveals Nd-rich phases only at grain boundary junctions. In order to improve coercive field, remanence and energy density, it is necessary to increase the volume fraction of the hard magnetic phase by decreasing the amount of oxygen, Nd and pores, and improving the degree of alignment of the $Nd_2Fe_{14}B$ grains. Oxygen is partly dissolved in the Nd-rich intergranular phase. On reducing the amount of oxygen (< 1000 ppm) the problem of abnormal grain growth during sintering becomes more severe.

The quantitative interaction between magnetisation and microstructure has been calculated by means of a micromagnetic finite element technique. Our micromagnetic simulations of the magnetisation reversal process revealed exact predictions of the influence of intergranular phases on the coercive field: Nd-rich phases at grain boundary junctions reduce the effective coupling between the grains and thus increase the coercive field up to 15 percent. Perfect grains without any reduction of the magnetocrystalline anisotropy show a behaviour which is expected from the Stoner-Wohlfarth theory. The coercive field decreases with increasing misalignment. Defects at the surface reduce the magnetocrystalline anisotropy locally and thus drastically decrease the coercive field. Then the coercive field shows only weak dependence on the degree of alignment. The defects change the local effective fields and thus reverse the behaviour observed for perfect grains: a better alignment slightly reduces the coercive field.

ACKNOWLEDGEMENT

This work is supported by the Austrian Science Fund projects FWF P13433 and Y132–PHY.

REFERENCES

1. K. J. Strnat, G. Hoffer, J. Oson and W. Ostertag, *J. Appl. Phys.*, 1967, **38**, 1001.
2. M. Sagawa, S. Fujimura, N. Togawa, H. Yamamoto and Y. Matsuura, *J. Appl. Phys.*, 1984, **55**, 2083.
3. J. J. Croat, J. F. Herbst, R. W. Lee and F. E. Pinkerton, *J. Appl. Phys.*, 1984, **55**, 2078.
4. J. F. Herbst and J. J. Croat, *J.M.M.M.*, 1991, **100**, 57.
5. J. Fidler and T. Schrefl, *JAP 79*, 1996, 5029.
6. T. Schrefl and J. Fidler, *J.M.M.M.*, 1998, 177–181, 970–975.
7. J. Bernardi, J. Fidler, and F. Födermayr, *IEEE Trans. Magn.*, 1992, **28**, 2127.
8. J Bernardi and J. Fidler, 1994, J. Appl. Phys. 76, 6241.
9. Y. Hirose, H. Hasegawa, S. Sasaki and M. Sagawa, Proc. XV Workshop on RareEarth Magnets and Applications, Vol. 1, Dresden, 1998, 77–96.
10. J. Fidler, 'Rare earth intermetallic magnets', Paper presented at the 11th Int. Conf. on Ternary & Multinary Compounds, ICTMC-11, Salford, UK, September 1997, *Inst. Phys. Conf. Ser. No. 152, Section G: Magnetic Materials*, 1998, 805–813.
11. I. R. Harris, in *Proceedings of the XII. International Workshop on Rare Earth Magnets and their Appl.*, B. Street ed., The University of Western Australia, 1992, 347.
12. R. Nakayama, T. Takeshita, M. Itakura, N. Kuwano and K. Oki, *JAP. 76*, 1994, 412.

13. M. Uehara, H. Tomizawa, S. Hirosawa, T. Tomida and Y. Maehara, *IEEE Trans. Magn.*, 1993, **29**, 2770.
14. J. J. Croat, *J. Less-Common Met.*, 1989, **148**, 7.
15. R. K. Mishra, *J. Appl. Phys.*, 1987, **62**, 967.
16. M. Sagawa, S. Hirosawa, H. Yamamoto, S. Fujimura and Y. Matsuura, *Jpn. J. Appl. Phys.*, 1987, **26**, 785.
17. E. Otsuki, T. Otsuka and T. Imai, *Proc. 11th Int. Workshop on rare earth magnets and their applications*, Pittsburgh, Shankar ed., Vol. 1, 1990, 328.
18. M. Endoh and M. Shindo, in *Proc. 13th Int. Workshop on rare earth magnets and their applications*, Manwaring *et al*. eds, Vol. 1, 1994, 397.
19. M. Sagawa and H. Nagata, *IEEE Trans Magn.*, 1993, **29**, 2747.
20. M. Sagawa, H. Nagata, O. Itatani and W. Watanabe, in *Proc. 13th Int. Workshop on rare earth magnets and their applications*, Manwaring *et al*. eds, Suppl., 1994, 13.
21. S. Sasaki, PhD Thesis, Vienna University of Technology, Austria, 1999.
22. W. Rodewald, B. Wall and W. Fernengel, *IEEE Trans. Magn.*, 1997, **33**, 3841.
23. A. S. Kim, F. E. Camp and H. H. Stadelmaier, *J. Appl. Phys.*, 1994, **76**, 6265.
24. T. S. Chin, M. P. Hung, D. S. Tsai, K. F. Wu and W. C. Chang, *J. Appl. Phys.*, 1988, **64**, 5531.
25. W. Fernengel, A. Lehnert, M. Katter, W. Rodewald and B. Wall, *J.M.M.M.*, 1996, **157/158**, 19.
26. W. F. Brown Jr., *Micromagnetics*, Wiley, 1963.
27. P. E. Gill, W. Murray and M. H. Wright, Practical Optimization, Academic Press, 1993.
28. D. R. Fredkin and T. R. Koehler, *IEEE Trans. Magn.*, 1990, **26**, 415–417.
29. S. Hirosawa and Y. Kaneko, in *Proceedings 15th Int. Workshop on Rare-Earth Magnets and their Application*, 1998, 43.
30. M. Bachmann, R. Fischer and H. Kronmueller, in Proc. X. *Symposium on magnetic anisotropy and coercivity in rare earth transition metal alloys*, Vol. 1, 1998, 217–236,
31. T. Schrefl and J. Fidler, 'Micromagnetics of nanocrystalline permanent magnets', *MRS-99 Proceedings of the Symposium I: Amorphous and Soft magnetic materials for Hard and Soft Magnetic Applications*, April 1999.

The Character and Role of Grain Boundaries in NdFeB-type Alloys and Magnets

I. R. HARRIS AND A. J. WILLIAMS

School of Metallurgy and Materials, University of Birmingham, UK

ABSTRACT

In this paper, the character and role of grain boundaries in NdFeB alloys and magnets are described and discussed. The role of the grain boundaries in the Hydrogen Decrepitation (HD) process is considered and the influence of the character and mobility of grain boundaries on the properties of fully dense sintered magnets is described. Oxygen at the grain boundaries is seen to play an important role, both positive and negative depending upon its concentration. Grain boundary chemistry is also shown to have an effect on the liquid phase sintering process and on the subsequent magnetic properties. The coercivity of sintered magnets are improved by annealing at 650°C for one hour and this has been attributed to changes in the grain boundary constitution and to the smoothing of the boundary by the liquid phase.

Grain boundaries also play an important role in the Hydrogenation Disproportionation Desorption and Recombination (HDDR)-process whereby the coarse grain, cast material is converted to very fine grain particles. During the solid-HDDR process, the initial penetration of the hydrogen takes place along grain boundaries and the finer the grains the more rapid are the hydrogenation kinetics. During the R-stage, the liquid Nd-rich material is redistributed along the newly formed boundaries of the ultra-fine grains with the consequent formation of relatively large scale cavitation at the original grain boundaries with a resultant large fall in the density. Because of the liquid phase at the grain boundaries of the HDDR material at temperatures >650°C (or lower with certain alloy additions, such as Cu), rapid grain growth occurs at elevated temperatures (>800°C) and this results in a degradation of the demagnetisation loop shape and in the coercivity.

INTRODUCTION

Grains and grain boundaries play vital roles in determining the magnetic properties of both hard (permanent) and soft magnetic materials. This article deals with their influence on the former and the latter are considered elsewhere in these proceedings. Specifically, their roles in determining the magnetic properties of NdFeB-type permanent magnets will be described and discussed. Currently there are two main types of magnet based on the ferromagnetic compound $Nd_2Fe_{14}B$, namely the fully dense sintered magnets[1] and the magnets based on the melt spun material.[2] The latter were described and discussed by Davies in this workshop[3] and the present authors will deal with the former together with the HDDR-material,[4,5] which will probably form the basis of a new generation of NdFeB-type magnets.

SINTERED NdFeB-TYPE MAGNETS

The structure of the $Nd_2Fe_{14}B$ matrix phase is shown in Fig. 1 and this tetragonal compound is ferromagnetic with a uniaxial (*c*-axis) magnetocrystalline anisotropy. This is an essential pre-requisite for the production of permanent magnets of this type. The powder metallurgical route enables the production of a fine grain (5–10 μm) highly aligned (*c*-axis), fully dense magnet.

The bulk alloy is produced by induction melting and then reduced to a fine powder (~2 μm) which is aligned and pressed in a magnetic field. The aligned 'green compact' is then fully densified by sintering in a vacuum at around 1050°C. The complete process is summarised in Fig. 2.

The coercivity of these magnets is governed by the nucleation mechanism whereby the microstructure is designed to inhibit the nucleation of reverse, 180° magnetic domains. This necessitates a fine, uniform grain size with smooth grain

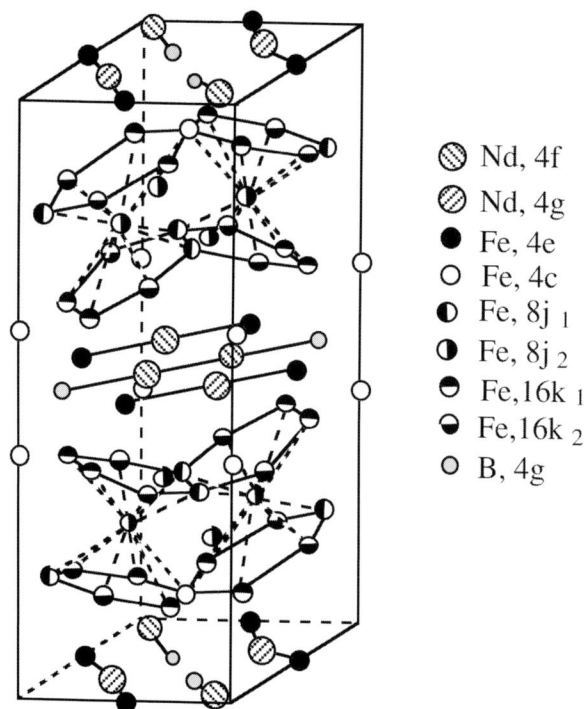

$\bigotimes$ Nd, 4f

$\oslash$ Nd, 4g

$\bullet$ Fe, 4e

$\circ$ Fe, 4c

$\obullet$ Fe, 8j $_1$

$\obullet$ Fe, 8j $_2$

$\obullet$ Fe, 16k $_1$

$\obullet$ Fe, 16k $_2$

$\circ$ B, 4g

Fig. 1 Tetragonal crystal structure of the $Nd_2Fe_{14}B$ phase.

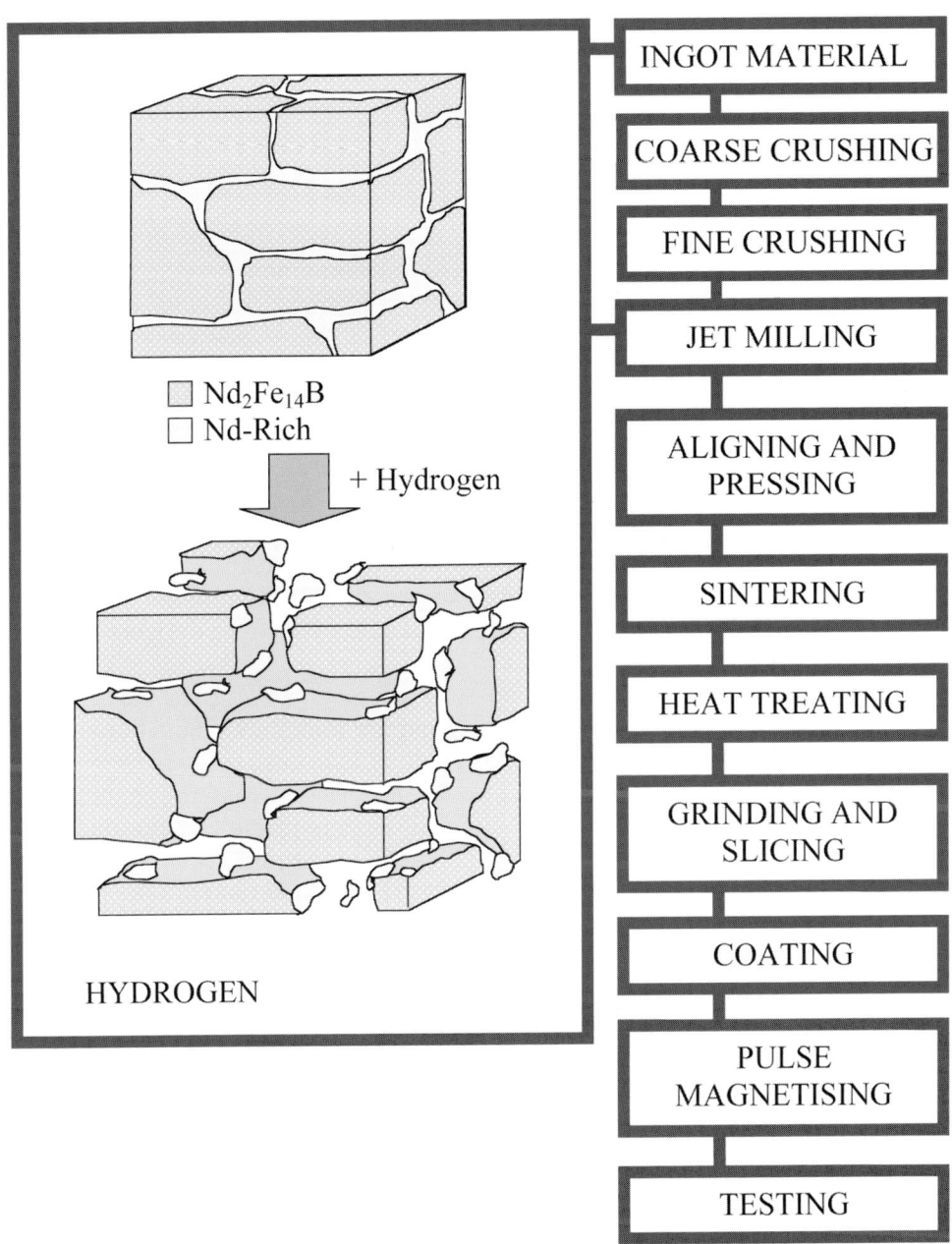

Fig. 2 Powder metallurgy processing route for NdFeB fully dense sintered magnets.

boundaries and the coercivity is further enhanced by the presence of a non-ferromagnetic film at the boundaries to produce magnetic isolation of the individual grains. It is also necessary to maximise the remanence (Br) and this requires a high degree of *c*-axis alignment for the grains, which results in a square hysteresis loop. This lowers the coercivity when compared with that of the isotropic material. The overall coercivity (H_c) can be expressed as

$$H_c = \alpha(T) \frac{2K_1}{M_s} - NM_s \tag{1}$$

Where $\alpha < 1.0$ and is a temperature dependent microstructural parameter, which is enhanced by a fine uniform grain size, K_1 is the anisotropy constant, M_s the saturation magnetisation and N is the demagnetisation factor. M_s should be as large as possible in order to maximise the remanence Br (and hence maximise the energy product $(BH)_{max}$). Thus, K_1 should be even larger to give a large value for the theoretical maximum coercivity of $2K_1/M_s$. K_1 is an intrinsic property of the ferromagnetic compound and can be increased substantially by substituting Nd with Dy. However, this results in a significant increase in the cost of the raw materials and a decrease in M_s (and hence Br) since the large Dy-moment is anti-parallel to those of the Fe and Nd atoms. The N-factor is increased by any microstructural features which concentrate the local demagnetisation fields such as surface damage, interphase boundaries, grain boundary steps. The microstructure of the magnets should be designed to minimise N.

The production route for NdFeB-type sintered magnets (Fig. 2), incorporates the Hydrogen Decrepitation (HD) route[6] whereby the bulk, induction melted material is reduced to a moderately fine powder by exposure to hydrogen (~1bar) at room temperature. A typical microstructure of the cast material is shown in Fig. 3 and this consists of the $Nd_2Fe_{14}B$-type (Ø) matrix phase (A), the intergranular Nd-rich (B) and $Nd_{1.1}Fe_4B_4$ (C) phases. There are regions of ∝-Fe dendrites (D) which form as a result of the peritectic nature of the Ø-phase.[7] The presence of these ductile regions toughens the ingot and makes conventional crushing and milling more difficult but these problems are obviated by the use of the HD-process.

The hydrogen is absorbed initially by the Nd-rich material to form NdH_x (x ~2.7) and subsequently by the Ø phase to form a $Nd_2Fe_{14}BH_x$ solution (x ~3). Under these conditions the $Nd_{1.1}Fe_4B_4$ phase does not absorb hydrogen. These processes lead to extensive differential expansion of the bulk ingot and to the decrepitation process. The failure process is predominantly intergranular in nature so that the HD-powder consists mainly of single grains of the original cast ingot and these are often cracked (as shown in Fig. 4). Thus, the finer the grains of the cast ingot the finer is the HD powder. Failure also occurs at interphase boundaries and some of the various HD-processes are summarised in Table 1.

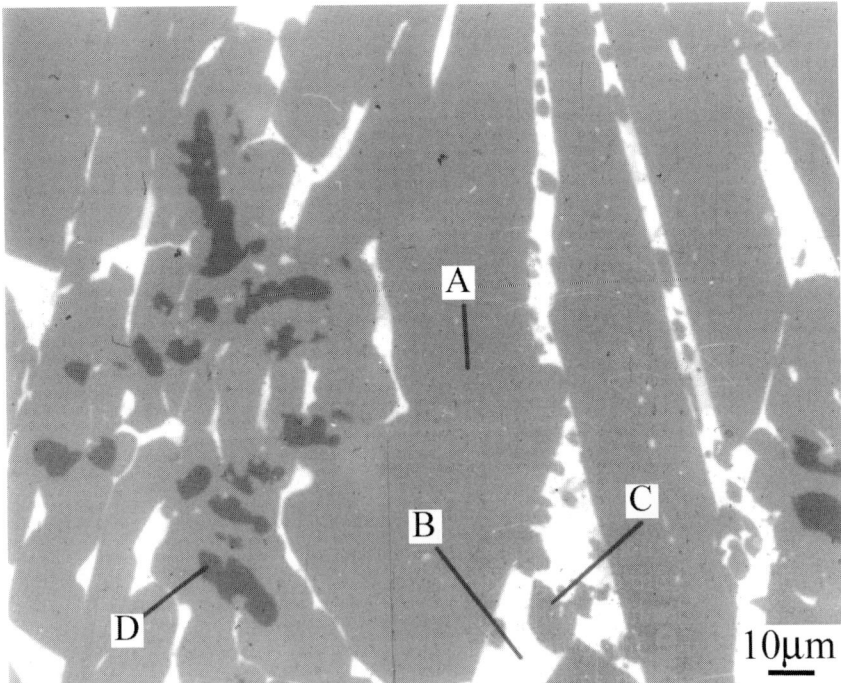

Fig. 3 Back-scattered electron micrograph of cast $Nd_{15}Fe_{77}B_8$: A is $Nd_2Fe_{14}B$, B is Nd-rich, C is $Nd_{1.1}Fe_4B_4$ and D is α-Fe.

Table 1 Some possible types of hydrogen decrepitation process.

Type	Appearance
(1) Intergranular fracture	Granular form might be equiaxed crystals, columnar or variations thereof. Smooth surface with grain boundary debris sticking to surface.
(2) Trangranular fracture	
(i) Random	Irregular shapes with sharp features. Smooth surfaces if brittle failure.
(ii) Cleavage planes	Regular crystallographic shapes. Smooth surfaces.
(iii) Second-phase interface	Depends on the nature of the interface, could result in needles, platelets etc.
(3) Ductile fracture	'Onion skin' effect. Flake-like particles with irregular surface. Poor reflectivity, high surface area.

It has been shown[8] that the HD-powder is extremely friable so that, in the subsequent jet milling process to produce finer powder, the effective capacity of the jet mill is increased by around a factor of 4.0 when compared with that using conventional powder. This spectacular difference in behaviour is illustrated in Fig. 5 where the particle size of the powder is plotted against the powder feed rate into the jet mill. A

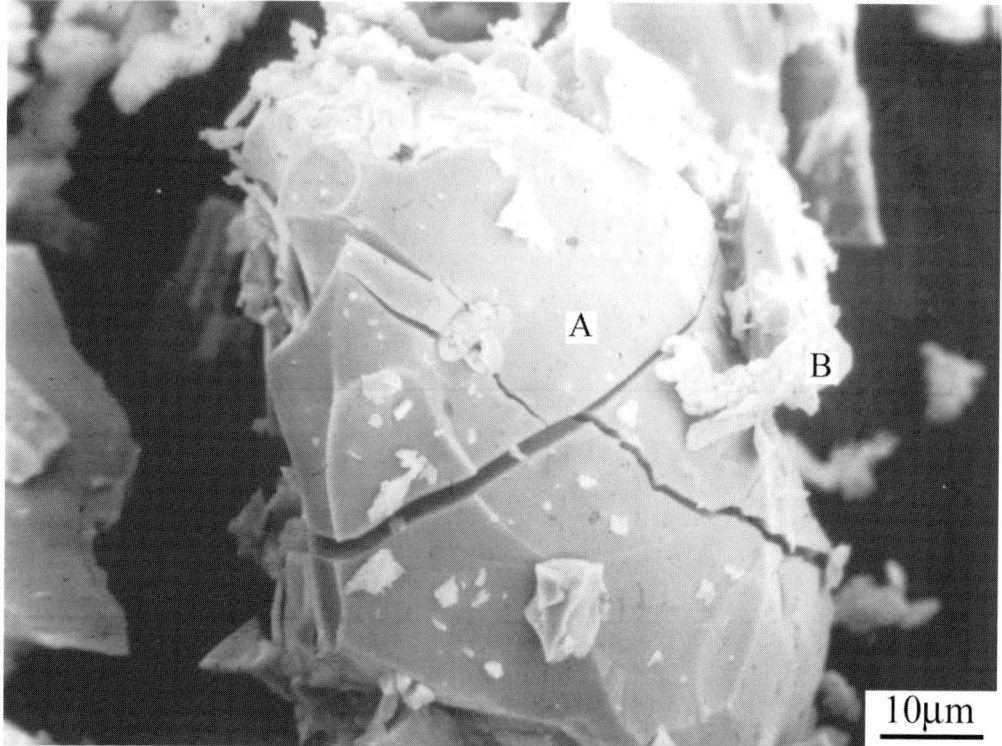

Fig. 4 Hydrogen decrepitated NdFeB alloy; A is particle of $Nd_2Fe_{14}BH_X$ showing transgranular cracks and B is fine particle of Nd-rich hydride.

much steeper dependence on feed rate is obtained for the non-hydrided powder. The finer the particles and hence the grain size of the sintered magnet the higher the coercivity (see above) and hence the advantage of using the HD-powder is clearly evident.

The aligned green compacts are sintered in a vacuum furnace and during the heating process the hydrogen is desorbed from the Ø-phase and from the Nd-rich material. The desorption processes are shown in Fig. 6 and the sintering process begins above the final desorption event at ~650°C, when the desorbed Nd-rich material melts and the new grain boundaries are formed by the process of liquid phase sintering.[9] The various stages of this process are shown in Fig. 7 which shows the sintering behaviour of an aligned $Nd_{16}Fe_{76}B_8$-type powder compact[10] and Fig. 8 shows the fracture surface on the fully sintered magnets.

The variations of density, remanence (Br) and intrinsic coercivity (iH_c) with the sintering temperature (for 1 hour duration) for a $Nd_{16}Fe_{76}B_8$ magnet[6] are shown in Fig. 9. It can be seen that the maximum density and remanence can be achieved by ~1000°C and remains approximately constant at temperatures up to 1080°C. There is, however, a marked fall in the coercivity at temperatures above ~1050°C and this corre-

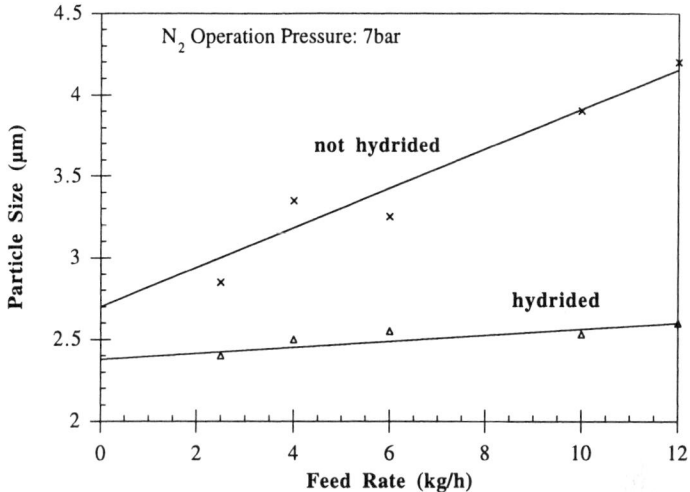

Fig. 5 Average particle size (measured by Fischer sub sieve sizer) plotted against feed rate into jet mill.

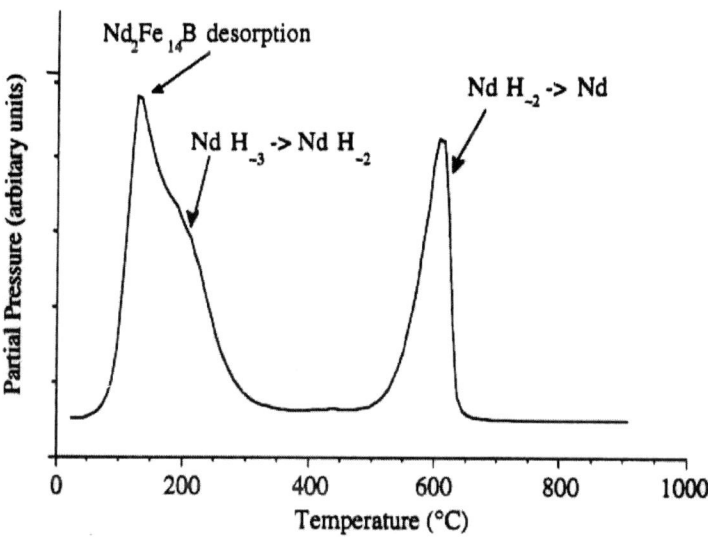

Fig. 6 Mass spectrometer hydrogen desorption trace from hydrided $Nd_{16}Fe_{76}B_8$ alloy.

171

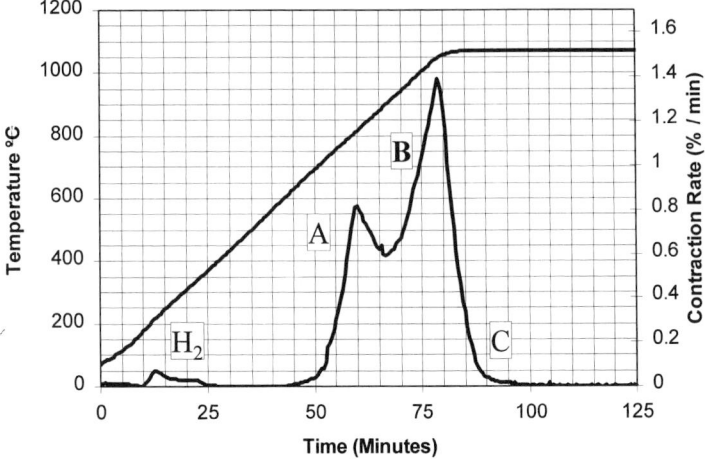

Fig. 7 Dilatometer trace of temperature and vacuum sintering (contraction) rate *v.* time; A is particle rearrangement, B is solution reprecipitation and C is solid state densification.

Fig. 8 Secondary electron micrograph of the fracture surface of a sintered fully dense $Nd_{16}Fe_{76}B_8$ magnet.

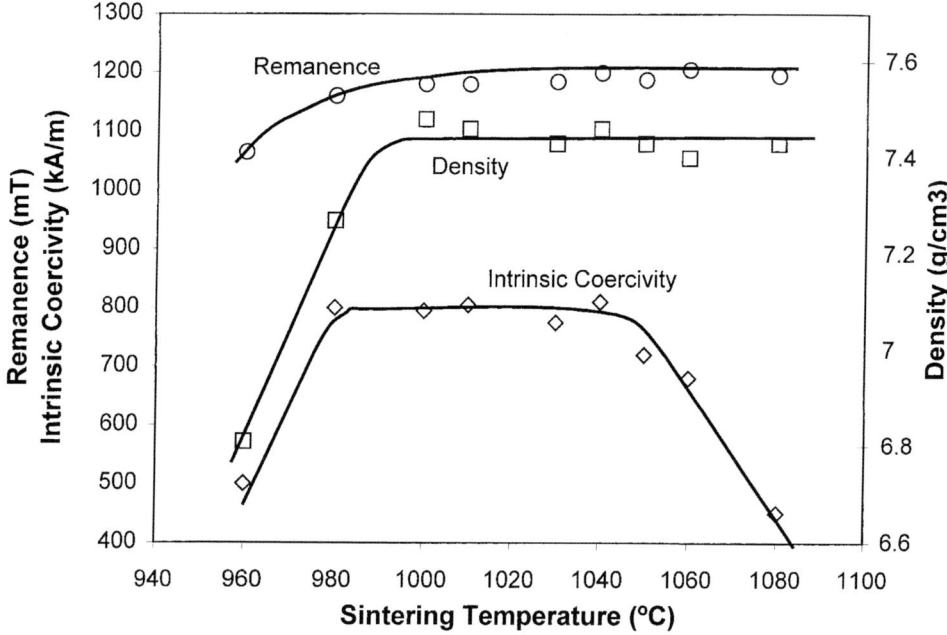

Fig. 9 Density, remanence and intrinsic coercivity plotted against sintering temperature (sintering time = 1 hour).

sponds with a marked increase in the grain size distribution. Microstructural studies[11] show that this corresponds to anomalous grain growth of a limited number of grains [see Fig. 10], which grow favourably in the basal plane of the tetragonal structure.[12]

The loop shape and iH_c are improved by an anneal at 650°C for one hour and this has been ascribed to the smoothing of the grain boundaries, changes in the magnetic nature of the grain boundary phases and improved magnetic isolation.[13] These observations emphasise the importance of the character of the grain boundaries in determining the loop shape and coercivity. This character can also be influenced by minor alloy additions such as aluminium and copper which improve the wetting of the grain boundaries by the liquid phase during the sintering process and subsequent annealing.[14, 15] These additions (particularly Cu) give rise to significant improvements in the coercivity. Amounts of Cu as low as 0.25 at .-% can have a significant effect on the coercivity but it should be noted that all the Cu is concentrated at the grain boundaries so its local concentration is much greater than that indicated by the overall figure.

Earlier, it was mentioned that dysprosium increases markedly the value of iH_c by increasing the value of K_1 in equation (1) but it also reduces the value of Br because of the anti-parallel alignment of its large magnetic moment to that of Fe and Nd. Another negative feature of Dy additions is that they increase the raw materials cost

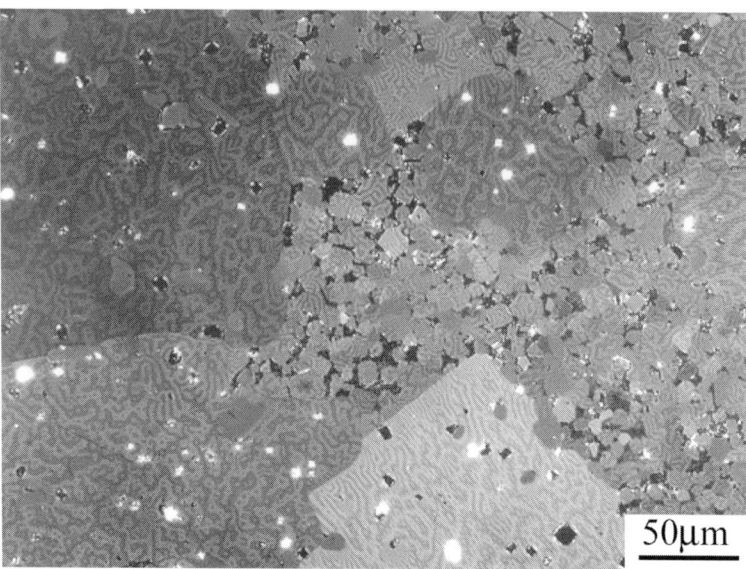

Fig. 10 Optical Kerr micrograph of a sintered $Nd_{16}Fe_{76}B_8$ magnet, annealed at 1000°C for 24 hours to enhance the anomalous grain growth.

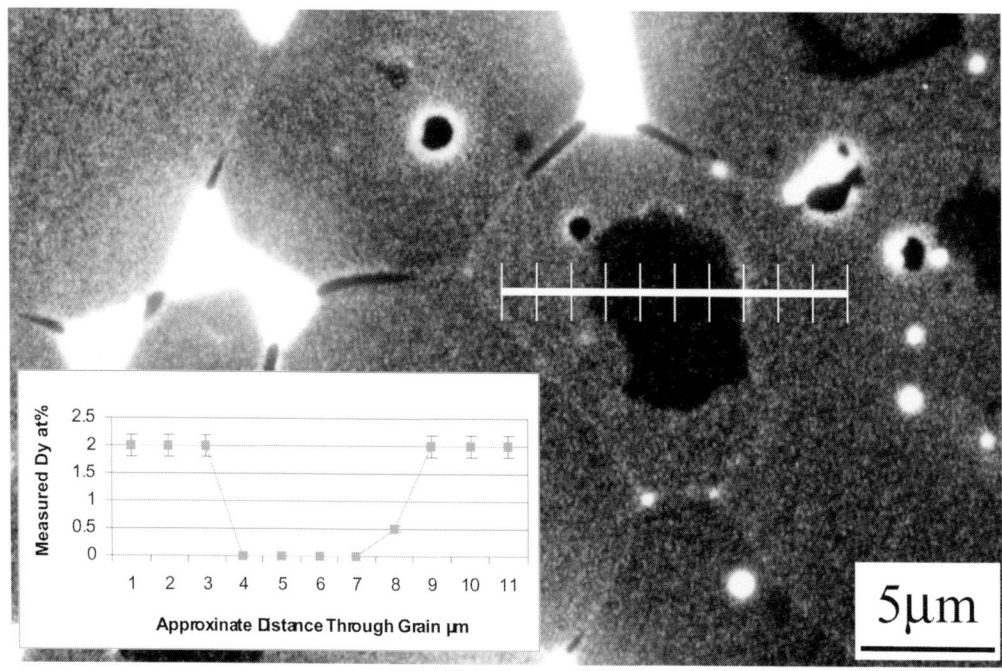

Fig. 11 Back-scattered electron micrograph of a sintered $Nd_{16}Fe_{76}B_8$ (blended with 2 at.-% Dy hydride) magnet; inset EDX analysis of Dy concentration across marked grain.

quite appreciably. Taking all these matters into consideration therefore, there would be a distinct advantage in concentrating the Dy at the grain boundaries where it would have the maximum positive effect on the coercivity and minimum negative effect both on the Br and on the raw material cost, as much smaller amounts of Dy could be employed. Progress has been made in this direction by blending the jet milled powder of a NdFeB alloy with powder of DyH_2 and then aligning and sintering the compacted mixture.[16] The microstructure of the sintered magnet is shown in Fig. 11 and electron back scattered contrast and EDX analysis shows that the Dy is absent from the centre of the grains. This microstructure gives rise to improved coercivity without appreciable losses in the remanence (see Fig. 12). A more rapid sintering treatment should result in an even more concentrated distribution of Dy at the grain boundary.

The levels of oxygen at the grain boundaries also play an important role in influencing the behaviour of NdFeB-type magnets. The oxygen is concentrated in the Nd-rich material at the grain boundary and Kim and Camp[17] have reported that an optimum level of oxygen actually improves the $(BH)_{max}$ value by producing a squarer

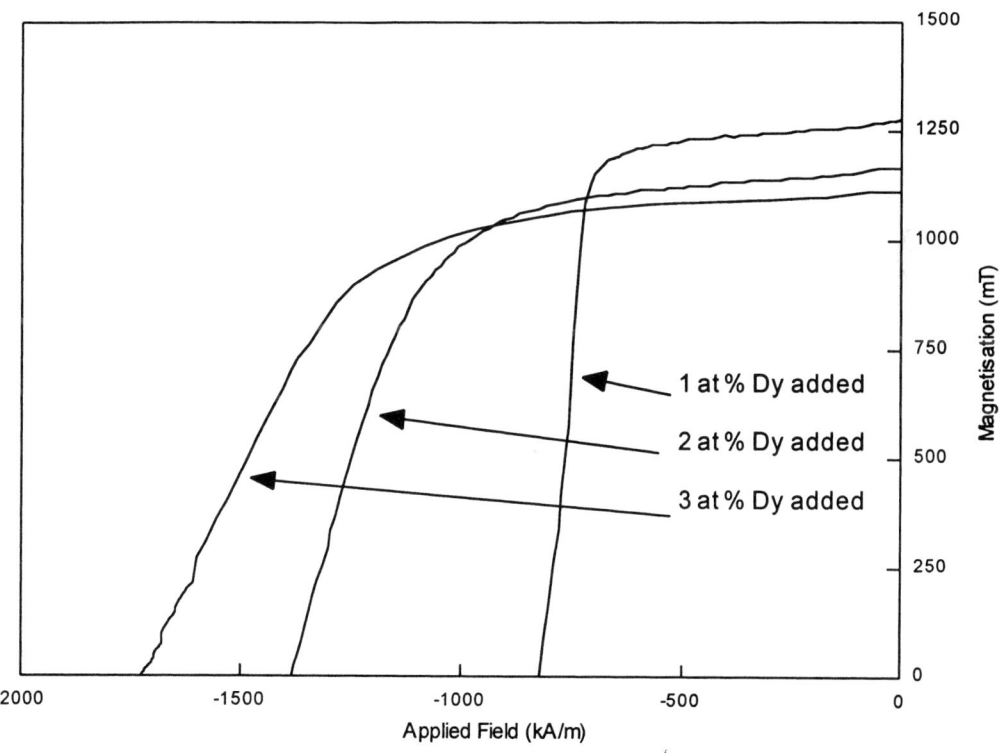

Fig. 12 Demagnetisation curves for $Nd_{16}Fe_{76}B_8$ magnets blended with various amounts Dy hydride.

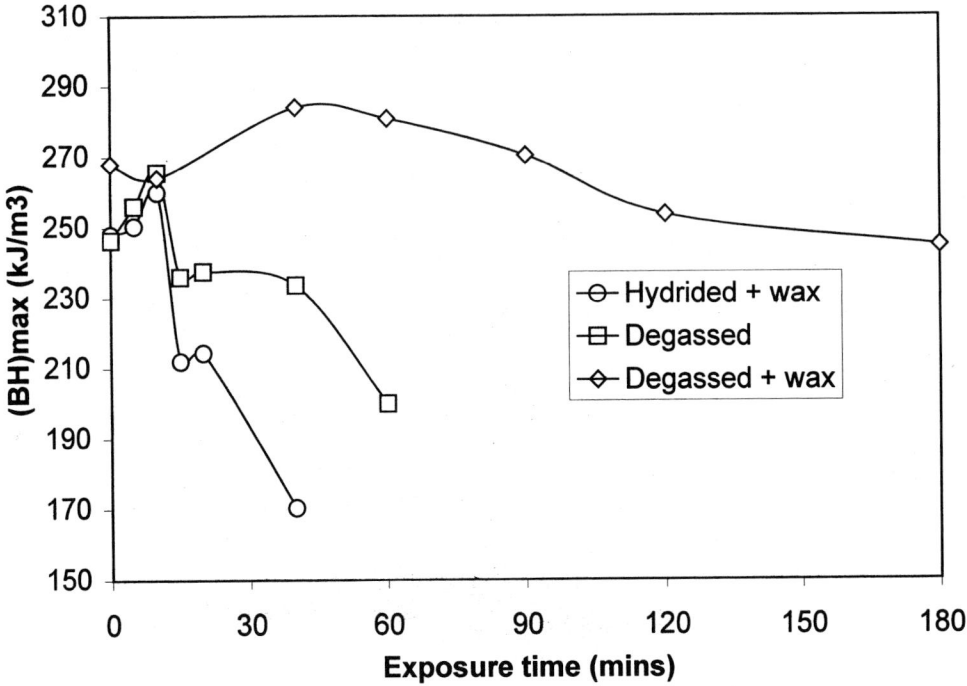

Fig. 13 Maximum energy product, $(BH)_{max}$, plotted against exposure time for fully hydrided or partially degassed powder either with or without pressing lubricant.

hysteresis loop. This has been confirmed by Verdier *et al.*[18] and the beneficial influence of oxygen is shown in Fig. 13. Too much oxygen however leads to excessive oxidation of the Nd-rich material and hence to the loss of liquid phase for the sintering process. This reaction is particularly critical for the high $(BH)_{max}$ magnets (>400 kJ m^{-3}) which have compositions close to that of stoichiometry. The improved loop shapes with optimum oxygen contents is probably related to the inhibition of grain growth by the partially oxidised grain boundaries. This effect can be observed in sintered NdFeB-type magnets which have been annealed at 1000°C for prolonged periods. These studies[19] show that grain growth has occurred within the bulk of the magnet but that much finer grains are observed in the outer region where the oxygen content is higher (see Fig. 14).

HDDR-TYPE MAGNETS

The HDDR-process is summarised in Fig. 15 and is an extremely effective means of dramatically reducing the grain size of the induction melted material (by up to 3 orders of magnitude). If the bulk ingot is heated in hydrogen (~1 bar) from room tem-

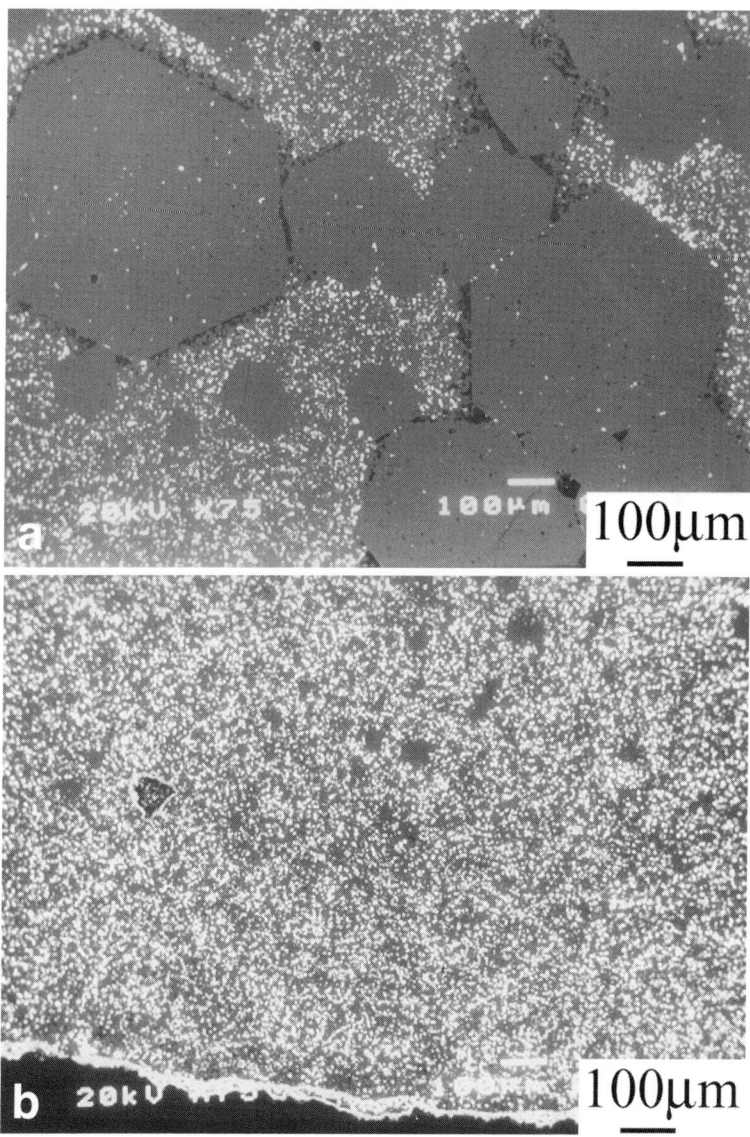

Fig. 14 Back scattered electron micrographs of a $Nd_{16}Fe_{76}B_8$ sintered magnet annealed at 1000°C for 24 hours taken from (a) the centre and (b) the edge of the magnet.

perature then the first stage is the normal decrepitation process described earlier and, as this is predominantly an intergranular fracture process, then the individual particles are predominantly single grains of the original bulk ingot. If the hydrogen is introduced at elevated temperatures ($\geqslant 650°C$) then the bulk alloy undergoes the Solid-HDDR process[20, 21] where the decrepitation process is avoided. In this case, the

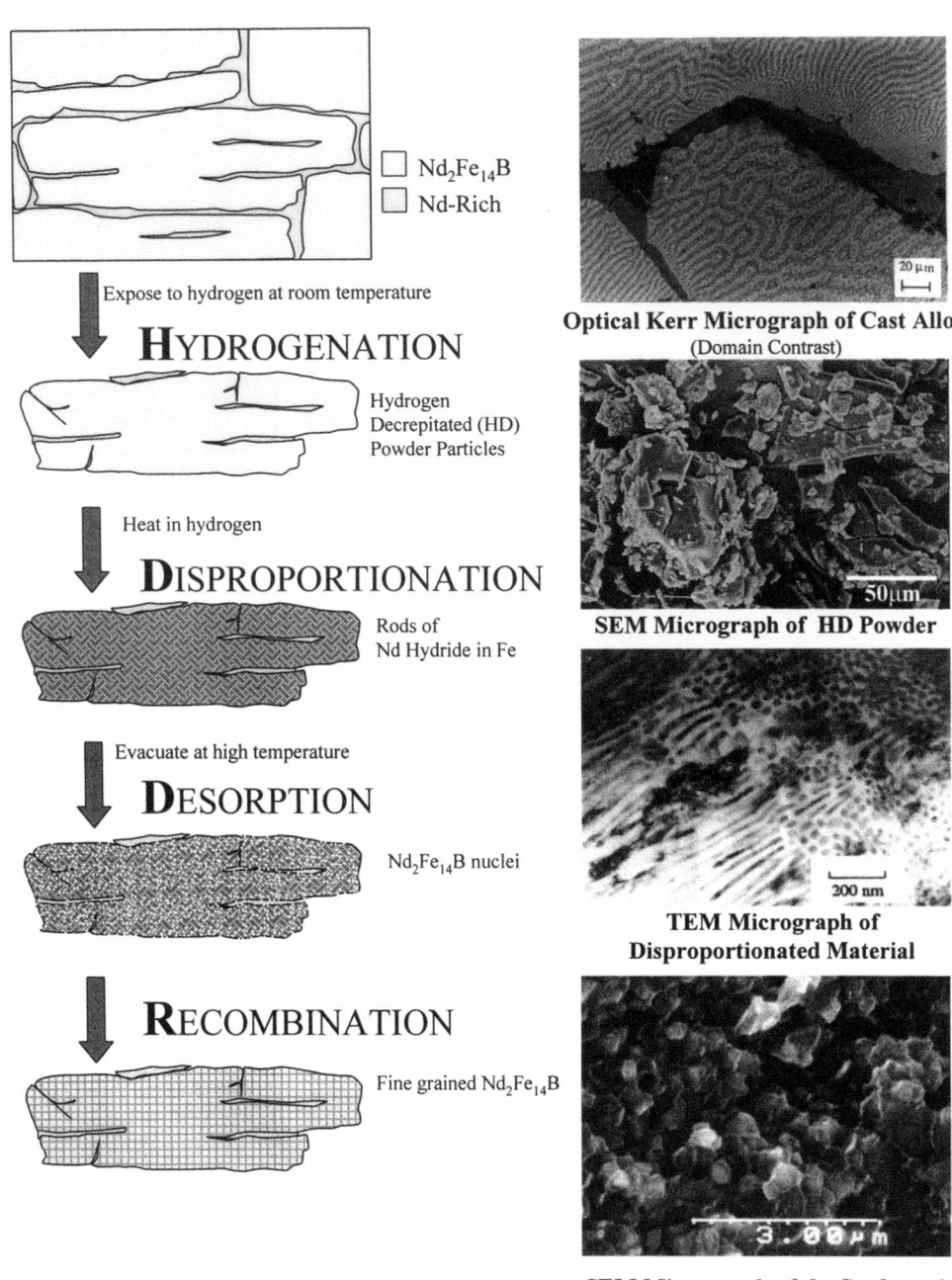

HYDROGENATION

Expose to hydrogen at room temperature

$Nd_2Fe_{14}B$
Nd-Rich

Hydrogen Decrepitated (HD) Powder Particles

Heat in hydrogen

DISPROPORTIONATION

Rods of Nd Hydride in Fe

Evacuate at high temperature

DESORPTION

$Nd_2Fe_{14}B$ nuclei

RECOMBINATION

Fine grained $Nd_2Fe_{14}B$

Optical Kerr Micrograph of Cast Alloy
(Domain Contrast)

20 μm

SEM Micrograph of HD Powder

50μm

TEM Micrograph of Disproportionated Material

200 nm

SEM Micrograph of the Surface of an HDDR Powder Particle

3.00 μm

Fig. 15 Summary of the HDDR processing route.

hydrogen quickly penetrates the grain boundaries and solidifies the previously liquid Nd-rich phase[22] by converting it to the hydride. After this initial, very rapid stage, the disproportionation reaction proceeds into the grains from the grain boundary regions (see Fig. 16).

The Solid-HDDR process can be followed very sensitively by means of electrical resistance measurements (see Fig. 17) and the initial spike (A) corresponds to the rapid hydriding of the grain boundaries as described earlier and the subsequent fall (B) corresponds to the disproportionation of the matrix phase into $\propto$-Fe, NdH_2 and Fe_2B. The process is complete when the resistance reaches the plateau value (C). On applying a vacuum then there is a further fall in resistance (D) resulting from the formation of an Fe, Nd and Fe_2B mixture prior to recombination to form $Nd_2Fe_{14}B$ with a steep rise in the resistance (E) and the formation of the fine (~0.3 μm) grain structure. The resistance of the recombined material exceeds that of the initial state and this discrepancy has been ascribed to cavitation[23] in bulk material. Here, the original Nd-rich intergranular material becomes molten on desorption of the hydrogen and subsequently disperses to the new grain boundaries leaving behind large scale cavitation at the original grain boundaries which causes a drop in the density (Fig. 18) and an increase in the electrical resistance (Fig. 17). The redistribution of the Nd-rich

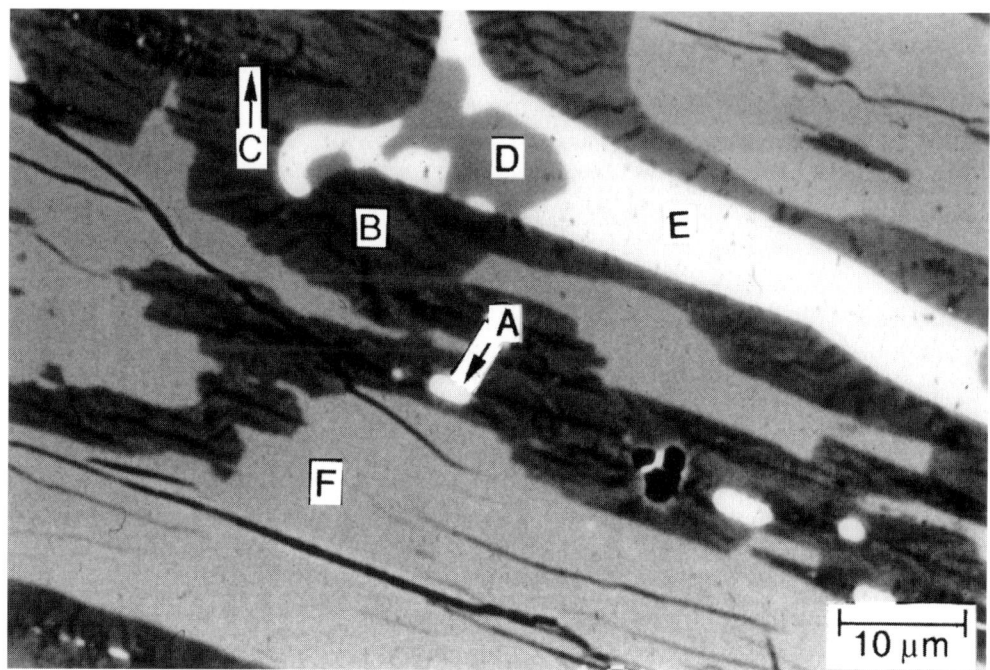

Fig. 16 Back scattered electron micrograph of partially solid-disproportionated $Nd_{16}Fe_{76}B_8$: A is Nd-rich precipitate, B is fine disproportionated mixture, C is coarsened disproportionated mixture, D is $Nd_{1.1}Fe_4B_4$, E is Nd-rich grain boundary phase and F is $Nd_2Fe_{14}B$.

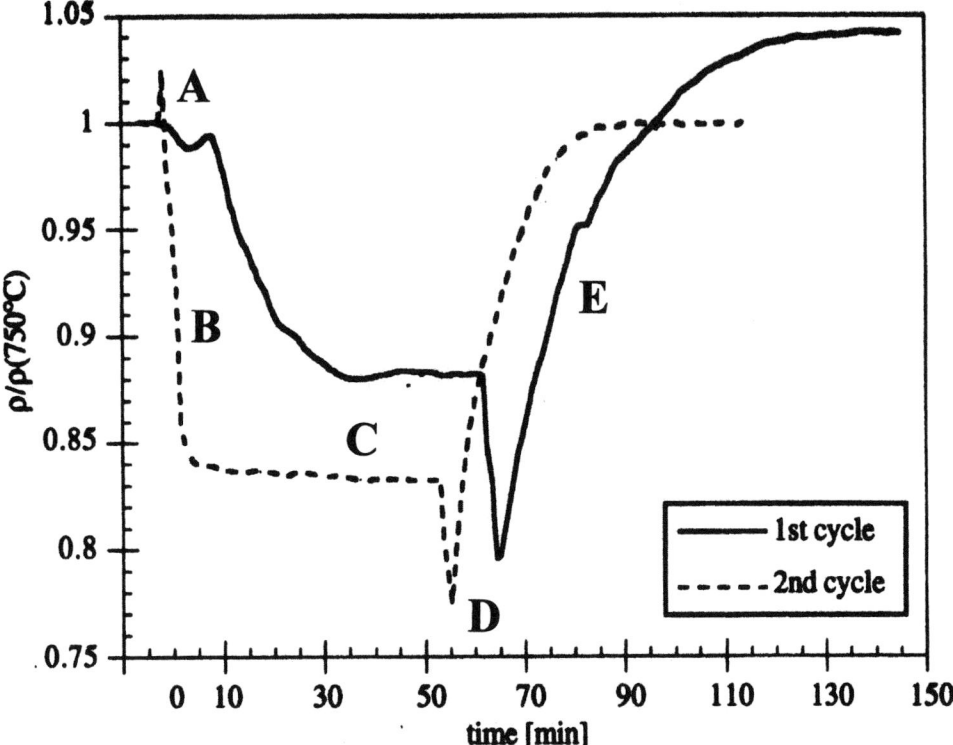

Fig. 17 Change in electrical resistivity of $Nd_{16}Fe_{76}B_8$ plotted against time for the first and second S-HDDR cycles at 750°C.

material is clearly evident from Fig. 19 which reveals the Nd-rich phase at the new, sub-micron grain boundaries. The extensive cavitation (A) can be seen in the fracture surfaces shown Fig. 20.

The subsequent HDDR cycle occurs at a much more rapid rate (Fig. 17) and this is consistent with the much reduced grain size and much increased grain boundary area. As there is no further cavitation on the second cycle then the electrical resistance of the recombined state is identical with that of the initial material and there is no further change in the density.

Because of the presence of the redistributed Nd-rich material at the new grain boundaries, there is a strong tendency for rapid grain growth in the HDDR material once this phase becomes liquid at temperatures >650°C. This is usually in the form of facetted grain growth as can be seen from Fig. 21 and these very angular grain boundaries and large grains result in a fall in the coercivity on annealing at elevated temperatures. Small additions of Zr have been found[24] to inhibit this grain growth and this has been attributed to the formation of Zr-rich preciptates at the grain boundaries which pin the grains.[25]

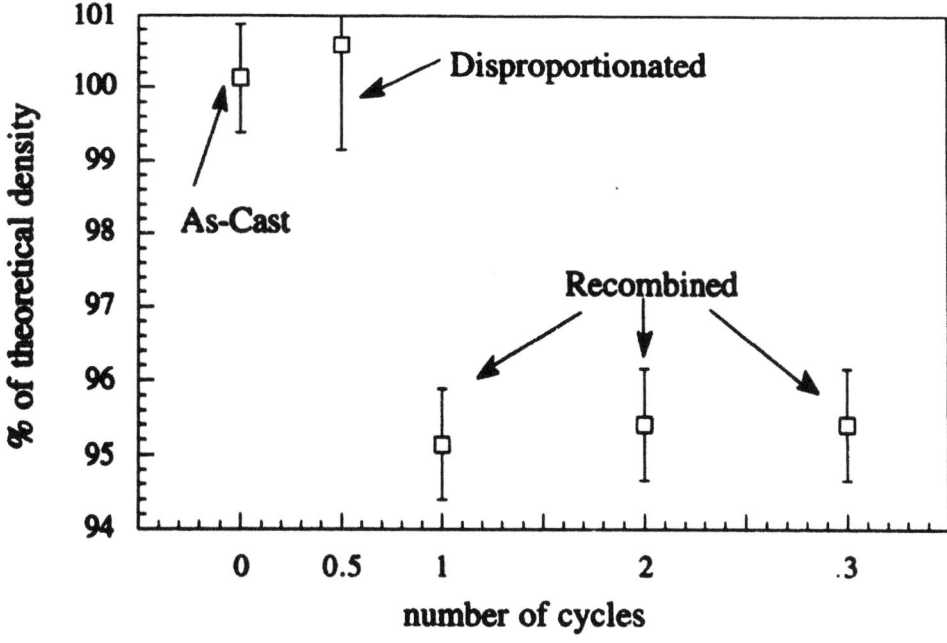

Fig. 18 Density of $Nd_{16}Fe_{76}B_8$ plotted against number of S-HDDR cycles.

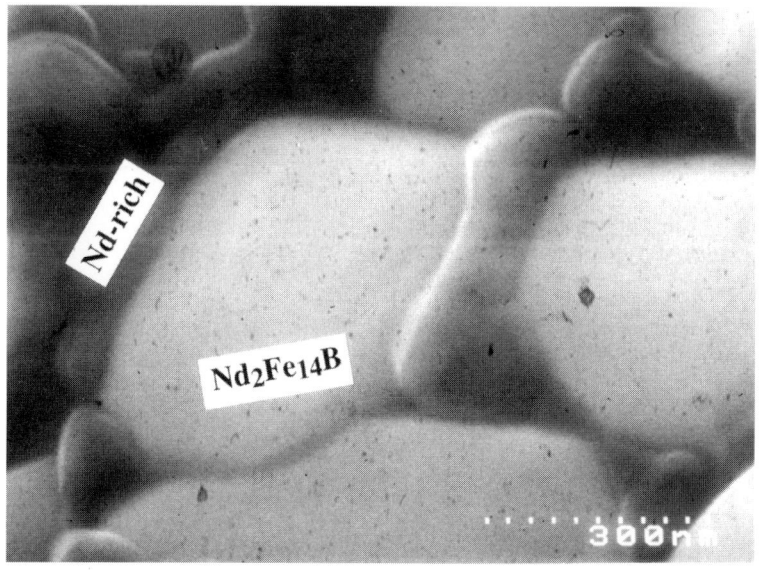

Fig. 19 Secondary electron micrograph of $Nd_{16}Fe_{76}B_8$ HDDR powder showing the redistribution of the Nd-rich phase to the fine $Nd_2Fe_{14}B$ grain boundaries.

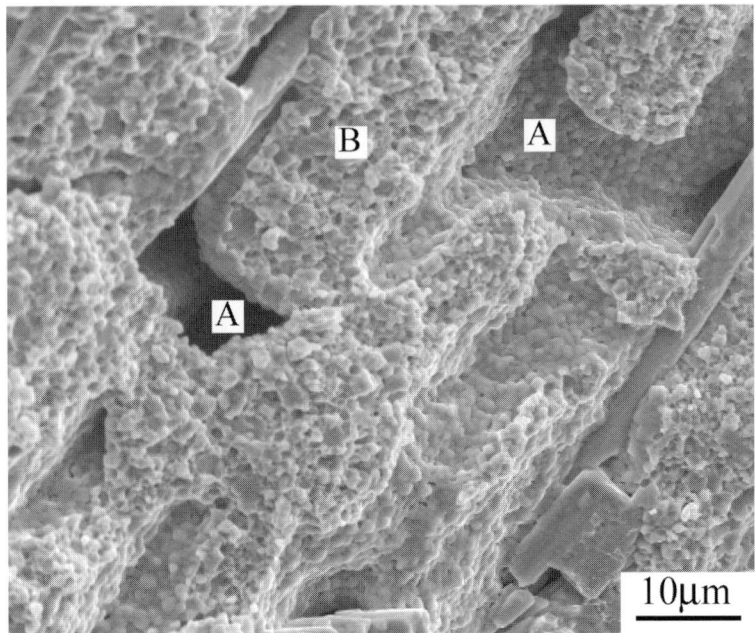

Fig. 20 Secondary electron micrograph of the fracture surface of S-HDDR treated $Nd_{16}Fe_{76}B_8$: A – cavitation and B – fracture surface.

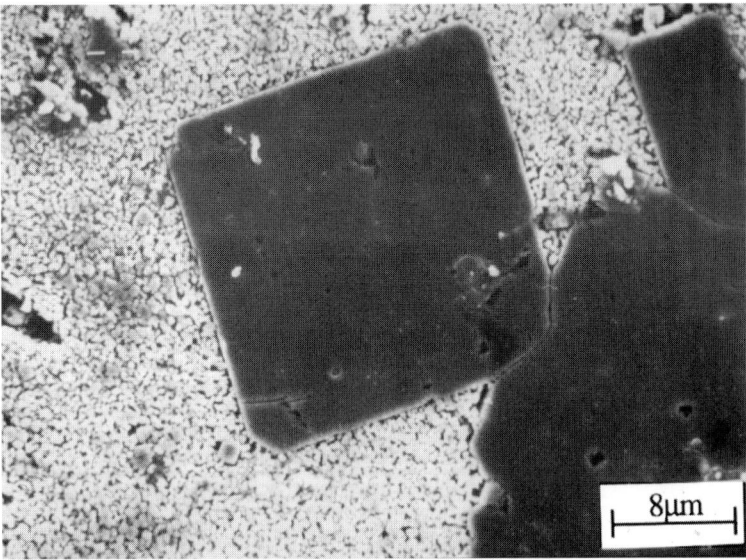

Fig. 21 Back scattered electron micrograph of over processed HDDR $Nd_{16}Fe_{76}B_8$, showing anomalous grain growth.

Fig. 22 Optical ferrofluid micrograph of annealed anisotropic HDDR powder; ferrofluid particles decorate the magnetic domain walls.

It is possible to produce anisotropic HDDR powder [see, for example, Refs. 26 and 27] by appropriate alloy additions and/or by particular HDDR treatments. This powder exhibits much enhanced values of Br and $(BH)_{max}$ compared with those of the isotropic powder. An anisotropic HDDR particle[28] is shown in Fig. 22 and here the grains have been deliberately grown by annealing in order to reveal the domain structure and hence orientations by ferrofluid. It can be seen that the grains have approximately the same orientation but it is by no means perfect. It has been proposed[29] that the metastable phase Fe_3B is a product of the disproportionation reaction and plays a decisive role in developing anisotropy during the recombination stage. The particular nature of the grain boundaries and their mobilities when compared with those of the standard isotropic material will be of particular interest.

CONCLUSIONS

The previous considerations have shown that grain boundaries play a crucial role in both the processing and properties of NdFeB-type sintered and HDDR-type magnets. The following conclusions can be made.

(1) A high proportion of single grain particles are formed by the HD-process because of the intergranular nature of the fracture process.
(2) The shape of the HD-particles reflect the shape of the grains in the original cast alloy.

(3) The rapid fall in coercivity at temperatures in excess of 1050°C for the $Nd_{16}Fe_{76}B_8$ composition is a result of anomalous grain growth of a few grains and this is accelerated by the presence of a liquid phase at the grain boundaries.

(4) Annealing for one hour at 650°C smoothes the grain boundaries and improves the magnetic isolation thus enhancing the demagnetisation loop shape and the coercivity.

(5) An optimum level of oxygen at the grain boundaries improves the loop squareness possibly by inhibiting grain growth. This is also manifest in grain size differences between the outside and inside of magnets annealed for prolonged periods at 1000°C.

(6) Excessive amounts of oxygen at the grain boundaries causes solidification of the previously liquid phase and thus inhibits sintering.

(7) Small additions of certain elements, such as Al or Cu, are concentrated at the boundaries and improve the grain boundary wetting of the liquid phase, hence increasing the loop squareness and the coercivity. In the case of Cu in particular it is sited exclusively at the grain boundaries so that even additions as low as 0.25 at.-% can have a marked effect on the coercivity.

(8) In the case of blending with DyH_2 it is possible to avoid solution of Dy in the central regions of the $\varnothing$-phase. By appropiate processing conditions it should be possible to concentrate this element exclusively at the grain boundaries and such a distribution would have very significant advantages.

(9) Grain boundaries play a crucial role in the S-HDDR process and above the melting point of the Nd-rich material they act as a fast diffusion track for the hydrogen and this is evident from the electrical resistance measurements.

(10) On the first Solid-HDDR cycle, the original Nd-rich liquid phase is redistributed around the new sub-micrometre grain boundaries with the consequent formation of extensive cavitation at the original grain boundaries. This gives rise to an increased electrical resistance and a decreased density.

(11) The much refined grain size and greater grain boundary area gives rise to much faster hydrogenation kinetics on the second cycle and the absence of any further cavitation results in the same electrical resistance at the end as at the beginning of the S-HDDR cycle.

(12) Because of the redistributed liquid phase, there is rapid grain growth in the HDDR material at temperatures >650°C and large facetted grains are formed which lower the coercivity. This grain growth can be inhibited by small additions of Zr to the basic alloy and this can be attributed to the presence of Zr-rich precipitates at the grain boundaries which pin the grains.

(13) Appropriate alloy additions and/or particular HDDR treatment can produce anisotropic HDDR powder. It has been proposed that a product of the disproportionation reaction in these series is the metastable phase Fe_3B, which plays a decisive role in developing anisotropy (preferred orientation) on subsequent recombination.

REFERENCES

1. M. Sagawa, S. Fujimura, M. Togawa, H. Yamamoto and Y. Matsuura: *J. Appl. Phys.*, 1984, **55**, 2083.
2. J. J. Croat, J. F. Herbst, R. W. Lee and F. E. Pinkerton: *J. Appl. Phys.*, 1984, **55**, 2078.
3. H. A. Davies: Presented at Workshop but not submitted for these proceedings.
4. T. Takeshita and K. Nakayama: *Proc. 10th Int. Workshop on Rare-Earth Magnets and Their Applications*, Kyoto, Japan, 1989, 551.
5. P. J. McGuiness, X. J. Zhang, H. Forsyth and I. R. Harris: *J. Less-Common Met.*, 1990, **162**, 379.
6. P. J. McGuiness, I. R. Harris, E. Rozendaal, J. Ormerod and M. Ward: *J. Mater. Sci.* 1986, **21**, 4107.
7. Y. Matsuura, S. Hirosawa, H. Yamamoto, S. Fujimura, M. Sagawa and K. Osamura: *Jpn J. Appl. Phys.*, 1985, **24**, L635.
8. H. Nagel and I. R. Harris: CEAM Value Report, SC 141/91 – UK, MAGPRO-1 1994.
9. R. M. German: *Powder Metallurgy Science* Metal Powder Industries Federation, Princeton, NJ, 1984.
10. R. S. Mottram and I. R. Harris: *Proc. 15th Int. Workshop on Rare-Earth Magnets and Their Applications*, Dresden, Germany, 1998, 473.
11. M. R. Corfield, A. J. Williams and I. R. Harris: *Proc. 15th Int. Workshop on Rare-Earth Magnets and Their Applications*, Dresden, Germany, 1998, 463.
12. C. D. Graham and L. Li: *Proc. 10th Int. Workshop on Rare-Earth Magnets and Their Applications*, Kyoto, Japan, 1989, 427.
13. D. Eckert, D. Hinz, A. Handstein and J. Schneider: *Phys. Status Solidi A*, 1987, **101**, 563.
14. K. G. Knoch, C. Schneider, J. Fidler, E. Th. Henig and H. Kronmüller: *IEEE Trans. Magn.* 1989, **25**, 3426.
15. O. M. Ragg and I. R. Harris: *IEEE Trans. Magn.* 1993, **29**, 2758.
16. R. S. Mottram, A. Kianvash and I. R. Harris: *J. Alloy. Comps*, 1999, **283**, 282.
17. A. S. Kim and F. E. Camp: *J. Mater. Eng.*, 1991, **13**, 175.
18. M. Verdier, J. Morros, D. Pere, N. Shell and I. R. Harris: *IEEE Trans. Magn.* 1994, **30**, 657.
19. M. R. Corfield, A. J. Williams and I. R. Harris: *J. Alloy. Compd*, 2000, **296**, 138.
20. X. J. Zhang, P. J. McGuiness and I. R. Harris: *J. Appl. Phys.*, 1991, **69**, 5838.
21. O. Gutfleisch, M. Verdier, I. R. Harris and A. E. Ray: *IEEE Trans. Magn.* 1993, **29**, 2872.
22. O. Gutfleisch, N. Martinez, M. Verdier and I. R. Harris: *J. Alloy Compd*, 1994, **204**, L21.
23. A. J. Williams, O. Gutfleisch and I. R. Harris: *J. Alloy Compd*, 1996, **232**, L22.
24. D. N. Brown, A. J. Williams, M. Strangwood and I. R. Harris: *MRS Conf.* San Francisco, CA, 1999.
25. G. Yi, D. N. Brown, J. N. Chapman and I. R. Harris: to be published.
26. T. Takeshita and K. Nakayama: *Proc. 11th Int. Workshop on Rare-Earth Magnets and Their Applications*, Pittsburgh, PA, USA, 1990, 29.
27. H. Nakamura, R. Suefuji, S. Sugimoto, M. Okada and M. Homma: *J. Appl. Phys.*, 1994, **76**, 6828.
28. P. J. McGuiness, C. L. Short and I. R. Harris: *IEEE Trans. Magn.* 1992, **28**, 2160.
29. T. Tomida, N. Sano, K. Hanafusa, H. Tomizawa and S. Hirosawa: *Acta Mater.*, 1999, **47**, 875.

Detection and Importance of Magnetic Surface Fields on Steels

A. J. MOSES

Wolfson Centre for Magnetics Technology, Cardiff School of Engineering, Cardiff University, UK

ABSTRACT

This paper focuses on aspects of magnetic fields detected on the surface of steels. The role of grain boundaries is specifically addressed in soft materials, particularly electrical steels used as magnetic core materials in motors and transformers. In these devices magnetic losses, namely hysteresis and eddy current losses, absorb almost 5% of all electrical energy generated, so it is important to understand factors which control the loss mechanisms. The internal performance of these steels can be predicted to some extent from measurement of the magnetic field on their surfaces when magnetised. The point to point variation of magnitude and direction of magnetic field can be detected by various types of sensor so that characterisation of such parameters as power loss, permeability and texture can be made. In this paper the detection methods and typical field distributions are reviewed and the role of grain size, orientation and boundaries in influencing the surface field and ultimately bulk magnetic properties are reviewed.

INTRODUCTION

Electrical steels used for concentrating and directing the magnetic flux in motors and transformers operating at power frequencies under a.c. magnetising condition are used in the form of thin sheets typically 0.23–0.65 mm thick. It is essential to build magnetic cores from such thin steels in order to restrict the eddy current loss and to make the material more magnetisable, ie to have a suitably high permeability. Some global statistics can be used to indicate the importance of electrical steels in large scale energy conservation. Around 60% of all energy generated is consumed in motors and within these motors, core losses consume around 45 million kWhr in the USA alone. In transformers, 3–5% of all energy generated is continuously wasted as eddy current and hysteresis loss in the magnetic core where, although a large machine might be more than 99% efficient, its core losses over its lifetime can exceed its initial capital cost. In the UK an improvement in an overall motor efficiency, which could be partly brought about by improved usage of core magnetic material, could reduce carbon dioxide emission by 1.5 million tons per year.

Electrical steels are broadly divided into two classes according to their grain structure. Non-oriented steels typically contain 0–3.5% silicon (balance iron) and are produced in sheet form 0.35–0.65 mm thick up to 1 m wide. They are regarded as being magnetically isotropic having mainly a random structure of mean grain diameter

187

typically 100 μm. They are used mainly in motors and other applications where magnetisation takes place in most directions in the plane of the sheets. Grain-oriented (or Goss-oriented), electrical steels have far better magnetic performance because of complex processing which produces a strong [001] (110) texture in the sheet typically 0.23–0.50 mm thick. It has a composition around 97%Fe, 3%Si and has large grains up to 10 mm in diameter and is highly anisotropic so it is used in transformers, etc where the magnetisation is mainly in one direction in the plane of the sheet. Its surfaces are coated with complex non-metallic layers, shown schematically in Fig. 1, which have a major influence on the electrical and magnetic properties of the steel. The important magnetic properties of both classes of material are power loss, permeability and *BH* characteristics which are strongly dependent on thickness, composition, texture, grain size and impurities.[1]

MAGNETISATION PROCESSES

The electrical steels are generally magnetised under alternating (a.c.) conditions at power frequency. The external magnetising field, H_a, produces an internal field H, (normally smaller than H_a because of demagnetising effects), which in turn produces a change in magnetic domain structure. This causes the internal magnetisation M and flux density B to increase according to the basic relationship

$$B = \mu_0 (H + M) \tag{1}$$

where μ_0 is the magnetic constant. The flux density is the parameter which machine designers need to optimise and carefully control to get best electrical performance.[2] Changes in domain structure under the influence of the magnetic field cause the magnetic losses. The ease of magnetisation, quantified as permeability, is measured as the relationship between B and H.

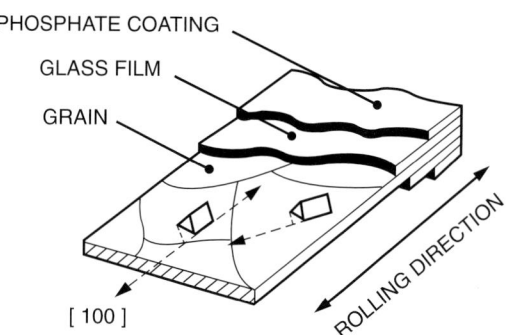

Fig. 1 Diagramatic representation of coatings on grain oriented SiFe sheet.

The power loss, P, is conveniently found to be given by

$$P = \int_0^{2\pi/\omega} H_s \frac{dB}{dt} \, dt \qquad (2)$$

where H_s is the instantaneous value of the tangential component of magnetic field at the steel surface, $\frac{dB}{dt}$ is the rate of change of the spacial, averaged internal flux density and ω is the magnetising frequency. As will be seen later, the surface field varies in magnitude and direction in a complex way, but nevertheless it can be measured using various types of sensor. Internally B and H vary with time and position in a very complex manner, but this need not be considered in loss and permeability assessment.

BASIC MAGNETIC DOMAIN STRUCTURE IN ELECTRICAL STEEL

The mobility of magnetic domains under the influence of a magnetic field determines the characteristics of the steel. These can be studied by optical or other means in grain oriented steels without too much difficulty because they are relatively large and regular.[2] In non-oriented steels it is far more difficult to study the domain structure because the surface structure, which is the region where domain observations are normally made, is not representative of the internal domain distribution.

In grain oriented steel, so-called 180° domain walls separate 'rectangular' domains oriented in the rolling, or [001] direction when the material is not magnetised or mechanically stressed. Domain patterns can be used to give a good indication of the position of grain boundaries in this material as shown in Ref. 2. The iron loss or power loss defined in equation (2) is dependent on the velocity and number of domain walls in motion at any time. As will be seen later, there are complex relationships between domain structure, surface field and other magnetic properties.

EFFECT OF GRAIN SIZE ON MAGNETIC PROPERTIES

Figure 2 shows typical variation of core loss with grain diameter in electrical steel at two magnetising frequencies.[3] The reasons for the occurrence of an optimum grain size are complex, but a simplified account is given here. In relatively thick material it has been found in many alloys, including NiFe and CoFe, that the a.c. magnetic properties deteriorate with increasing grain size.[4] One reason for this is that the bar domains referred to earlier become wider as grain size increases in order to reduce the magnetic free energy of the material. When such material is magnetised, the domain walls need to move further than if there were more domains, thus meaning that their mean velocity has to increase, hence losses and other magnetic properties

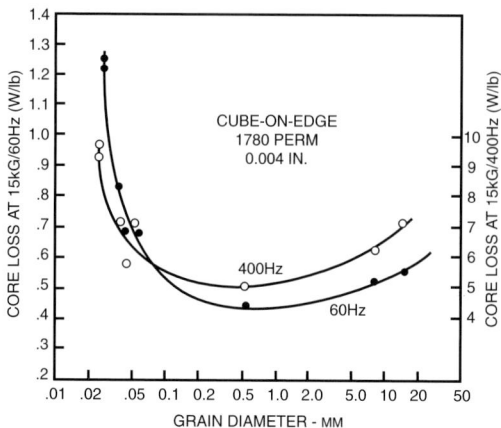

Fig. 2 Typical variation of loss with sheet thickness in grain oriented silicon iron at 60 and 400 Hz.[3]

deteriorate. When more domain walls are present, as in smaller grains, they do not need to move so far on average during the magnetisation process hence losses will be lower.

The increase in loss with decreasing grain size shown in Fig. 2 is due to an effect opposing the mechanism referred to above. Grain boundaries are sites of demagnetising fields[5] associated with so-called closure domains and spike domains occurring at boundaries between mis-oriented grains. This causes domain wall pinning at the boundaries which degrades the magnetic properties partly by causing the main domains to become relatively wider again.

In grain oriented steel a particular problem has been overcome in recent years by so-called domain refinement techniques. The [001] (110) orientation is produced in a complex manner during a high temperature secondary recrystallisation process.[1] When the best texture is obtained large grains are present hence they have wide domain wall spacing which is not desirable. The steel is coated with a stress inducing phosphate layer which has a beneficial effect of reducing the wall spacing. This can be reduced further to reduce losses by 5–10% by surface treatment in which controlled stressed areas are produced on the surface by mechanical, chemical or laser treatment.[6] The final result is that the large grain, well oriented material operates magnetically more like a small grain material hence improving the magnetic performance.

Some reference should be made to two other important soft magnetic alloy groups. Nanocrystalline materials are of growing importance in high frequency devices. They have grain sizes in the range of 10–300 nm, far lower than the electrical steels, and have very good magnetic properties at high frequencies. The mechanisms which control the effect of grain size are complex since the influence of magnetics exchange interactions dominate over the magnetocrystalline anisotropy phenomena which dominate the processes in materials within the electrical steels thickness range.[7]

Amorphous magnetic material has replaced electrical steels in some applications because of its very low losses, as little as a quarter of that of the best electrical steels. These materials have no grain boundaries, but their excellent properties are mainly due to their high electrical resistivity, low thickness and low anisotropy.

SURFACE MAGNETIC FIELD MEASUREMENT

Surface magnetic fields can be measured in various ways depending on the application and requirement. In the laboratory small, sensitive sensors can be placed very close to the surface of steels to obtain very accurate profiles. When surface fields are to be measured on steel, while moving on a strip production line, it is impossible to mount sensors close to the surface and detailed studies, calibration and interpretation of results are needed.

The earliest forms of surface field detects were magnetic potentiometers and H coils. These cannot be made very small, they are sometimes not robust and they detect $\dfrac{dH}{dt}$ so their outputs must be integrated. They are, however, used in some on-line applications where relatively large sensors remote from the surface are acceptable.

The Hall-effect sensor has been used by many workers with considerable success. Figure 3 shows a field profile measured using orthogorical Hall plates scanned over the surface of silicon steel in 0.3 mm steps.[8] The active length of the sensors here was 3 mm. The Hall effect sensor was also used for detection of grain boundaries in coated steels at high speed.[9]

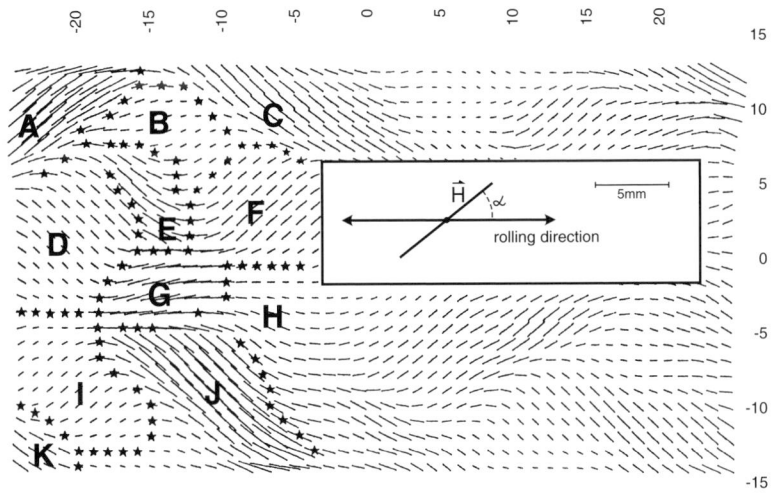

Fig. 3 Field distribution on large grain silicon iron obtained using tangential Hall plates.[8]

The most promising method of surface field measurement is using magneto resistive sensors.[10] It has a relatively high sensitivity, operates over a wide frequency range from d.c. to MHz, its output is proportional to H in the linear region of the resistance-field characteristic, it can be made smaller than 1 mm in overall dimensions and it can easily be integrated with electronic processing or computing systems.

Surface fields can be measured when a sample is magnetised or unmagnetised. Figure 4 shows a typical arrangement.[11] Here the yoke is used to concentrate the flux over a small region of the sample. The Permalloy sensor has a measurement region of 1.5 mm $\times$ 10 µm and it is scanned across a 30 mm $\times$ 20 mm measurement zone where orthogonal components of tangential field are recorded at 5000 points. In this case the loss is also computed making use of equation 2. Figure 5 shows a typical loss variation across a grain as in a Goss-oriented sheet.[11] The static domain structure shows the presence of spike domains originating at the grain boundary. The loss is high towards the right hand side of the grain because of the wider domain wall spacing in the region.

SURFACE FIELD PATTERNS ON GRAIN-ORIENTED STEEL

Field distributions similar to that shown in Fig. 3 are common on surfaces of silicon steel sheet. The field is seen to change in magnitude and direction. At first sight it might be expected that the field directions correspond to [001] directions in particular grains, but as will be shown later, this is not the case. If the detector is not close to the surface the field detected will include stray components due to the applied field

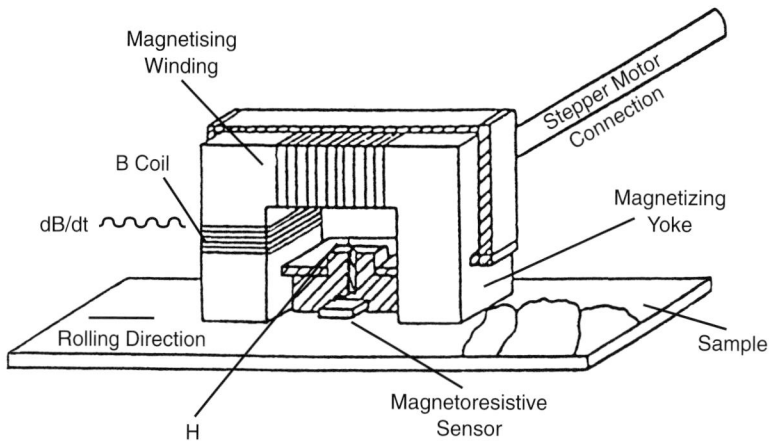

Fig. 4 Basic system for magnetising and field detection using a.c. magnetising winding and magnetoresistive sensor respectively.

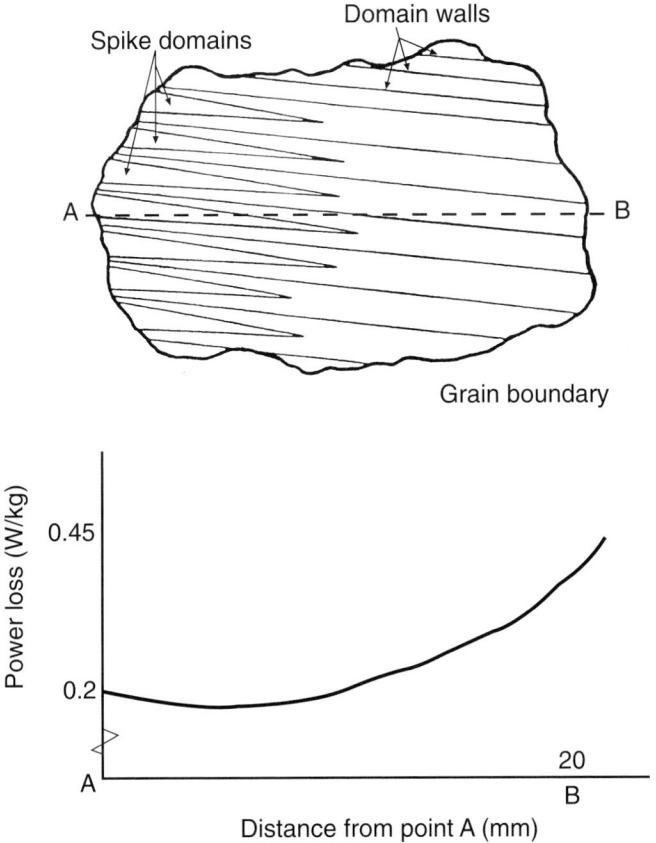

Fig. 5 Loss variation along the line AB in a grain of [001] (110) oriented steel obtained using an MR sensor and B coil.

or other near sources so for accurate quantitative measurements the field must be as close to the surface as possible.

It might be correctly assumed that the regions in Fig. 3, where abrupt changes of field direction or magnitude are observed, grain boundaries are present. Figure 6 shows a typical field pattern on grain-oriented steel superimposing the position of grain boundaries. In practice all boundaries are not always detected and some pseudo-boundaries appear presumably caused by stressed regions. Figure 7 shows a typical relationship between actual grain structure and measured tangential field magnitude contours obtained using a so-called magnetovision system.[12]

Figure 8 shows how the local demagnetising fields in individual grains affect the surface field direction. The [001] directions of a few grains are shown in Fig. 8a. The effect of an applied field *Hâ* is shown in Fig. 8b. In each grain the local magnetisation will create a demagnetising field Hd in its M direction such that the surface field is

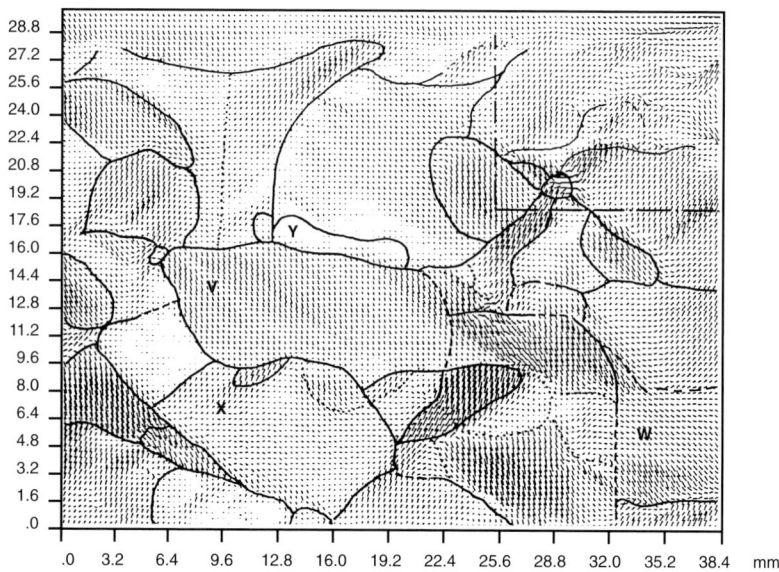

Fig. 6 Field vector plot showing variation of *H* direction and position of grain boundaries over a 3 × 3 cm area of grain oriented 3%SiFe.

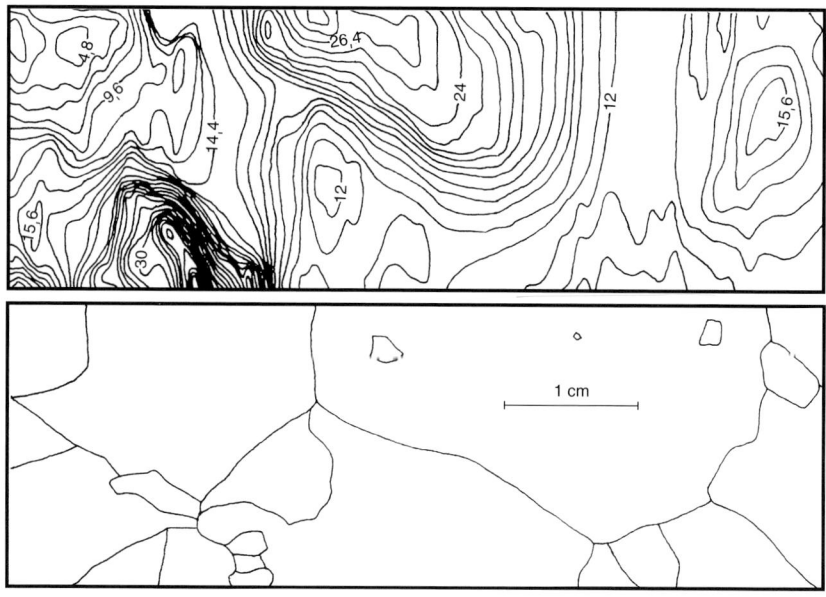

Fig. 7 Field distribution obtained using magnetovision system and corresponding grain structure.[12]

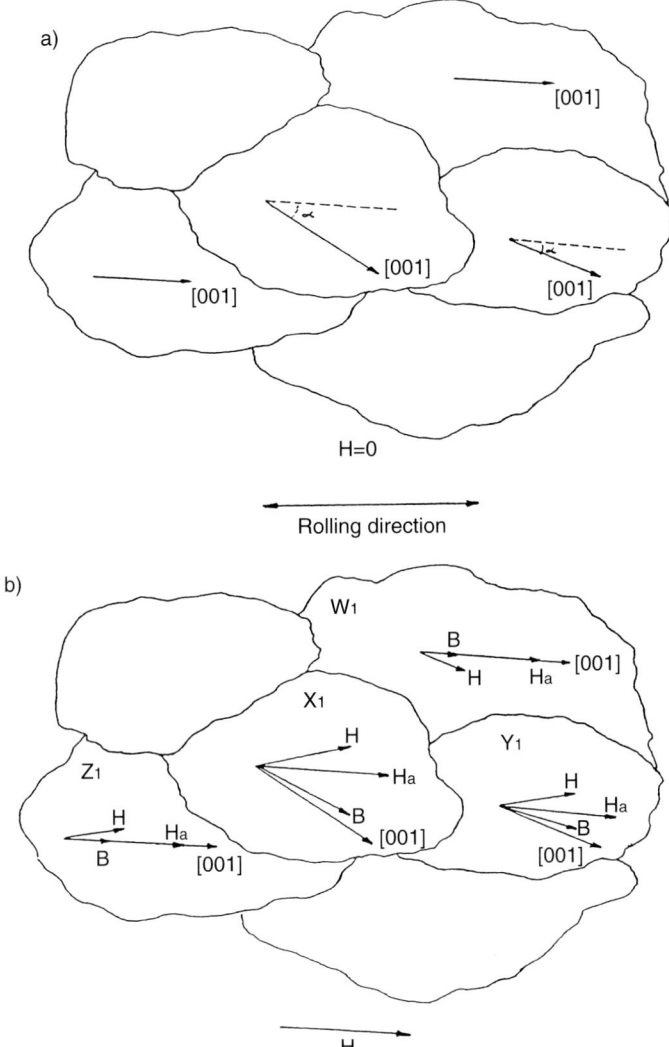

Fig. 8 Representation of grain misorientation and demagnetising field effects in grain oriented 3%SiFe.

the resultant of Ha and Hd, this will not be along the [001] or rolling direction in general and its magnitude will vary according to the degree of magnetisation in individual grains. Further work is necessary to quantify the surface effects and ultimately it may be possible to use the measurement to rapidly and remotely assess grain size and direction in electrical steels.

Attempts have been made to use surface field measurements to map domain structures on surfaces of grain oriented steels. The goal here is to detect the domain wall

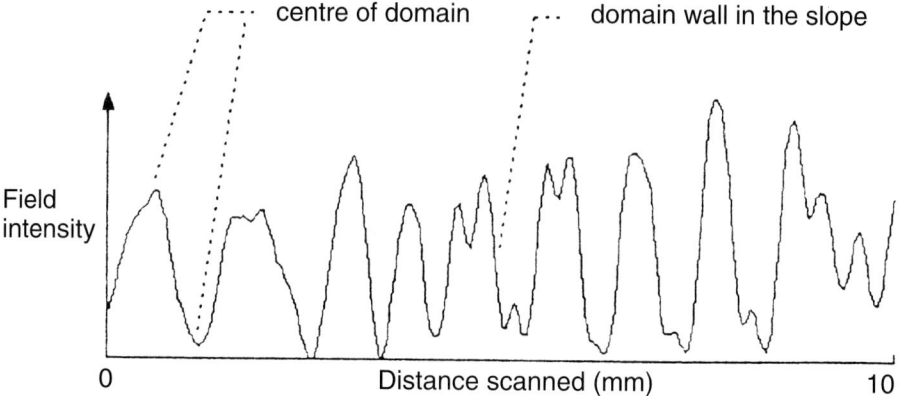

Fig. 9 Field distribution obtained by one scan across 10 mm of grain oriented SiFe over a grain with less than 5° misorientation obtained using a 2 mm long MR element.[13]

motion under the a.c. condition without removing the phosphate coating which changes the magnetic characteristics of the steel. This cannot be done easily using other domain observation techniques. Figure 9 shows a typical distribution of normal component of field measured using a 2 mm × 10 μm magnetoresistive sensor 40 Å thick, stepped at 50 μm intervals.[13] The vertical component of strain fields is generated at grain boundaries, at the 180° block domain walls and at the surfaces of domains whose [001] direction tilts out of the steel surface. This latter component is greatest and forms the distribution Fig. 9. By relating the field intensity with a grey colour scale, an image representing the domain structure can be obtained, as shown in Fig. 10. The static domain pattern measured in the same region using a commercial domain viewer is shown for comparison. The field generated image is not so good, but refined image processing and improved sensor design will improve this with the potential for a.c. studies, which the other technique cannot be used for.

SUMMARY

This paper attempts to show how magnetic fields on the surface of electrical steel can be used for assessing grain structure and magnetic properties. Localised measurements are not possible on non-oriented steels, but it may become feasible in the future as it becomes possible to make sensors smaller and more sensitive. There is a wealth of potential not covered in this paper for use of surface field measurements in NDT applications. This is already a proven principal which may be developed in the future. Also there is a huge potential for making use of surface field measurements for detection of mechanical properties such as hardness or residual stress on a local or global basis. This also has not been covered in this text, but it is hoped that the reader

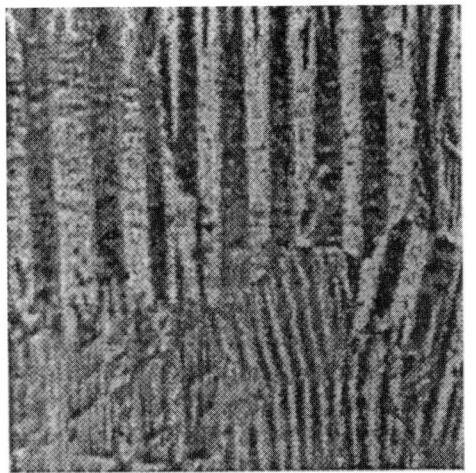
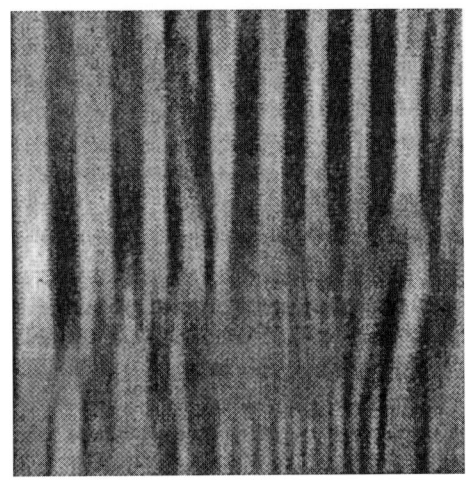

Fig. 10 Domain image over the area referred to in Fig. 9: (a) obtained using the Bitter technique; (b) using the 2 mm MR sensor over a 1 cm × 1 cm area.[13]

will see from what has been discussed that exciting possibilities exist for exploiting such measurements in wide areas.

REFERENCES

1. A. J. Moses: *Proc. Inst. Electr. Eng.*, 1990, **137A** (5), 233–245.
2. P. Beckley: *Power Eng. J.* Aug. 1999, 190–200.
3. D. Jiles: *Introduction to Magnetism and Magnetic Materials*, Chapman and Hall, London, 1989.
4. F. Pfeiffer and C. Radeloff: *J. Magn. Magn. Mater.* 1980, 19, 190–207.
5. H. Schafer: *Phys. Scr.*, 1989, **39**, 531–534.
6. J. W. Schoen: *IIT Conf. on Properties and Applications of Magnetic Materials*, 1985.
7. G. Herzer: *IEEE Trans. Mag.*, 1990, **26** (5), 1397–1402.
8. H. Pfutzner: *J. Magn. Magn. Mater.*, 1980, **19**, 27–30.
9. K. Mohri and T. Fujimoto: *Mem. Kyushu Inst. Technol. Eng.*, 1977, **7**, 33–37.
10. B. B. Mohd Ali and A. J. Moses: *J. Fiz. Mater.* 1988, **9**, (75), 77–80.
11. A. J. Moghaddam and A. J. Moses: *IEEE Trans. Mag.* 1993, **29** (6), 2998–3000.
12. S. Tumanski and M. Stabrowski: *J. Magn. Magn. Mater* 1996, 160, 165–166.
13. M. H. So, P. I. Nicholson, T. Meydan and A. J. Moses: *IEEE Trans. Mag.*, 1995, **31** (6), 3370–3372.

Grain Boundary Segregation in Thin Film Media

K. O'GRADY

School of Electronic Engineering & Computer Systems, University of Wales Bangor, Dean Street, Bangor, Gwynedd, LL57 1UT, UK

H. LAIDLER

School of Physics, University of Exeter, Stocker Road, Exeter, EX4 4QL, UK

ABSTRACT

The majority of the active materials involved in magnetic recording technology consist of sputtered magnetic thin films. As such, these films are granular in nature and hence are subject to similar grain boundary effects to other such materials. Of particular interest is the nature of the thin film coatings that form the magnetic disk upon which information is stored. This is because for these materials grain boundary effects are crucial in determining both the bulk magnetic properties and thereby the recording performance of the system itself. In this paper an introduction to magnetic recording technology is presented followed by a description of the effect of grain boundary phenomena on the performance of this medium. The crucial parameter here is the effect of grain segregation on intergranular magnetic exchange coupling between grains. A technique is described by which such coupling and thereby the grain segregation can be measured from a bulk magnetic measurement and thereafter, using the same technique, the effects of high Cr content on the Co alloys used is explored. The magnetic data is compared with results associated with high resolution Electron Energy Loss Spectroscopy (EELS) and the resulting magnetic structure determined by Magnetic Force Microscopy (MFM).

INTRODUCTION

Magnetic recording technology is probably the most advanced materials technology in the world today. This statement derives not only from the nature of the materials themselves where complex multilayer structures with up to thirteen individual sub-nanometre strata are laid down, but also because of the complex interactions that occur in such materials. Magnetics technology is unique in this respect since, in addition to the normal interatomic forces, there is also the possibility of the magnetic dipole–dipole interaction and more importantly the complex quantum mechanical exchange interaction that occurs between all ordered magnetic materials. No other class of materials has these additional levels of complexity and therefore the management of this complex, intergranular coupling matrix requires extraordinary materials control. To compound these difficulties the nature of the computer industry is such that remarkable economic demands are made upon the technologists who produce the materials commonly used particularly in hard disk drive (HDD) applications. For example currently product lifecycles are down to nine months with the launch of a new product each quarter. R&D departments are currently operating a three year development cycle and hence the engineers working in the industry are continuously developing not less than twelve new products simultaneously!

This remarkable rate of research and development is resulting in an astonishing rate of technological advance. Recently the figure of merit which is the areal density at which information is stored, i.e. the number of bits per unit area has accelerated so that current progress is at the rate of approximately 100% per annum. Indeed during the preparation of this manuscript it has been announced that IBM have recorded data at 35.3 Gbit in^{-2} compared with the world record of 10 Gbit in^{-2} which they set last year. This is the textbook definition of exponential growth. To achieve this rate of growth new materials with improved properties are required every three months. Hence, a detailed understanding of the complex behaviour of such materials is required on an ongoing urgent basis. Figure 1 shows the areal density curve which dominates the industry and shows data realised up to 1997. The shift in the gradient of the curve from the 60% per annum shown to 100% per annum occurred in the last quarter of 1998.

Almost all the materials used, both in the rigid disk itself and in particular in the read head, are sputtered alloys. Such materials are granular in nature and invariably multilayered. The grain size typical in both the head and the disk are in the order of

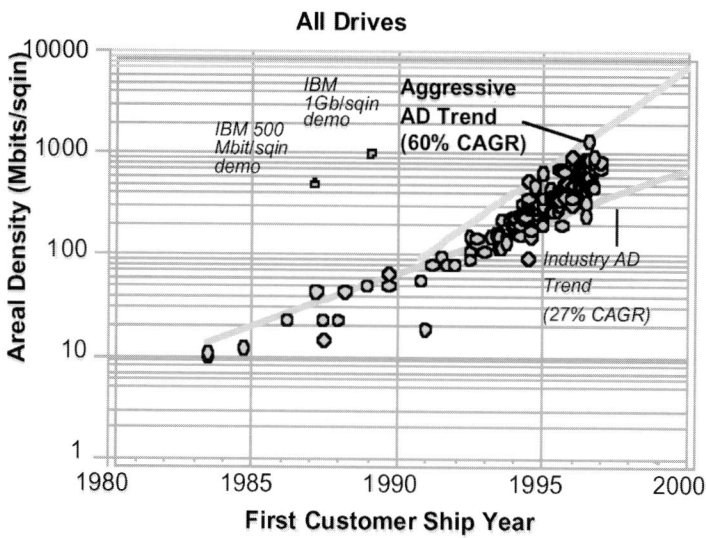

Fig. 1 The rate of evolution of rigid disk technology represented as a graph of areal density versus first customer shipment year. The solid lines mark the trends commonly referred to in terms of Compound Annual Growth Rate (CAGR) (after Ref. 1).

100–200Å and hence are truly nanophase. Given that the grains are of this size it is clear that a significant proportion of the atoms in the materials lie at or very near to, grain boundaries and thus, particularly in the light of the exchange interaction, boundary effects are often dominant.

Given that the subject of this paper is associated with the materials technology associated with such grain boundaries, it is necessary to introduce the reader to the basic physics of the magnetic recording process viewed from the perspective of disk drives. Of course, in the limited space available in this article, it is not possible to give a comprehensive review of magnetic recording technology and thus the interested reader is referred to the recently issued second edition of the text by Mee and Daniel.[2]

THE PHYSICS OF MAGNETIC RECORDING

The basic recording system is as shown in Fig. 2. Fundamentally, a ring magnet with a gap contained within it passes over a disk which has previously been saturated in a direction indicated to the left in the figure. Pulses of current are sent through the coil resulting in the writing of areas of magnetisation aligned to the right as shown in the figure. Such a system gives rise to an output signal if a similar head passes over the written bits of information as shown in the lower half of the figure. In this instance, the change in magnetisation at the edges of the domains produce pulses of voltage. For each domain written in the reverse direction, two voltage pulses occur at either

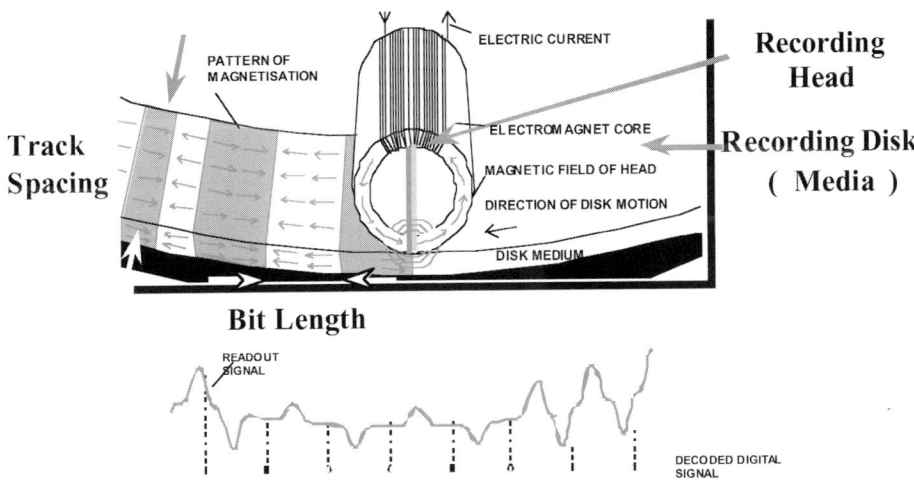

Fig. 2 Schematic diagram showing a) the basic principle of a magnetic recording system and b) a representation of the di-bit output signal. Further explanation is given in the text.[1]

end of the domain, giving rise to what is known as a dibit transition. Such a system means that each bit of information has two chances to be read and hence leads to in-built error correction. Also shown in the figure is the areal density which is given by:

$$\text{Areal density} = (\text{Tracks per inch}) \times (\text{Bits per inch}) \qquad [1]$$

The size of a magnetic bit as historically used and as projected for the future, is as shown in Fig. 3. Clearly as magnetic recording technology advances a significant reduction, not only in the bit length but also in the track width must occur. At the time of writing, current technology available commercially, particularly as used in laptop computers has an areal density of the order of 5 Gbit in^{-2}. Also it should be noted that recent announcements have confirmed laboratory demonstrations at 10 Gbit in^{-2} and beyond.[3, 4] Thus, examining Fig. 2 for the case of 10 Gbit in^{-2} shows that the track width is now down to a figure of the order of 50 nm and thus the bit sizes are also nanophase. In this context it should be borne in mind that the grain size used to achieve these densities was in the order of 10 nm and thus the track width is now only five grains wide. The reason why the bit length is significantly greater than the track width is associated with sample shape demagnetising effects which also further complicate the use of magnetic materials in this application.

The basic read-back process is as shown in Fig. 4. On the left of the figure is shown the concept of an inductive read head. This technology was used consistently until about 1991 when the alternative technology, also shown in the figure, using magnetoresistive sensors was launched commercially. Significantly, reference back to Fig. 1 shows a sig-

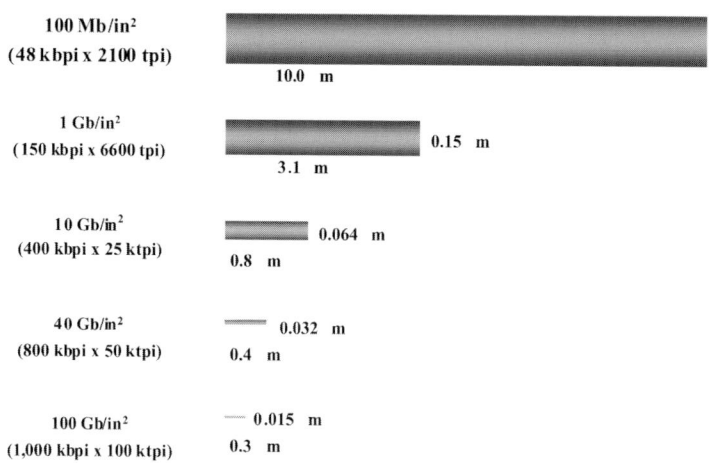

Fig. 3 Schematic representation of the evolution of actual bit size in terms of areal density. The bit width is given in kilobits per square inch (kbpi) and the bit length is given in terms of tracks per inch (tpi).[1]

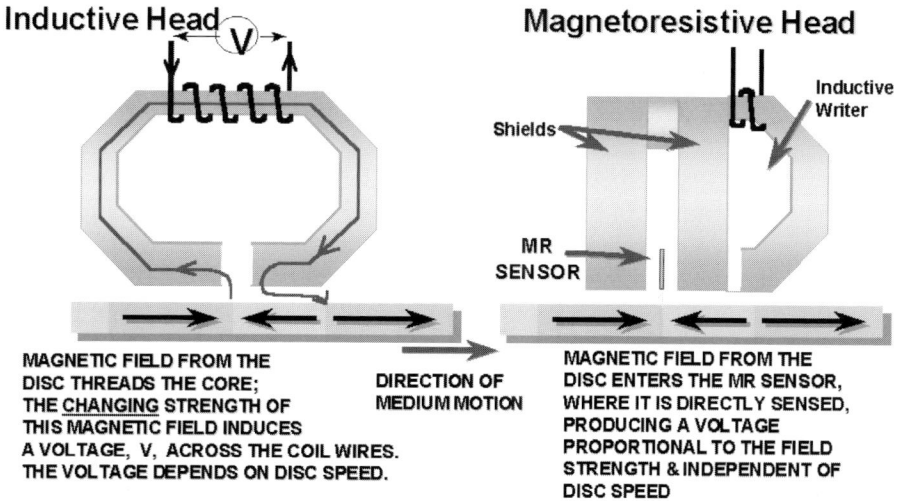

Inductive Head

Magnetoresistive Head

Shields

Inductive
Writer

MR
SENSOR

MAGNETIC FIELD FROM THE
DISC THREADS THE CORE;
THE <u>CHANGING</u> STRENGTH OF
THIS MAGNETIC FIELD INDUCES
A VOLTAGE, V, ACROSS THE COIL WIRES.
THE VOLTAGE DEPENDS ON DISC SPEED.

DIRECTION OF
MEDIUM MOTION

MAGNETIC FIELD FROM THE
DISC ENTERS THE MR SENSOR,
WHERE IT IS DIRECTLY SENSED,
PRODUCING A VOLTAGE
PROPORTIONAL TO THE FIELD
STRENGTH & INDEPENDENT OF
DISC SPEED

Fig. 4 Schematic diagram illustrating the read-back process using an inductive head (as shown on the left) and a magnetoresistive head (as shown on the right).[1]

nificant acceleration in the areal density trend with the arrival of magnetoresistive (MR) sensors.

The technology associated with the use of MR sensors is extremely complex. First the inductive write head which is a complex structure in its own right, is produced by electrodeposition, and onto this the MR sensor is sputtered. Of course, to prevent crosstalk between neighbouring bits of information, magnetic shields are incorporated which absorb the flux from neighbouring bits thus preventing interference with the sensor itself. Additionally, MR sensors require permanent magnet layers to bias them to their operating region and of course there is a need for connections to be made to the MR sensor. Hence a massively complex electrodeposited and sputtered structure is required to produce a combined thin film read-write head.

In modern high density disk-drives the magnetoresistive sensor itself is even more complex due to the use of so-called spin-valve technology. Most bulk magnetic materials exhibit some form of magnetoresistance which is typically of the order of 1–2% but this occurs over a very large field range. However, in 1988 the phenomenon of Giant Magnetoresistance (GMR) was first discovered.[5] In this system a multilayer of two magnetic films of Fe separated by a non-magnetic Cr spacer was shown to exhibit a very large magnetoresistance which in practical devices at room temperature can be as high as 20%. However, the field range over which such a system operated was again very large and hence the technology as such was not suitable for incorporation in a read head. However subsequent developments using an antiferromagnet to pin one of the magnetic layers so that the other magnetic layer alone rotates, has resulted in the development of an MR sensor that exhibits typically 8%

MR for a changing field of 1–2 Oe. Given that one of the layers rotates to allow this to occur, such a system is analogous to a valve. Given that the origin of magnetoresistance lies in spin dependent scattering, such a system is described correctly as a spin-valve. Hence in order to use such sensors in magnetic recording systems, it is clear that material science is required that is truly quantum mechanical engineering.

The structure of a typical spin-valve is shown in Fig. 5 where the incredible complexity of such a system can be observed. It should be noted that the circular region in the figure is drawn to scale and by way of calibration the Ta cap is of order 10 nm and the NiMn antiferromagnetic layer is typically 60 nm. From the figure it can be seen that the upper NiFe layer is pinned by the antiferromagnetic and the purpose of the Co layer is to improve the magnetoresistance. A subsequent copper layer followed by a further layer of Co and a second NiFe layer is the one that is free to rotate. Please note the presence also of the conductors which make electrical connection to the spin-valve and also the permanent magnet biasing layers which are necessary to bring even a spin-valve head to its operating region. For review of the operation of spin-valve recording heads see Ref. 6.

For the purpose of discussing grain boundary effects it is much simpler to consider thin film recording disks rather than the massively complex spin-valve structures where complex epitaxial effects and quantum mechanical phenomena such as spin frustration lead to great complexity. As such, the remainder of this article will be concerned with the structure and behaviour of disks and by way of background an introduction is now provided. A reader interested in the evolution and the requirements for disk technology is referred to an excellent review by Johnson.[7] The basic structure of a thin film disk is as shown in Fig. 6. Today the substrate is typically glass upon which is sputtered a seed layer followed by the underlayer shown in the figure. The magnetic layer is typically a Co alloy and the overcoat consists of a sputtered layer of carbon to prevent the build up of static charge. The whole structure is then dipped in

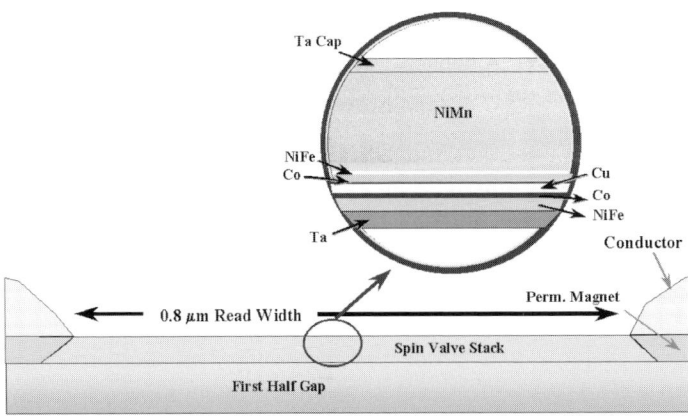

Fig. 5 Structure of a typical spin-valve read head.[1]

a PFPE lubricant so that the head structure is sufficiently lubricated that it will take off and fly above the surface of the disk when the disk is rotated. An often quoted analogy to the behaviour of the head gimbel above the surface of the disk is that it is equivalent to the flying a Boeing 747 Jumbo Jet at a height of 0.1" (2mm) above the ground. In the case of thin film technology the head medium spacing is now of the order of 10 nm.

Over the last two decades there has been a steady evolution of the type of alloys used in the magnetic layer. Fig. 7 shows the alloy evolution with the associated dates. Initially thin film media consisted of Co alloyed with Cr, Ni or Pt to improve the resistance to corrosion of the thin film. Subsequently it was found that by additions of tantalum, Cr or Ni films with low noise and high coercivity were possible. A significant development was the inclusion of Pt in such alloys which enabled the systematic control of the medium performance to be obtained. The physics of this evolution is that Pt causes a stretching of the *c*-axis in a sputtered Co film which traditionally grows with a reduced c/a ratio. In bulk hcp Co the c/a ratio is √8/√3 however, when films are produced by sputtering, this is reduced. The presence of a large atom such as Pt or Ni increases the c/a ratio whilst also improving corrosion resistance. However, should the c/a ratio reach the ideal value it is possible and indeed quite often common that the Co reverts to the fcc structure. Hence such additives must be used with care since fcc Co has a very low anisotropy and hence low coercivity. The inclusion of Cr in these alloys was the first indication of the importance of grain boundary effects. Cr has limited solubility in Co and its segregation to the grain boundary results in the formation of a paramagnetic phase at concentrations typically above about 25%. Such a phase of the grain boundary results in exchange decoupling between grains and hence to a low noise medium.

In subsequent developments it was found that the inclusion of Ta and Cr in a CoPt alloy resulted in improved performance by a further grain boundary segregation. This

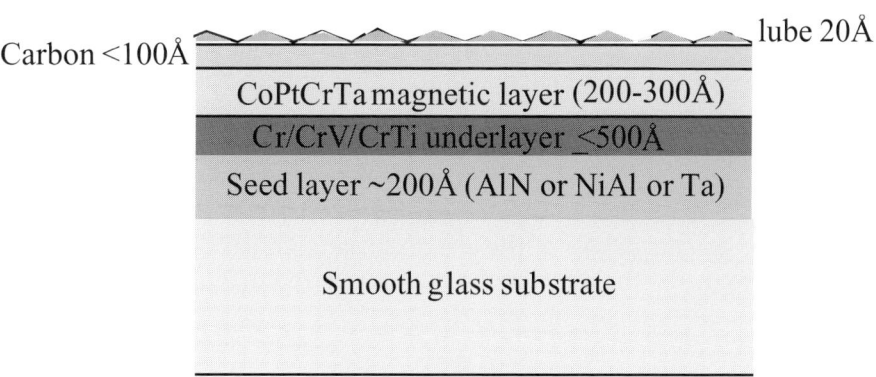

Fig. 6 Schematic diagram of the typical structure of advanced thin film media produced on glass substrates.

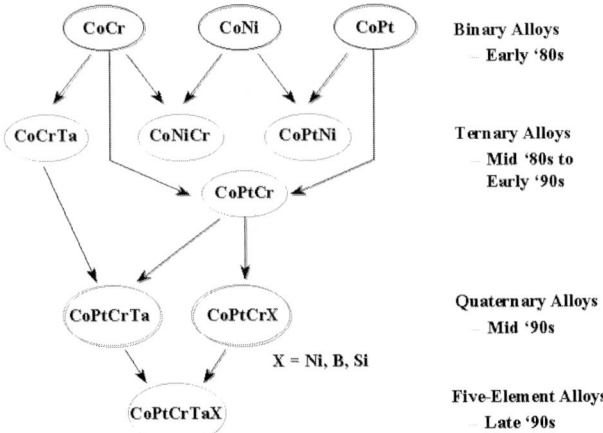

Fig. 7 Sputtered magnetic alloy evolution. The associated dates that the alloys were used in production are given on the right of the diagram.[1]

topic is further discussed in the experimental results section. Other additives are now commonly used which include Ni, B or even Si. Very recently quinary alloys have also appeared, particularly in research laboratories associated with media with many complex additions, the role of which is not as yet fully understood. The reader specifically interested in the evolution of the types of alloy used is again referred to the excellent review by Johnson.[7] However, for more recent developments the interested reader is referred to very recent papers appearing in the literature associated with data densities in excess of 10 Gbits in^{-2}.[3]

The mechanism of media noise reduction is shown clearly in Fig. 8. In this figure it is seen that an attempt to write a reversal in the direction of magnetisation into a medium requires a step change in the direction of magnetic moments in neighbouring grains. Such a step change is not favoured in terms of the micro-magnetics and in early media intergranular exchange coupling resulted in highly irregular so called zig-zag domains.[8] Significant micromagnetic modelling of such systems has shown that the critical parameter in enabling a sharp reversal of magnetisation direction to be obtained is associated with the control of inter granular exchange between the grains. In this way the critical importance of the inclusion of Cr in thin film media is clearly demonstrated since its segregation to grain boundaries can result in exchange de-coupling. For the specific details of the effect of intergranular decoupling and dipolar effects in thin film media the interested reader is referred to a recent book chapter by Zhu.[9]

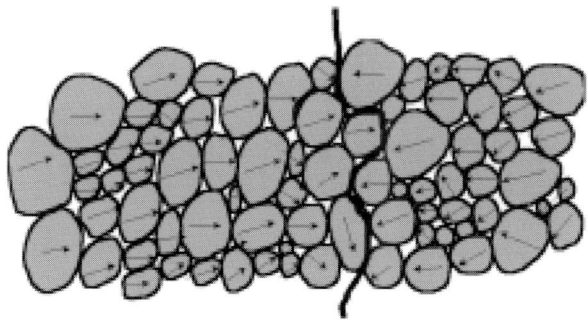

Fig. 8 Schematic diagram of a granular magnetic medium where an attempt has been made to write a bit reversal. The red line marks the region where a step change would be required in the direction of the magnetic moments in each grain (depicted by the arrows).[1]

THE DETERMINATION OF COUPLING AND GRAIN BOUNDARY EFFECTS

We have seen that exchange coupling between grains leads to co-operative reversal which can result in the movement of transitions and generally leads to an irregular shape for the regions of magnetisation reversal hence giving rise to noise in a thin film recording medium. In order to improve the performance of such a system it is not only necessary to determine mechanisms by which exchange decoupling can occur but also to determine mechanisms by which such effects can be measured. Fundamentally, grains can be exchanged decoupled via two mechanisms: the first of these is spacial segregation. There are techniques available by which, through the careful control of growth, the grains in a thin film can be separated from each other spatially. For the case of Co alloy thin films it is usual that a Cr underlayer is used. It has been found that the thickness of the Cr underlayer determines not only the size of the grains that result but also their physical location. By use of very thick Cr underlayers it has been established that total exchange decoupling can result leading to the generation of a very low noise thin film medium. Such films were originally grown by and reported on in detail in terms of both the spacial segregation and the resulting performance by Yogi and co-workers from IBM.[10] Detailed magnetic measurements on these same films are reported later in this paper.

The second process that can be used to exchange decouple films is compositional segregation. This is the mechanism whereby the migration of Cr or any other element to grain boundaries results in a discontinuity in the spin order within the thin film, giving rise to a breaking of the effective exchange interaction and hence lowering the noise in a similar manner to spacial segregation. Such compositional segregation can be achieved by using high levels of Cr in the alloy typically of the order 15 to 25%. Again the nature of coupling in such films is discussed later in this article.

Of course, it is of critical importance to determine the effect of such coupling effects on the recording performance and some direct measure of correlation is therefore required. In the case of a magnetic recording medium it is the case that the output of the recorded signal is in some way a direct reflection of the micromagnetic configuration of the thin film. The micromagnetic configuration is in turn a reflection of the underlying physical microstructure which gives rise to it. Hence in principle a magnetic measurement can be used as an indicator of the physical microstructure and also at the same time, a potential indicator of the likely recording performance of the resulting medium.

From a range of studies on both thin film and bulk magnetic materials such as permanent magnets it has been established through experiments and large scale numerical modelling that exchange coupling affects the magnetic properties of materials in a number of well defined ways. It is generally found that an increase in intergranular exchange coupling decreases the coercivity of materials.[11, 12] It can also result in an increased remanence as observed for example in two phase permanent magnet materials where a hard magnetic phase is coupled to a soft magnetic phase giving rise to an increased remanence.[13] A general description of the effect of exchange coupling would be that it causes the occurrence of a square loop.

The difficulty with all the above indicators of the effects of exchange coupling is that they are purely qualitative in nature and for such a sophisticated technology as information storage, it is desirable if not necessary to have a quantitative indicator of what is occurring to allow for the effects of a large number of quantities such as process variables to be assessed in a relatively short period of time.

The manifestation of such exchange coupling variations on the hysteresis loop is seen in Fig. 9 for a set of CoNiCr thin films grown on a range of thickness of Cr underlayers. These are the original films grown and reported on by Yogi *et al*.[10] In these samples the film grown on a 100 Å layer was found to be essentially continuous with very close physical proximity between the grains leading to strong exchange coupling. As can be seen from the figure, it appears that, as the Cr layer thickness is increased, a progressive increase in coercivity is observed. However, by the time the underlayer thickness reaches 1000Å it appears that a further increase in the Cr underlayer makes no significant different to the hysteresis loop. This is essentially classic behaviour as described above and as determined from the hysteresis loop. However it will be shown in subsequent sections that this impression is somewhat misleading since the effect on the interaction configuration by a further increase in the thickness of the Cr underlayer are in fact dramatic. It should also be noted that[10] it was shown that the increasing Cr underlayer thickness resulted in an increasing spacial segregation. For example, for the film with 500 Å of Cr an average spacial segregation of approximately 1 Å resulted. However, for the thickest Cr underlayer spacial segregation of the order of 4 Å resulted. Clearly such changes in spacial segregation would be expected to have some effect on intergranular coupling but this is not apparent from the hysteresis loop. Hence improved measurement techniques are needed.

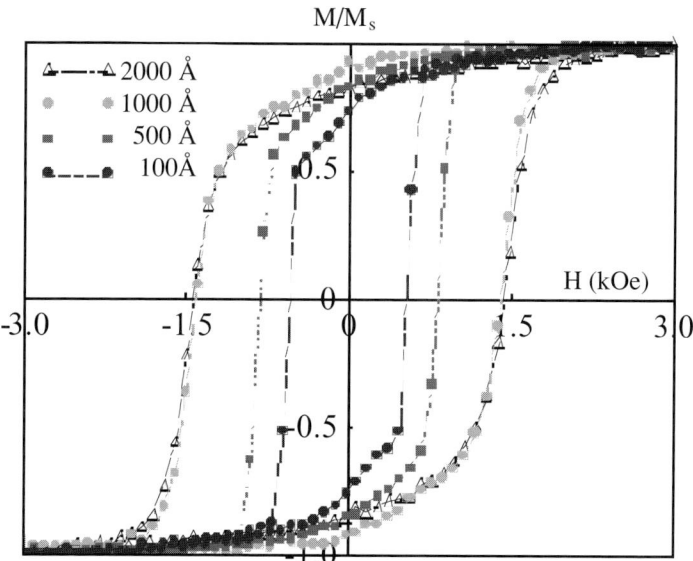

Fig. 9 Hysteresis loops of identical CoNiCr films grown on different thicknesses of Cr under-layer. The thickness of the Cr underlayer for each sample is given in the legend.

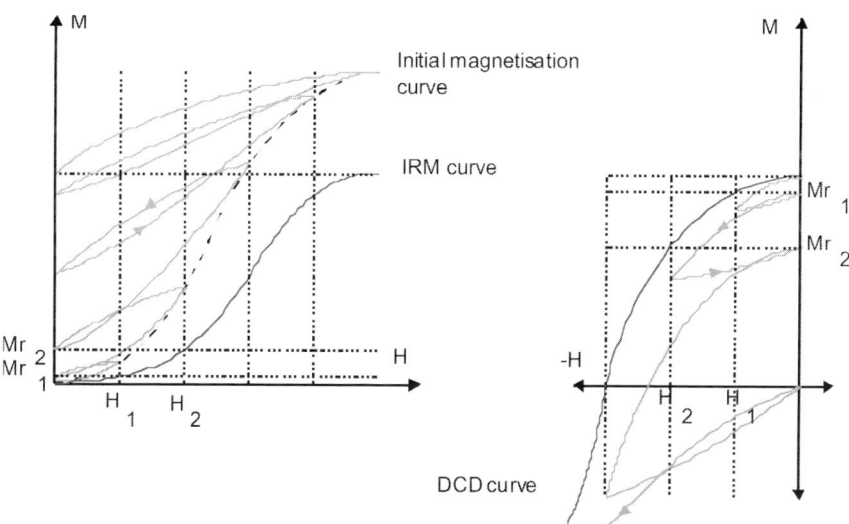

Fig. 10 The two principal remanence curves: the isothermal remanence magnetisation (IRM) curve and the dc demagnetisation remanence (DCD) curve. H_r is the remanent coercivity and H'_r is the half saturation point on the IRM curve.

REMANENCE MEASUREMENTS AND DELTA-M CURVE

In magnetic information storage technology it is the switch in the direction of magnetisation in the grain that results in the storage of information. On a bulk hysteresis loop the changes in magnetisation observed under the influence of an applied field have two distinct origins. These origins are switches in the direction of magnetisation of a grain as described above also known as irreversible changes, but reversible changes in magnetisation can also occur whereby the moment of a single domain grain rotates slightly out of its easy direction but on removal of the field reverts back to that original direction. Hence the full magnetisation curve of such a system consists of reversible and irreversible effects.

In order to examine the viability of a given material as a recording medium, it is clearly preferable to study only irreversible changes. Such changes can be measured via a change in the remanent magnetisation along a minor hysteresis loop. Such minor loops can be measured from two different micro-magnetic configurations being the demagnetised state and the saturated state. In both cases small fields are then applied in the direction in which magnetisation is to be oriented and then removed and the resulting change in the remanence measured. Such a change in remanence is then a reflection only of the irreversible changes in the loop. The Isothermal Remanence Magnetisation (IRM) curve and the DC demagnetising remanence (DCD) curve are shown schematically in Fig. 10. In the case of magnetic recording, the recording process is undertaken from the saturated state and hence the DCD magnetising remanence curve is a reflection of those processes which actually occur when a bit of information is written. The differential of such curves is a highly useful parameter since it shows in detail the switching behaviour of such systems. A good example of this detail switching analysis is shown in Fig. 11 where the differential of both the magnetising and demagnetising remanence curve for the same set of CoNiCr thin films as discussed previously is shown. In these figures it is clear that reversal from the saturated state occurs over a much narrower range of fields than is the case for the magnetising curve. It is also clear that as the Cr film thickness increases the breadth of the switching field distribution measured from the saturated case also broadens. Furthermore, it should be noted that for the strongly exchanged coupled film the position of the demagnetising remanence curve switching film distribution lies to the left of that for the magnetising case. This is in contrast to the sample grown on the thickest curve where the reverse is true. Given that these films were all prepared from exactly the same material and sputtered under exactly the same conditions, giving rise to exactly the same grain size distribution, it is clear that the only possible origin of these differences in switching behaviour, which is also partly manifest on the hysteresis loop, arises due to the effects of intergranular magnetic coupling.

In principle if there were no magnetic coupling effects between grains then the two remanence curves would be identical with the exception that the demagnetising curve would run from positive saturated remanence to negative whereas the magnetising curve would run from zero to the saturated remanence state, ie a difference

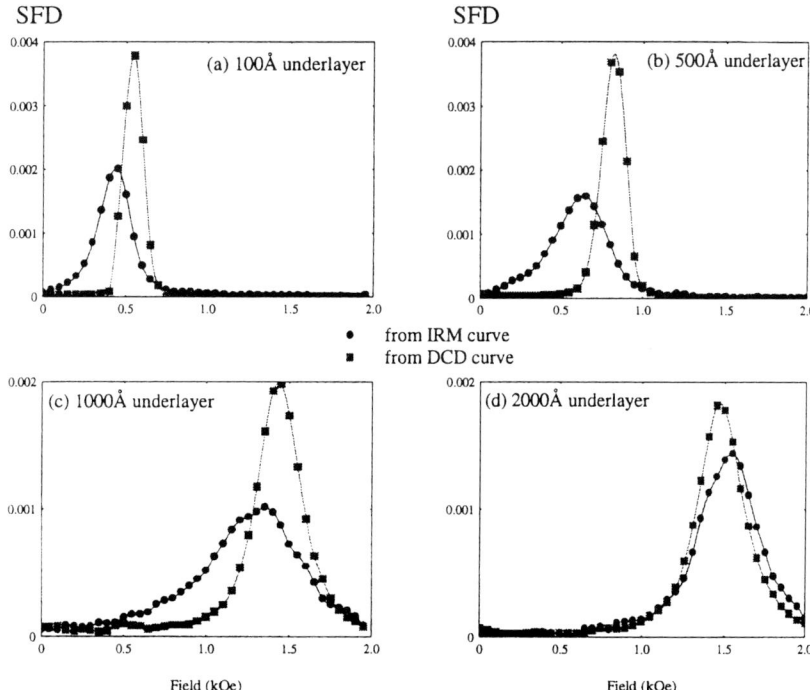

Fig. 11 Switching field distributions of CoNiCr films. The underlayer thickness for each film is shown in the legend. Data for the differential of the IRM and DCD curves are denoted by circles and squares, respectively.

of a factor 2. This fact was realised by E. P. Wohlfarth who in a defining paper wrote down what is now known as the Wohlfarth equation:[14]

$$\overline{M}_d (H) = 1 - 2 \overline{M}_r (H) \tag{2}$$

This equation in principle allows a measurement of the effective coupling by its effect on remanence to be determined. In a subsequent work many years later it was realised that the parameter of greatest significance was in fact the numerical difference between the two remanence states which is now known as the Delta-M (ΔM) curve.[15] This curve uniquely provides a relatively simply and accessible measurement of the effects of intergranular coupling in a quantifiable manner and therefore allows an insight into the effects of a number of parameters such a spacial or compositional segregation on the coupling and therefore the behaviour of the magnetic grains in a thin film medium. Figure 12 shows the ΔM curves for the four samples that have been discussed thus far. Clearly the curves exhibit dramatically different behaviour with the sample grown on the thinnest underlayer exhibiting very strong co-operative reversal with a much lower coercivity than is the case for samples on thicker layers.

211

Also from the figure it can be seen that the sample grown on the 2000 Å Cr underlayer exhibits behaviour which is clearly anomalous compared with the other samples.

The interpretation of such curves has been modelled using large scale theoretical and computational models by Zhu and Bertram[16] who have determined unambiguously that a positive ΔM curve implies exchange coupling with any decrease in the strength of exchange resulting in an almost exact mimicking in the trend seen in Fig. 12. The curve for the sample on the thickest Cr underlayer could only be generated in this model by reducing the exchange coupling to zero and this implies that when a negative ΔM profile is observed, total segregation of the grains has occurred such that the exchange interaction is effectively negligible.

Importantly, Zhu and Bertram also examine the potential correlation between such curves and the noise exhibited in a thin film medium. They found that the maximum slope of the ΔM curve through the reversal region was a quantitative indicator of the noise that would be generated in an information storage device. The importance of this result cannot be overstated since it allows for the rapid evaluation of alloys and process conditions without the need to fabricate a disk capable of supporting a flying head. The validity of this result has been confirmed in a number of experimental studies that were carried out subsequently.[17, 18]

Of course, such advances in an understanding of exchange decoupling of grains is of vital importance to the thin film media industry. However the growth of layers of

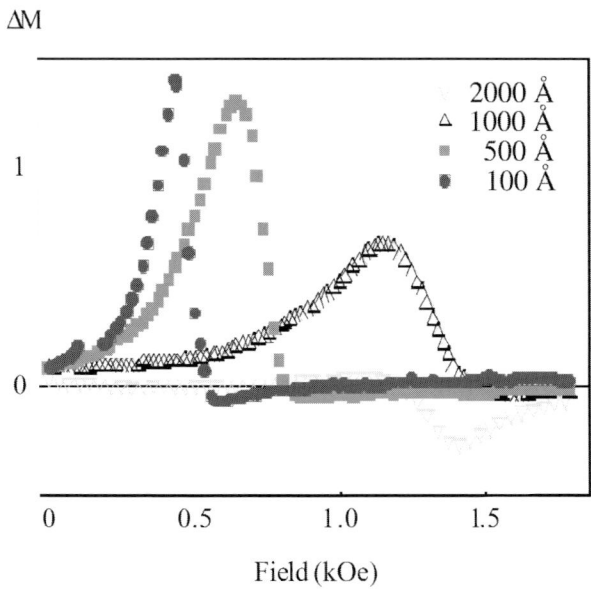

Fig. 12 ΔM curves for the four CoNiCr films with different Cr underlayer thickness, as shown in the legend.

Cr as thick as 2000 Å are not practical in an economic sense since much more rapid production times are needed. Thus the trend is to look at other techniques for grain segregation through grain boundary effects. The generation of such samples has been greatly aided by the use of measurement techniques such as ΔM and is described in the subsequent section.

CONTROL OF GRAIN SEGREGATION

The limited solubility of Cr in Co-alloy films enables grain segregation to be achieved via migration of the Cr to the grain boundaries. In this way a paramagnetic or very weakly ferromagnetic phase results at the grain boundary that limits exchange coupling. Typically films of a Co, Cr, Pt alloy containing about 5% Pt and 15–25% Cr have been grown on the Cr underlayers. These underlayers are typically 200Å thick and are grown onto a heated substrate formed by an AlMg disk substrate coated with a NiP electrolessly deposited smoothing layer. Such Cr layers grow with an in-plane (200) texture. This leads to a strong epitaxy for the growth of the Co alloy layer which is subsequently sputtered with c-axes in two orthogonal directions. This leads to the formation of so-called bi-crystal films.[19] The crystallography and the nature of the epitaxy that leads to this bi-crystallinity is shown in Fig. 13. The occurrence of bi-crystals can mean that given grains can grow with the c-axis orthogonal within the plane of the film or there can be sub-grains within a given Co alloy grain having orthogonal c-axis directions due to the polynucleation of the growth of the Co grain. Such effects lead to very complex magnetic behaviour but it has been shown via large scale computer modelling that the occurrence of such bi-crystal films could in principle lead to a medium exhibiting very low levels of recording noise. Hence a significant study on grain segregation in bi-crystal media has been undertaken.[20, 21]

In our work we studied samples with varying Cr levels and with two different grain sizes produced by heating during the sputtering process. The samples we examined, labelled D and F, had typical grain sizes of 174 Å whereas samples E and G had typical grain sizes of 415 Å. Samples D and E were sputtered with a low Cr alloy (13%) whereas F and G had a high Cr level (22%).

We have again studied the properties of these films via the ΔM technique described previously in this work and the results for the four films are shown in Fig. 14. At first glance, it appears that the ΔM curve for each of these samples is of similar amplitude and shape and thus in principle the coupling would be the same. However, careful analysis of the data shown in Fig. 14 has revealed that the slope of the ΔM curve does vary in a systematic way, exhibiting a smaller slope for samples F and G, which have the high Cr level as compared with D and E. Inevitably, this indication of lower coupling resulted in an improved recording performance for these samples.

Data such as that shown in Fig. 14 has lead to a substantial effort to examine the role of Cr segregation in thin film media since it is clear that these effects dominate over grain size effects as indicated by our results. In particular, Futomoto and co-

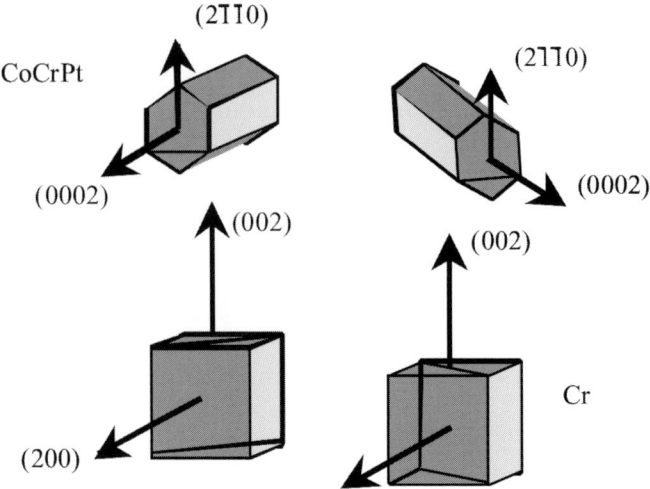

Fig. 13 Epitaxy of the bicrystal CoCrPt structure with respect to the Cr underlayer.

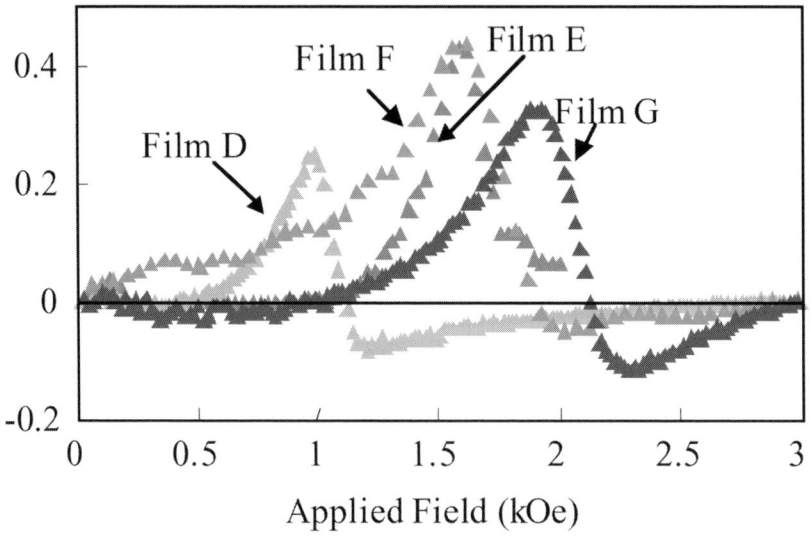

Fig. 14 Delta M curves for a set of CoCrPt bi-crystal films grown on Cr underlayers on NiP plated Al substrates. Samples D and E have 13% Cr whereas samples F and G have 22% Cr. The slopes of the ΔM curves are higher for films F and G with higher Cr concentration and this was found to correlate with a higher signal to noise ratio.

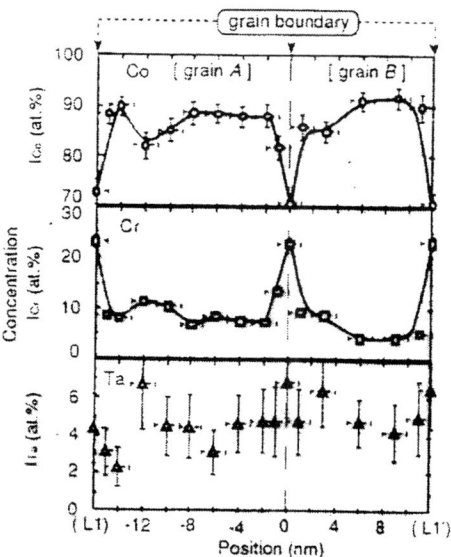

Fig. 15 EELS data of the concentration of Co, Cr and Ta across two grains in a CoCr$_{15}$Ta$_4$ thin film.[22]

workers[22] have examined not only Cr segregation but also the role of other components in the alloys in promoting such segregation. Figure 15 shows one example of their results taken via high resolution EELS studies. In the scan shown the level of Co, Cr and, in this case Ta, across a grain and at grain boundaries has been examined. Clearly the Ta level, whilst difficult to determine, remains essentially constant across the grain and the grain boundary. However, there are significant peaks in the Cr levels at grain boundaries which correspond almost monotonically to the depletion in Co at the same positions. This is clear evidence of the role of Cr in grain boundary segregation leading to the exchange decoupling that is required. Whilst this data is really quite stunning in its confirmation of the behaviour which is expected, it begs the question as to the role of the Ta in the alloy. It is well known that Ta produces low noise media and studies on alloys without the inclusion of Ta has failed to provide evidence for the level of Cr segregation shown in Fig. 15. It is therefore believed that the role of Ta is to occupy those sites within the crystal lattice that would be occupied by Cr thereby reducing the overall solubility of Cr in the Co alloy. Hence the role of Ta is in promoting the Cr segregation and not in any intrinsic behaviour in its own right.

The micromagnetics of the four films under discussion have been examined via the use of a magnetic force microscope. For the purposes of examining the near equilibrium micromagnetic state the films were ac-erased by exposure to a ramped-down ac field which originally saturated the samples. These images are shown in Fig. 16. From these images it is clear that films D and E exhibit much larger 'domain' sizes than the samples F and G. Here of course the apparent domains are in essence artificial and the term correlated regions of magnetisation is more appropriate. This is the case because these materials are truly granular and hence do not have an intrinsic domain structure as such. Such domains have been described previously as interaction domains since they exist by virtue of the coupling between the grains. From these data it is clear that those samples having a high Cr level exhibit a significantly enhanced reduction in the equilibrium domain size compared with those having a low Cr concentration. There is no obvious correlation between the micromagnetic state and grain size in these films. This data was entirely consistent with the bulk magnetic measurements and with the recording performance of these films which was reported elsewhere.[20]

CONCLUSIONS

In this paper we have described the outline structure of a magnetic recording medium for use in a high density rigid disk system. Such a material clearly is a most complex metallurgical structure and since it is composed of a sputtered thin film alloy where the grain size is typically a few tens of nanometers, grain segregation and grain boundary effects are clearly of major importance. In fact it could be argued that the grain boundary is almost the entire material.

We have described the effects of exchange coupling in determining the ultimate

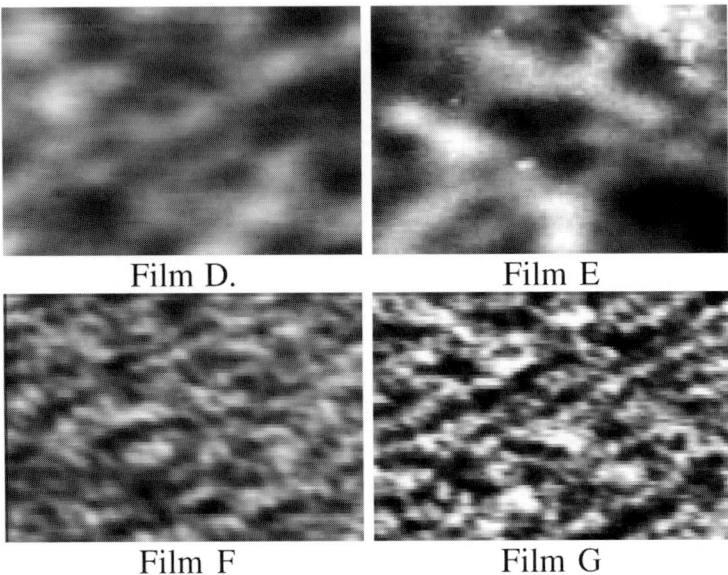

Film D.   Film E

Film F   Film G

Fig. 16 MFM images of the ac demagnetisated state for the four bi-crystal films D-F described in Fig. 14.

quality (noise) of such a recording system and obviously the nature of the grain boundaries is of major importance in determining the degree of coupling. For an ideal system a high degree of uniformity in the grain boundary is essential to maintain uniformity of exchange and thereby of recording performance from bit to bit. In order to achieve such grain boundary control it is essential to have quantitative measurement techniques that enable grain boundary quality to be determined rapidly. We have shown that bulk magnetic measurements and in particular the ΔM curve can provide a quantitative indicator of inter-granular exchange coupling, particularly for the case of spatial segregation. However the interpretation of data for the more complex bi-crystal systems where Cr segregation is dominant is somewhat more difficult but nonetheless the technique can still be used. The interpretation of Cr segregation as determined by ΔM has been validated by reference to published work on EELS data and also by reference to equilibrium domain structures that exist in the material.

Thus, in conclusion, it is clear that a wide range of techniques are need to understand and fully characterise such highly complex materials and as the degree of complexity increases in line with the dramatic growth in the technology the development of further and more sophisticated techniques will inevitably be required.

ACKNOWLEDGEMENTS

The authors wish to acknowledge the financial support of IBM Storage Systems Division, San Jóse and Seagate Media Research Center, Fremont, California towards this work. Many of the students and research assistants who undertook the original measurements.

REFERENCES

1. Seagate Corporation, private communication.
2. C. D. Mee and E. D. Daniel, *Magnetic Recording Technology*, McGraw Hill, 1996.
3. J. Li, M. Mirzamaani, X. Bian and M. Doerner, *J. Appl. Phys.*, 1999, **85** (8), 4286.
4. See press releases on data density breakthroughs at the relevant websites: http://www.storage.ibm.com; http://www.fujitsu.co.jp; http://www.readrite.com; http://www.seagate.com.
5. M. N. Baibich, J. M. Broto, A. Fert, F. N. Vandau, F. Petroff, P. Etienne, G. Creuzet, A. Friederich and J. Chazelas, *Physical Review Letters*, 1988, **61**, 2472.
6. J. Nogués and I. K. Schuller, *J. Mag. Magn. Mater.*, **192**, 1999, 203.
7. K. E. Johnson, J. A Merz, R. L. White and A. Wu, *IBM J. Res. Develop.*, 1996, **40** (5), 511.
8. B. K. Middleton, M. Aziz, M. Wdowin and J. J. Miles, *IEEE Trans. Magn.*, **34**, 1998, 2339.
9. J-G. Zhu, in Ref 2 above.
10. T. Yogi, G. L. Gormann, C. Hwang, M. A. Kakalec and S. E. Lambert, *IEEE Trans. Magn.*, 1988, **MAG-24**, 2727.
11. G. Hughes, *J. Appl. Phys.*, 1983, **54**, 5306.
12. M. el Hilo, K. O'Grady, R. W. Chantrell, I. L. Sanders, M. M. Yang and J. K. Howard, *IEEE Trans. Magn.*, 1991, **27** (6), 5061.

13. J. Fidler, T. Schrefl, *J. Appl. Phys.*, 1996, **79** (8), 5029.
14. E. P. Wohlfarth, *J. Appl. Phys.*, 1958, **29,** 595.
15. P. E. Kelly, K. O'Grady, P. I. Mayo and R. W. Chantrell, *IEEE Trans. Magn.*, 1989, **25** (5), 3449.
16. J. Zhu and H. N. Bertram, *J. Appl. Phys.*, 1988, **63**, 3248.
17. P. I. Mayo, K. O'Grady, P. E. Kelly, J. A. Cambridge, I. L. Sanders, T. Yogi and R. W. Chantrell, *J. Appl. Phys.*, **69** (8), 1991, 4733.
18. P. Dova, K. O'Grady, M. P. Morales and M. F. Doerner, *J. Appl. Phys.*, 1997, **81** (8), 3949.
19. M. Mirzamaani, C. V. Jahnes and M. A. Russak, *J. Appl. Phys.*, 1991, **69**, 5169.
20. P. Dova, K. O'Grady, M. P. Morales and M. F. Doerner, *J. Appl. Phys.*, 1997, **81** (8), 3949.
21. J-G. Zhu, X-G. Ye and T. C. Arnoldussen, *IEEE Trans. Magn.*, 1993, **29**, 324.
22. M. Futamoto, N. Inaba, Y. Hirayama, K. Ito and Y. Honda, *Mat. Res. Soc. Symp. Proc.* 1998, **517**, 589.

Grain Boundary Segregation and Hardness in Ni₃Al

R. E. SMALLMAN

School of Metallurgy and Materials, The University of Birmingham, Edgbaston, Birmingham B15 2TT, UK

C. S. LEE

Department of Materials Science, City University of Hong Kong, Kowloon, Hong Kong

ABSTRACT

Grain boundary brittleness of intermetallics is briefly reviewed with special reference to Ni_3Al. The improvement in ductility as a result of small ternary additions and the various explanatory models are outlined. Recent work on grain boundary segregation and on microhardness is presented showing that boron doping of Ni_3Al can (i) reduce the grain boundary hardening contribution, (ii) improve the transfer efficiency of shear stress across the boundary (iii) reduce dislocation interaction hardening and (iv) cause a significant solution hardening.

INTRODUCTION

There is considerable world-wide interest in the properties of intermetallic compounds. In terms of their properties, intermetallics are generally regarded as a class of material between metals and ceramics, arising from the bonding being a mixture of metallic and covalent. Intermetallics are intrinsically strong (and in the Ll_2-ordered compounds the strength increases with temperature up to about 600°C) with high elastic modulus. The strong bonding and ordered structure also gives rise to lower diffusion coefficients and hence greater stability of diffusion-controlled properties. Intermetallics containing aluminium or silicon exhibit a resistance to oxidation and corrosion because of their adherent surface oxides. Those based on light elements have attractive low densities giving rise to high specific properties particularly important in weight-saving applications. Like ceramics, however, the biggest disadvantage of intermetallics is their low ductility, particularly at low and intermediate temperatures.

In recent years, interest in these materials has increased with the observation that many compounds may be made more ductile by the addition of a third element. This is particularly spectacular in Ni_3Al with B addition, Ti_3Al with Nb and TiAl with Mn.

LIMITED DUCTILITY

Over the years various proposals have been made to explain the lack of ductility in intermetallics. Whilst recognising the overall commonality of these materials, the reason for the lack of ductility varies from compound to compound and includes (i) a limited number of easy deformation modes to satisfy the von Mises criterion, (ii) segregation of deleterious solutes to the grain-boundaries (iii) intrinsic grain boundary weakness (iv) environmental susceptibility and (v) co-valent bonding and a high Peierls-Nabarro stress.

Ball and Smallman[1] as early as 1966 pointed out that the B2 structure of NiAl did not possess enough independent deformation modes to provide polycrystalline plasticity. Single crystals can exhibit some ductility (~2%) depending on orientation and increases significantly at somewhat elevated temperatures under glide-climb conditions.[2] Small additions of Fe or Mo can also give slightly improved ductility.

Polycrystallne NiAl can show a little ductility (~1%) if it is stoichiometric but any deviation from stoichiometry results in intergranular brittleness. Generally, the grain boundaries are free from segregation and the behaviour arises from a basic, inherent lack of ductility. Additions such as boron and carbon have no effect on the ductility nor substitutional elements, such as iron and molybdenum, which can improve the ductility of single crystals. It would appear that any small improvement in grain boundary strength is outweighed by the significant solid solution strengthening these elements produce.

From early work it has also been recognised that environment, particularly processing environment, markedly influence the ductility of the compound. A spectacular example is the 'pest' phenomena investigated by Westbrook and co-workers who demonstrated that oxygen can penetrate the grain boundaries during heat treatment causing the sample to decrepitate and turn to powder. The testing environment can also influence the ductility, as discussed later for Ni_3Al.

In recent years deformation studies and the enhancement of ductility has been most extensively studied in the Ll_2 intermetallics and, of these, Ni_3Al has received by far the greatest attention as a model materials for all intermetallics. This Ll_2 compound has sufficient slip systems for polycrystalline plasticity, exhibits intergranular fracture and is ductilised with the addition of specific third element additions. Whereas single crystals exhibit some ductility, most binary A_3B compounds in polycrystalline form are intergranularly brittle. Since Auger studies do not reveal any particular grain boundary segregation in these undoped compounds (within the limits of detection), it is generally concluded that the grain boundaries in these compounds are intrinsically weak.

From early work it has long been recognised that the environment, not only during processing but also during mechanical testing, can influence the brittleness behaviour. Carefully prepared polycrystalline Ni_3Al exhibits[3] a room temperature ductility of 3.1 to 4.8% in air but 12.6 to 15.8% in dry oxygen. The environmental embrittlement is considered to result from the chemical reaction

$$2Al + 3H_2O \rightarrow Al_2O_3 + 6H$$

Aluminium atoms in the aluminide react with moisture in the air resulting in the generation of atomic hydrogen that penetrates the crack tip, in a similar manner to hydrogen embrittlement. The best ductility behaviour is obtained by testing in dry oxygen when it is considered that the reaction of the aluminium atoms with oxygen

$$4Al + 3O_2 \rightarrow 2Al_2O_3$$

competes with the moisture reaction, thereby reducing the generation of atomic hydrogen. Unlike hydrogen embrittlement in steels which occurs by slow stable crack growth the environmental cracking observed in tensile tests of Ni₃Al occurs quite rapidly. It is therefore concluded that crack nucleation is the process being influenced by the environment and once formed the crack propagates rapidly, whatever the environment. Even testing in dry oxygen the specimens exhibit intergranular fracture pointing to possible inherent grain boundary weakness.

DUCTILISATION

The improvement in the ductility of Ni₃Al by third element addition has been recognised for almost two decades, but the exact mechanism is still not fully understood. Although boron has the most spectacular effect in ductilising Ni₃Al other elements of a substitutional nature have been shown to be beneficial. These elements, eg, Fe, Mn, Be, Pd, Zr, Hf must be added in substantial concentration, usually greater than 1%, and are thought generally to improve the ductility of the matrix, allowing easier yielding across grain boundaries and reducing the tendency for crack nucleation. They are not considered to influence ductility by segregating to grain boundaries. They may also exert an influence on the environmental sensitivity. Zirconium and hafnium, which do not segregate to grain boundaries, are observed to ductilise Ni₃Al, but titanium does not. It is noted that Zr and Hf have a large misfit (~11%) and are found to substitute for Al and therefore possibly trap H from the environmental reaction in contrast to Ti which has a small misfit and does not improve the ductility.[3]

By contrast the spectacular improvement in ductility by the addition of boron requires only a trace-element composition, with an optimum around 0.1%. The improvement produced is, however, very dependent on the non-stoichiometric composition being below 25% aluminium; for the stoichiometric composition and aluminium-rich alloys boron additions produce little improvement in properties. Boron is a strong interstitial solid solution strengthener and with increasing boron the strength of the alloy counteracts the improvement in ductility. The reason why Ni-rich alloys show an improvement in ductility with boron addition but stoichiometric and Al-rich alloys do not, is still unclear. Although there is same uncertainly about the validity of analytical electron microscopy studies of grain boundary microchemistry there is little or no evidence to indicate that boron alters the grain boundary chemistry. In stoichiometric Ni_3Al the grain boundaries have essentially the same composition as the bulk grain in both boron containing and boron-free material. In Ni-rich alloys the microchemistry shows that aluminium is reduced in the grain boundary region but this is uninfluenced by the presence of boron. The fact that boron may form strong bonds with nickel atoms on the boundary and the resultant reduction in directional Ni-Al bonding could well improve the boundary cohesion.

It has been well demonstrated[4] that boron is effective in eliminating the sensitivity of Ni_3Al to environmental embrittlement. Boron-free $Ni_{26}Al_{24}$ has a ductility of 8% in dry oxygen whereas the boron-doped alloy exhibits more than 40%. Since boron and hydrogen both occupy similar sites it is possible that segregation of boron to grain boundaries reduces the diffusivity of hydrogen along these boundaries.

A further influence of boron proposed is that it enhances slip transmission across grain boundaries. From the conventional picture of yielding, with dislocations piled up at a grain boundary, stress relief by plastic deformation in the next grain occurs before fracture as a result of boron in the boundary. Since k_y in the Hall–Petch relation, Fig. 1 is a measure of this slip transmission, a reduction in the slope of the plot of (Yield Stress) versus $d^{-1/2}$ as pointed out by Baker *et al.*[5] would seem to point to such an effect.

Overall, most of the attention in the literature attributes ductilisation to increasing the cohesive strength of the grain boundary, coupled possibly with easing slip transfer across grain boundaries. Little attention has been given to the effect on bulk properties, eg B making polyslip easier such that plastic deformation can be accommodated more easily. The most significant support for the hypothesis of an improvement in bulk properties is the observation[6] that superlattice intrinsic stacking faults (SISFs) are removed by B. This effect is unlikely to result from any influence of B on SISF or antiphane boundary (APB) energy, although γ_{100} does seem to be reduced. Possibly more likely is a locking effect preventing the reaction which creates SISFs. However a reduction in fault density must lead to a reduced friction stress in the bulk grain and ease polyslip by

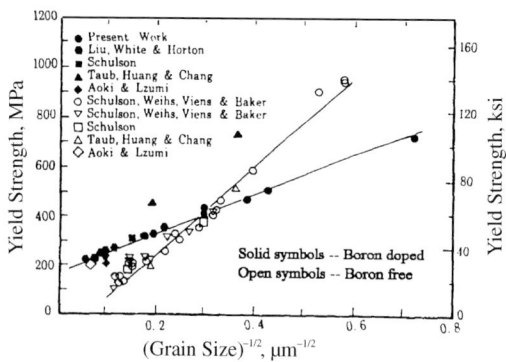

Fig. 1 Hall–Petch relationships for boron free and boron doped polycrystalline Ni_3Al.[5]

removing obstacles to dislocations. Other evidence supporting a bulk effect follows from the observation that [001] oriented single crystals oriented for duplex slip are ductilised more than those oriented for single slip ie 55% compared to 25%.

GRAIN BOUNDARY SEGREGATION AND HARDENING IN NI₃AL

Lee and co-workers[7] have recently investigated the suppressed hardening effect from the interaction of dislocations moving on different slip systems and the improved transfer efficiency of shear stress across a grain boundary as a result of boron segregation. <100> single crystals and bicrystals were each prepared from Ni–23.3at.%Al and Ni–23.3at.% Al–0.27at.%B respectively. After homogenisation the binary alloy crystals were furnace cooled whereas the boron-doped Ni₃Al were subjected to various cooling rates, including water quenching, air cooling, furnace cooling and step annealing i.e. furnace cooling held at 800°C and 600°C for 6 hours. The various cooling rates were expected to produce different amount of segregation at the grain boundary with the greater amount of segregation existing for the slower cooling rate.

Table 1 gives the measured compression yield strength and microhardness values of the binary and boron-doped Ni₃Al single crystals. The hardening observed is related to the increase in lattice friction due to the boron solid solution. The potent hardening arises from the asymmetrical strain field of the occupied interstitial site.

For the bicrystals, the microhardness profiles are shown in Fig. 2. The microhardness near the grain boundary is greater than in the grain interior and with boron addition the hardening effectiveness of the grain boundaries is reduced, decreasing with increasing boron segregation.

GRAIN BOUNDARY MODELLING

The transfer of shear stress τ_1 in grain 1 across a grain boundary to a neighbouring grain as shown in Fig. 3 is given by

$$\tau_2 = \frac{\alpha}{2} \frac{\pi(1-\nu)}{G_b} \cdot d\tau_1^2 \cos\varphi = \alpha T d\tau_1^2 \tag{1}$$

where d is the grain diameter, T a stress transmission factor and α a term to reflect

Table 1 The microhardness and compression yield strengths of binary and B-doped Ni₃Al single crystals.

Composition (at.%)	Ni–23.3Al	Ni–23.3Al–0.27B
Microhardness (HV), 10 gf	189.2	261.1
Yield Strength, MPa	159.2	448.5
HV10 gf/Yield strength	1.19	0.58

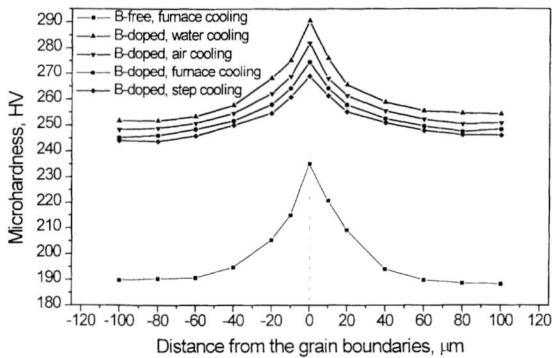

Fig. 2 Grain boundary microhardness profiles.

the efficiency of the stress transmission; defects in the boundary, such as microcavities, would reduce the term α and hence reduce the effective stress on the slip plane in the adjacent grain 2.

Now if τ_{ss} is the resolved shear stress for plastic deformation by the operation of single slip and τ_{ps} the resolved shear stress for polyslip when other slip systems operate, then $\tau_{ps} = H \tau_{ss}$ where H is a hardening factor due to dislocation interaction. Hence to operate a slip system in grain 2, τ_2 has to reach the value $\tau_{ps} = H \tau_{ss}$, so that the resolved shear stress in grain 1, τ_{1gb} to overcome the grain boundary hardening can, from equation 1, be expressed as

$$\tau_{1gb} = \sqrt{\left(\frac{H\tau_{ss}}{\alpha Td}\right)} = \sqrt{\left(\frac{H\tau_{ss}}{\alpha T}\right)} \cdot d^{-1/2} \tag{2}$$

The overall resolved shear stress needed to operate a slip system in grain 1, ie, overcome grain boundary resistance and intrinsic resistance, is $H \tau_{ss} + \tau_{1gb}$ and the flow stress σ equal to

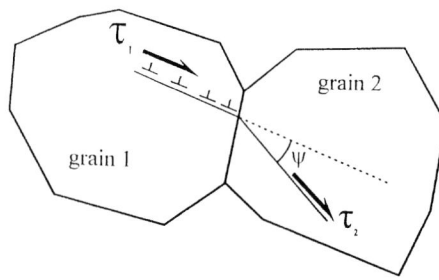

Fig. 3 Stress transfer across a grain boundary ahead of a dislocation pile-up.

$$\sigma_f = \frac{1}{m}\left[H\tau_{ss} + \sqrt{\left(\frac{H\tau_{ss}}{\alpha T}\right)} \cdot d^{-1/2} \right] \tag{3}$$

where $\overline{m}$ is the average modified Schmid factor for the polycrystal.

For single crystal behaviour the grain boundary effect can be ignored so that $\frac{\sigma}{f} = \tau_{ss}/m$ for single slip deformation and $\frac{\sigma}{f} = H\,\tau_{ss}/m$ for polyslip deformation. From the compressive yield strength of binary and doped <100> Ni$_3$Al single crystals given in table 1, $(\frac{\sigma}{f})$doped$/(\frac{\sigma}{f})$ binary = 2.82 indicating that the lattice resistance is increased by almost a factor of 3 upon boron doping. Furthermore, if it is assumed that the hardness is proportional to the flow strength, ie HV = $\xi \frac{\sigma}{f}$, then

$$\frac{\text{HV Hardness}}{\text{Compressive Yield Strength}} = \frac{\xi H\,\tau_{ss}/m}{\tau_{ss}/m} = \xi H$$

and the results of table 1 give $(H)_{\text{doped}}/(H)_{\text{binary}} = 0.49$, indicating that boron doping decreases the dislocation interaction hardening to about half that of the undoped Ni$_3$Al.

From equation (3) k_y in the Hall–Petch relationship is given by $k_y = \frac{1}{m}\left[\frac{H\tau_{ss}}{\alpha T}\right]^{1/2}$

and since T and $\overline{m}$ should be the same for both doped and undoped Ni$_3$Al the k_y ratio can be written as $k_y =$

$$\left[\frac{H\tau_{ss}}{\alpha}\right]^{1/2}_{\text{doped}} / k_y = \left[\frac{H\tau_{ss}}{\alpha}\right]^{1/2}_{\text{binary}}.$$

The grain size dependence plot, Fig. 1, gives the k_y ratio to be 0.427 and with the above values for $H\tau_{ss}$, then

$$\alpha_{\text{doped}} / \alpha_{\text{binary}} = 7.58$$

indicating a very significant increase in shear stress transfer across the grain boundaries.

CONCLUSION

Analysis of the doping results shows that boron (i) produces a considerable solution strengthening, (ii) reduces the hardening contribution from the grain boundaries, (iii) suppresses the hardening effect from the interaction of dislocations within the grain undergoing polyslip deformation and (iv) improves the transfer efficiency of shear stress across grain boundaries. These conclusions are consistent with the experimental observations reported in the literature.

REFERENCES

1. A. Ball and R. E. Smallman, *Acta Met.*, **14**, 1996, 1349.
2. H. L. Fraser, M. H. Loretto and R. E. Smallman, *Phil. Mag.*, **28**, 1973, 651.
3. E. P. George, C. T Liu and D. P. Pope, *Structural Intermetallics*, R. Daralio, D. V. Miracle and M. V. Nathal eds, TMS, 1993, 431.
4. C. T. Liu, *Structural Intermetallics*, R. Daralio, D. V. Miracle and M. V. Nathal eds, TMS, 1993, 365.
5. I. Baker, E. M. Schulson and J. A. Horton, *Acta Metall.*, **35**, 1987, 1533.
6. W. Yan, I. P. Jones and R. E. Smallman, *Scripta Metall.*, **21**, 1987, 1611.
7. C. S. Lee, G. W. Han, R. E. Smallman, D. Feng and J. K. L. Lai, *Acta Metall.*, **47**, 1999, 1823–1830.

The Mechanism of Diffusionless Transformations in Metallic and Protein Crystals

R. C. POND AND T. NIXON

Materials Science and Engineering, Department of Engineering, University of Liverpool, Liverpool L69 3GH

T. T. CHENG AND M. AINDOW

School of Metallurgy and Materials, University of Birmingham, Edgbaston, Birmingham B15 2TT

ABSTRACT

The mechanisms of diffusionless phase transformations are compared in metallic and protein crystals. Although the elastic moduli of the former crystals are greater than the latters' by a factor of about 100, it is proposed that the underlying mechanisms involve the motion of interfacial dislocations in both cases. In the case of martensitic transformations in metals the interface must be an invariant plane of the transformation in order to minimise the elastic strain. In the protein case, the interface orientation is not so prescribed, and large elastic strains may be present. It is demonstrated that the dislocation transformation mechanisms are diffusionless in examples of the two cases, namely NiTi and the tail-sheath of a bacteriophage.

I. INTRODUCTION

Diffusionless phase transformations are important in both technological and natural materials. For example, martensitic transformations are the basis of shape-memory and super-elastic materials,[1] and the T-even bacteriophage exploits such a transformation in order to inject its genetic material through a target cell membrane.[2] The object of the present article is to compare the transformation mechanisms in these two apparently disparate systems, and to show that the concepts developed in the theory of solid-state defects are useful tools in the analysis of both cases. Metals exhibit three-dimensional crystalline structures and elastic moduli of the order of 100 GPa, whereas the structural proteins that form the tail-sheath of the bacteriophage have two-dimensional crystallinity and elastic moduli more than 100 times smaller than for metals. Consequently, certain aspects of transformations in the two materials, such as the magnitude of elastic strains involved, are very different in magnitude. However, we propose that both transformations proceed through the motion of interfacial defects and that the topological properties of these can be identified using a unified framework.

In order that a transformation is diffusionless from the nucleation event through the growth stage it is necessary to show that transformation dislocations, referred to here as disconnections, move along the interface between the two chemically identical but structurally different phases without a flux of material being required. Moreover, if other defects are already present in this interface, it is necessary to demonstrate that the interactions of disconnections with these additional defects also do not require a diffusional flux, or that the sum of these two fluxes is zero overall. These issues have been clarified recently and are briefly reviewed in the next section.

Transformations in metals must involve relatively small elastic strains in the matrix surrounding a martensitic product. A particularly favourable situation arises when the transformation mechanism leads to the evolution of plate-like crystals with habit planes that are invariant planes of the transformation. A phenomenological theory of martensite crystallography (PTMC) has been developed[3, 4] that describes an algorithm for identifying such planes. The present authors demonstrate here for the case of NiTi, an important shape-memory and super-elastic alloy, that a disconnection mechanism can be identified that is consistent with the experimentally verified predictions of the PTMC.

In protein crystals, larger elastic strains can be tolerated, and hence the orientation of the interface between the two phases is not so prescribed as in the metallic case. Olsen and Hartman[5] have shown that the interface orientation for the T-even bacteriophage does not correspond to that expected on the basis of the PTMC. Their treatment is briefly reviewed here, but it is shown that a disconnection mechanism can operate in a diffusionless manner in the observed interface.

II. DIFFUSIONAL FLUXES DUE TO DEFECT MOTION AND INTERACTION

Interfacial line-defects such as transformation dislocations are known to effect phase transformation. The topological theory developed by Pond and co-workers[6] enables the characterisation of all admissible defects in a given interface. In particular, disconnections, which are interfacial line-defects exhibiting step and dislocation character,[7] are characterised topologically by their Burgers vector, $\mathbf{b}$, and step height, h. This partitioning of a disconnection into step and dislocation portions enables the glide/climb behaviour of such defects to be clarified and related to the transformation and deformation aspects of interfacial processes. Only those defects that exhibit finite h can lead to transformation of crystal structure, whereas the dislocation portion of a disconnection is related to deformation of crystal structure. The expression describing the incremental number of atomic sites, δN, due to motion of a disconnection of length L (lying parallel to the direction x) through a distance δy along an interface separating two crystals designated λ and μ, is given by[7]

$$\delta N = L(h\Delta X + b_z X(\lambda/\mu))\delta y \tag{1}$$

where $X(\lambda/\mu)$ represents the density of atomic sites in the λ or μ crystal, $\Delta X = X(\lambda) - X(\mu)$ and b_z is the component of $\mathbf{b}$ perpendicular to the interface. If the material is not monatomic, equation (1) must be written for each species present. Hirth and Pond[7] have shown that the only two circumstances under which disconnection motion along an interphase interface is conservative are when b_z and h are simultaneously zero, or when the interface is an invariant-plane (IP) since $h\Delta X = -b_z X(\lambda/\mu)$ for this case. Fluxes that arise as a result of the interaction of two interfacial defects have been discussed by Sarrazit and Pond[8] and Nixon and Pond.[9] For example, the former authors considered the motion of aligned defects (oriented parallel to x) and showed that if defect α moves in the direction $-y$ through defect β, the additional number of atoms required, owing to the climb of α and β, up/down each others steps, is given by

$$\delta N^i = L\left(b_y^\alpha h^\beta - b_y^\beta h^\alpha\right)X(\lambda/\mu) \qquad (2)$$

III. PHASE TRANSFORMATION MECHANISM IN NiTi

The transformation process is envisaged to begin with nucleation of a favourably oriented but homogeneously strained embryo that is taken as the reference state. Operation of a source then generates an array of mobile disconnections that simultaneously effects growth of the embryo and partially relieves its strain. Further relief of the strain is envisaged to arise by appropriate lattice-invariant deformation (LID), involving slip and/or twinning in one (or both) adjacent crystal(s). The optimum defect density arises when the interface is an IP of the transformation relating the fully relaxed crystals. This interface is referred to as being compatible, and it corresponds to that predicted by the PTMC. In the present model, the state of strain in the second phase changes continuously until the compatible state is reached. The habit plane then corresponds to regions of the initial terrace plane delineated by the arrays of disconnections and LID.

NiTi undergoes a martensitic transformation from the B2 to the B19′ structure, and observations of the product phase are in good agreement with the predictions of the PTMC.[10] The first step in this model is to choose a candidate nucleus exhibiting an IP and requiring a relatively small deformation to become the relaxed B19′ structure. A plausible choice is shown in Fig. 1; the invariant interface, or terrace plane, is $(110)_c$: $(001)_m$ and the martensite is extended by 4·18% and 3·37% along $[100]_m$ and $[010]_m$ respectively relative to its relaxed crystal structure. The crystallography of this nucleus predetermines the topological properties of the set of admissible disconnections that can arise.[6] Amongst this set, the defect illustrated schematically in Fig. 2, is thought to be the most likely physically since it exhibits small $|\mathbf{b}|$ and h ($\mathbf{b} = [0·08, \overline{0·08}, 0·18]_c$, $h = 2d_{(110)c}$), which imply stability and reasonable mobility,[11] and, according to equation (1), would move conservatively across the terrace.

Various authors have reported different modes of LID; in the present work type I

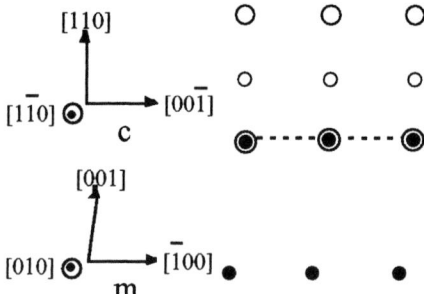

Fig. 1 Schematic illustration of the crystallographic orientation of the cubic (open symbols) and strained martensite nucleus (filled symbols) configuration exhibiting an invariant-plane interface.

$(011)_m$ twinning is assumed to be active.[12] The $\mathbf{b}^T$ of the twinning dislocations is then $[0{\cdot}17, 0{\cdot}07, \overline{0{\cdot}07}]_{m'}$ and substitution of this value with $h = 0$ into equation (2) shows that the intersections of the mobile disconnections defined above and these twinning defects can be conservative. The twins intersect the terrace plane along $[100]_{m'}$ and, according to calculations using the PTMC, the twinning fraction must be $0{\cdot}29$. To determine the line direction and spacing of the disconnections it is necessary to ident-ify the deformation necessary to transform the nucleus into the relaxed martensite in the compatible state. This deformation can then be substituted into the Frank–Bilby equation,[13] and, by using probe vectors defined in the compatible interface, the required parameters can be determined, as illustrated in Fig. 3.

As a final check on the consistency of this model of nucleation and growth with the PTMC description of the compatible state, the relative orientations of the two

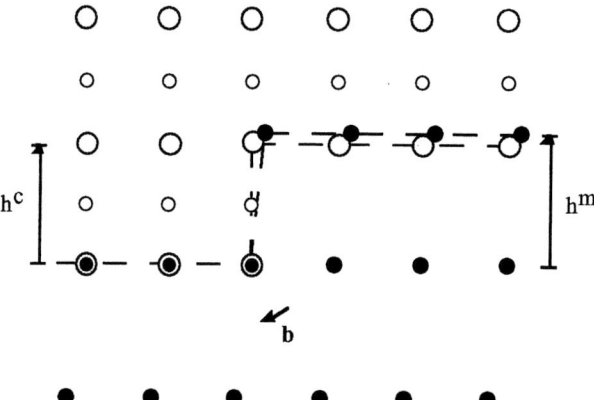

Fig. 2 Schematic illustration of the formation of a disconnection ($\mathbf{b}$, h^c) on the terrace plane of the nucleus.

crystals and the orientations of the habit plane can be compared using the two approaches. In terms of the line defect modelling, this is done using knowledge of the steps introduced to the terrace plane by the two defect arrays. Only one set of steps of height h_c = 4·26 nm (see Fig. 2) oriented parallel to ξ_d (see Fig. 3) with spacing d_d = 0·9 nm arises in the cubic phase corresponding to the interfacial plane being $(0·42,0·87,0·25)_c$. On the other hand, two sets of steps associated with the disconnections and the twins lead to the interface plane corresponding to $(0·18, \overline{0·44},1)_m$ in the martensite phase. These values are in good agreement with the PTMC.

IV. TRANSFORMATION MECHANISM FOR BACTERIOPHAGE

The structural change arising when the tail-sheath of the bacteriophage contracts is shown schematically in Figs. 4 and 5. In Fig. 4a the cylindrical arrangement of globular proteins is viewed along its axis, and Fig. 5a shows the two-dimensional array of molecular sites obtained if the cylinder is imagined to be cut parallel to its axis and then flattened out. The corresponding diagrams when the sheath shortens along its axis and increases in radius are depicted in Fig. 4b and 5b.

Following Olsen and Hartman,[5] the principal distortions involved in the transformation from the extended, λ, to the contracted, μ, crystalline forms can be determined using the methods developed in the PTMC. In addition, the undistorted and unrotated directions associated with this transformation, which delineate possible invariant-line interfaces between the two crystals can be found.[5] Neither of these corresponds to the observed interface line, although the latter is an Eigenvector of the transformation. It is also noted that the symmetry of the two crystal forms is the

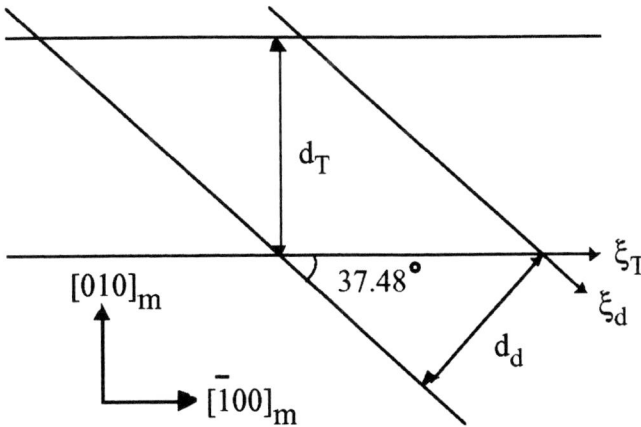

Fig. 3 Schematic illustration of the defect array in the compatible interface: disconnections ξ_d, d_d = 0·95 nm and twins with equivalent thickness d_T when projected onto the interface.

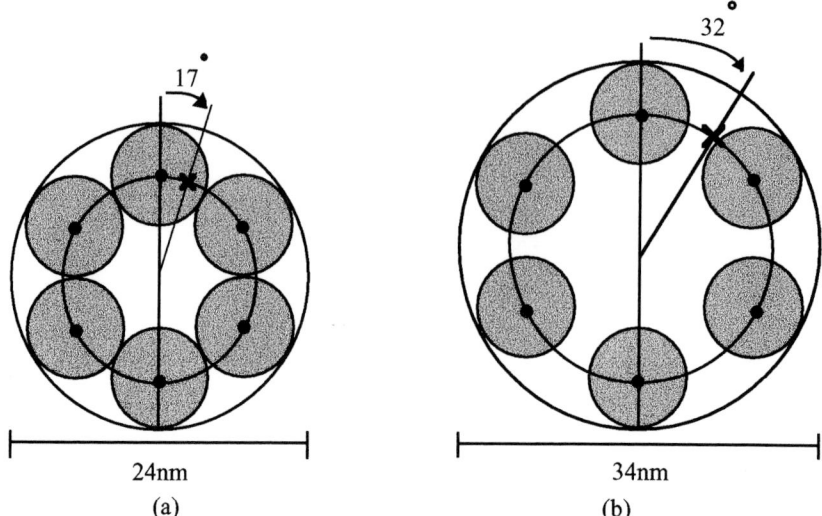

Fig. 4 Schematic view of protein molecules viewed along the axis of the virus tail sheath: (a) extended, λ, and (b) contracted, μ, axis arrangements. The angular rotation of molecular sites in the next annular ring are indicated.

same so that, unlike typical cases in metals, only one variant of the product phase arises.

The actual interface line between the extended and contracted forms, as shown schematically in Fig. 6, is perpendicular to the cylinder axis and corresponds to the interface with minimum length rather than an invariant-line. In Fig. 6a the λ and μ crystals are shown in their relaxed forms abutting along the interface and this defines the chosen reference state. The interface is depicted as an IP in Fig. 6b following appropriate elastic straining of the crystals. Using dislocation terminology, the interface in Fig. 6b exhibits a defect array that can be identified using the Frank–Bilby equation, and can be visualised as one edge dislocation with $\mathbf{b}^\beta = \mathbf{t}(\lambda) - \mathbf{t}(\mu)$ in each period of the interface, as depicted in Fig. 6b.

Now that the dislocation content of the observed interface has been identified, the motion of a disconnection along the interface can be considered. A plausible disconnection is shown in Fig. 6c; this has overlap step height $h^\alpha = \mathrm{d}(\mu)$ and $\mathbf{b}^\alpha = \mathbf{a}_2(\lambda) - \mathbf{a}_2(\mu)$ so that the component $b^\alpha_z = \mathrm{d}(\lambda) - \mathrm{d}(\mu)$. Since its motion would involve relatively small molecular displacements, it would probably exhibit reasonable mobility. Next, it is shown that motion of such a disconnection along the interface line would be conservative. To do this the interaction of the disconnection ($\mathbf{b}^\alpha, h^\alpha$) is considered with the array of pre-existing defects ($\mathbf{b}^\beta, 0$) identified above. Equation (1) is used with the defect line length taken to be unity for the two-dimensional crystal case, $X(\lambda) = (t(\lambda)\mathrm{d}(\lambda))^{-1}$, $X(\lambda) = (t(\mu)\mathrm{d}(\mu))^{-1}$ and y is taken to be parallel to the

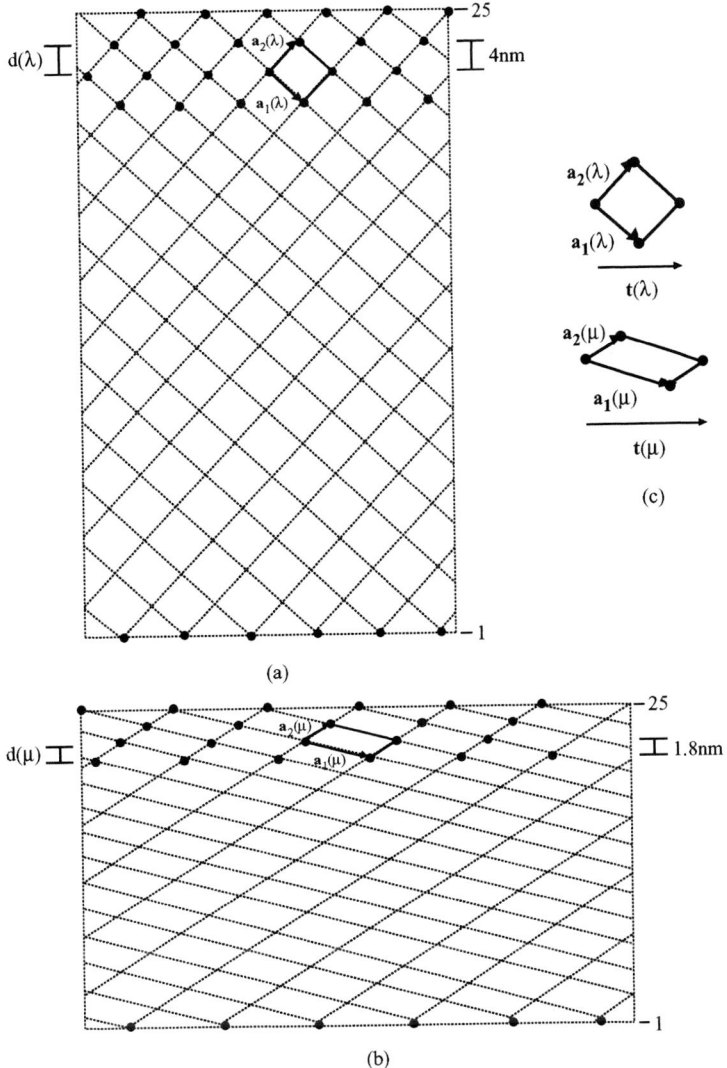

Fig. 5 Schematic view of protein sites in the virus tail sheath after cutting and flattening: (a) extended λ, (b) contracted μ axis arrangements, and (c) unit cells in the two forms.

interface line and z parallel to the cylinder axis (Fig. 6a). For motion of the disconnection in Fig. 6c to the left by the distance $\delta y = -\mathbf{t}(\lambda)$ we have, from equation (1), for this case

$$\delta N = \left(h^\alpha \Delta X + b^\alpha_z X(\lambda) \right) \delta y$$

$$= -\left[d(\mu) \left\{ (t(\lambda)d(\lambda))^{-1} - (t(\mu)d(\mu))^{-1} \right\} + (d(\lambda) - d(\mu))(t(\lambda)d(\lambda))^{-1} \right] t(\lambda)$$

$$= \left(\frac{t(\lambda)}{t(\mu)} - 1 \right) \tag{3}$$

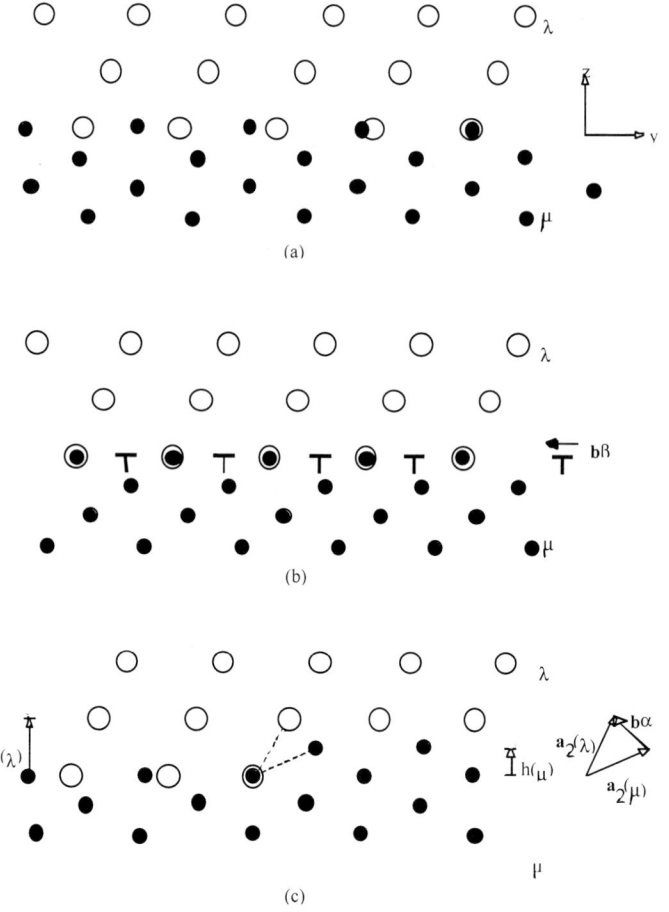

Fig. 6 Schematic illustration of the interface between the extended λ and contracted (μ) crystals showing: (a) as the reference state with relaxed crystals, (b) the elastically strained state modelled as a dislocation array, and (c) a candidate disconnection in the reference state.

This negative flux implies that material must flow away from the disconnection. However, in moving leftwards, disconnection α must pass through defect β, and the latter must climb up the former's step. According to equation (2), the number of molecules required for this is

$$\delta N^i = -b^{\beta}_{y} h^{\alpha} X(\mu) = -(t(\lambda) - t(\mu))d(\mu)(t(\mu)d(\mu))^{-1}$$

$$= \left(1 - \frac{t(\lambda)}{t(\mu)}\right) \tag{4}$$

Thus the total flux obtained by adding equations (3) and (4) is zero, and hence the mechanism described is conservative.

V. DISCUSSION

The present authors have compared diffusionless transformation in a metallic system, NiTi, and a protein crystal, and shown that disconnection motion along the interface separating adjacent phases is a plausible mechanism in both cases. Other aspects of the two transformations are quite different, but the framework of solid-state defect theory offers a unified approach to understanding transformation mechanisms. In the metallic case the transformation occurs by nucleation and growth. A nucleus with epitaxial crystallography exhibiting IP interfaces arises. At this stage the interfacial energy makes a significant contribution to the total energy, and may dominate the elastic strain energy of the nucleus. The nucleus then grows by means of the introduction of an array of mobile disconnections and simultaneously the increasing elastic strain is relieved by these same defects in combination with LID. By this means, the interface structure remains microscopically and macroscopically invariant, finally exhibiting the habit plane predicted by the PTMC. At this stage, the interfacial energy makes a relatively small contribution to the total energy, as is attested by the plate-like shape of the martensite. The dominant factor in minimising the total energy is the preservation of an IP interface, thereby enabling the two high modulus phases to fit together with very small elastic strains parallel to the interface.

In the case of the bacteriophage, the tail-sheath is self-assembled in the host cell, and is thought to form in a metastable crystalline arrangement through epitaxial constraints from the base-plate of the tail. When the bacteriophage alights on a suitable cell membrane, the base-plate becomes distorted and this may initiate the transformation to the contracted crystalline form. Unlike the metal case, the interface orientation corresponds to that with minimal length rather than being an invariant line since the latter cannot form a closed loop on a cylindrical surface. Consequently, large strains arise parallel to the interface, and these can be formally represented by an array of interfacial edge dislocations. In effect, the interface becomes an IP of the total transformation. Thus, in a defect mechanism of phase transformation in this bacteriophage, a mobile disconnection must be imagined to move through this pre-existing array. In other words this process can be visualised as the motion of disconnections along the interface, which requires a flux of material, and climb of the edge dislocations up the steps of the disconnections, which requires an equal and opposite flux of material.

The source of disconnections has not been identified in this work for either the metallic or protein case. In the latter case, a single mobile disconnection could bring about the transformation of the whole sheath by moving on a spiral path, whereas expansion of a disconnection dipole could only transform a single annular ring of protein molecules and hence repeated nucleation of dipoles would be required. For the metallic case, the density of disconnections introduced into the interface must reach a particular value in order to relieve the strain in the nucleus and create a macroscopically invariant habit plane.

REFERENCES

1. *Shape Memory Materials*, K. Otsuka and C. M. Wayman, eds. Cambridge University Press, Cambridge, 1998.
2. Y. Kikuchi and J. King: *J. Mol. Biol.*, 1975, **99**, 695.
3. M. S. Wechsler, D. S. Lieberman and T. A. Read: *Trans. AIME*, 1953, **197**, 1503.
4. J. S. Bowles and J. K. Mackenzie: *Acta Metall.*, 1954, **2**, 129, 224.
5. G. B. Olson and H. Hartman: *J. Phys. Colloque C4*, 1982, **43** (12), C4–855.
6. R. C. Pond: *Dislocations in Solids*, F. R. N. Nabarro ed. North Holland, Amsterdam, 1989, 1.
7. J. P. Hirth and R. C. Pond: *Acta Mater.*, 1996, **44**, 4749.
8. F. Sarrazit and R. C. Pond: *Interface Sci.*, 1996, **4**, 99.
9. T. Nixon and R. C. Pond: *Proc. 9th Int. Conf. Intergranular and Interphase Boundaries in Materials*, P. Lejcek and V. Paidar eds., Switzerland, Trans Tech Publications Ltd, *Mater. Sci. Forum*, 1999, **294–296**, 123.
10. O. Matsumoto, S. Miyazaki, K. Otsuka and H. Tamura: *Acta Metall.*, 1987, **35**, 2137.
11. J. W. Christian: *Metall. Trans.*, 1982, **A13**, 509.
12. M. Nishida, H. Ohgi, I. Itai, A. Chiba and K. Yamauchi: *Acta Metall. Mater.*, 1995, **43**, 1219.
13. B. A. Bilby, R. Bullough and E. Smith: *Proc. R. Soc.* London A, 1955, **231**, 1955, 263.

Grain Boundaries in Engineering Ceramics

M. H. LEWIS

Centre for Advanced Materials, Department of Physics, University of Warwick, UK

ABSTRACT

Grain boundary structure, local composition and related transport and cohesive properties have a critical influence on fabrication and mechanical or environmental behaviour of ceramics. These issues assume greater importance than for metallic polycrystals in view of the directional (covalent) contribution to bonding and the frequent use of liquid or solid state sintering additives. The presence of grain boundary segregated layers or residual phases, often in the glassy-state, provide the diffusional pathway operative during sintering, in creep deformation or for transport of cations to initially passive oxidation layers.

Examples of these phenomena and associated grain-boundary structure are presented via a survey of the development of silicon nitride based ceramics and the relation between properties and intercrystalline constitution, determined by electron imaging and spectroscopic techniques.

INTRODUCTION; THE KEY ROLE OF GRAIN BOUNDARIES IN CERAMICS

Grain boundaries in both structural and functional ceramics play a key role in the determination of properties, with an influence generally much greater than for metallic polycrystals. The underlying reason for this sensitivity to grain boundary structure and composition stems from a greater directionality of crystal bonding, frequently of a covalent origin, in many nitride, carbide and oxide systems. Hence there is a special problem in retaining normal coordination chemistry within the intercrystalline zone, which results in reduced cohesion and frequently intergranular failure. The ceramic grain boundary is the major mass transport pathway both during diffusional creep and in the sintering process necessary for fabrication from powder precursors. It is for this latter reason that most ceramics are prepared with 'sintering additives' which enhance diffusion or modify interfacial and surface energies to facilitate densification. Grain-boundary transport is often more efficient when the additives form intercrystalline liquids, providing the frequently used mechanism of 'liquid-phase-sintering' as an alternative to solid state diffusion. However, the liquid residues normally dominate the final grain-boundary microstructure and are frequently in the glassy state, with obvious influence on properties. In addition to determining creep and fracture behaviour the boundaries form rapid transport paths for out-diffusion of segregated atoms or liquid components to the free surface where they may modify the protective properties of surface films and coatings. A specific example is the reduction in viscosity of passive silica oxidation layers by out-diffusion of modifying cations, with consequent acceleration in oxidation kinetics due to inward oxygen transport. Within functional ceramics both ionic and electron transport, either parallel or normal to be grain boundary, are influenced by structure and segregated species, exemplified by varistors, thermally sensitive oxide switches and thermometers, fuel-cell membranes and electrodes and the cuprate superconductors.

In this paper a survey is presented of the evolution in the understanding of grain boundary constitution and its influence on properties of one prominent class of structural ceramic, the silicon nitrides and sialons. These ceramics exemplify a typical range of liquid-phase-sintered systems to which the generic techniques of electron imaging and electron-probe spectroscopies have been successfully applied in the definition of grain boundary characteristics.

EARLY Si_3N_4 CERAMICS; THE IDENTITY OF LIQUID PHASE SINTERING

The pioneering work of Deeley, Herbert and Moore in 1961[1] first demonstrated that silicon nitride powders could be densified by hot-pressing in the presence of minor oxide additives. Best results were achieved with MgO which was subsequently the basis for commercially produced HPSN from Norton (NC132) and Lucas (HS110) containing 1–2% of the additive, hot pressed at ~1700°C.

An understanding of the sintering mechanism was derived from X-ray diffraction at Newcastle University[2] and electron microscopy at Warwick[3] which demonstrated the occurrence of a glassy silicate intergranular phase (Fig. 1) derived from a eutectic liquid formed by MgO–SiO_2 reaction, the latter being an impurity oxidation layer on the α-Si_3N_4 starting powder. The silicate glass constitution was estimated from Auger electron spectra derived from high temperature intergranular fracture surfaces,[3,4] prior to the advent of nanoprobe EDAX spectroscopy in TEM. The marked reduction in fracture-stress above ~1000°C was linked to the glass softening point.[4] The characteristic hexagonal prism morphology of β–Si_3N_4 (Fig. 1) reflects the symmetry of crystals growing within the liquid sintering-aid by a solution-reprecipitation mechanism.[3] In retrospect, an understanding of the necessity for liquid-assisted sintering comes from an extrapolation of solid-state diffusion rates for Si_3N_4 to high temperatures. Rates which would permit pure solid-state sintering are only reached at temperatures of ~2000°C, which is above the decomposition temperature for Si_3N_4.

The subsequent development of Si_3N_4 ceramics concentrated on a refinement of sintering liquid composition, avoiding Ca impurities[5] which reduced the intergranular glass softening point and, later, using alternatives to the divalent metal oxide additives.[6] Hot isostatic pressing (HIP) was developed as an alternative to simple die-pressing and enabled more uniform densificiation and component shaping. However, the more significant developments in the 1970s were the tailoring of sintering liquid

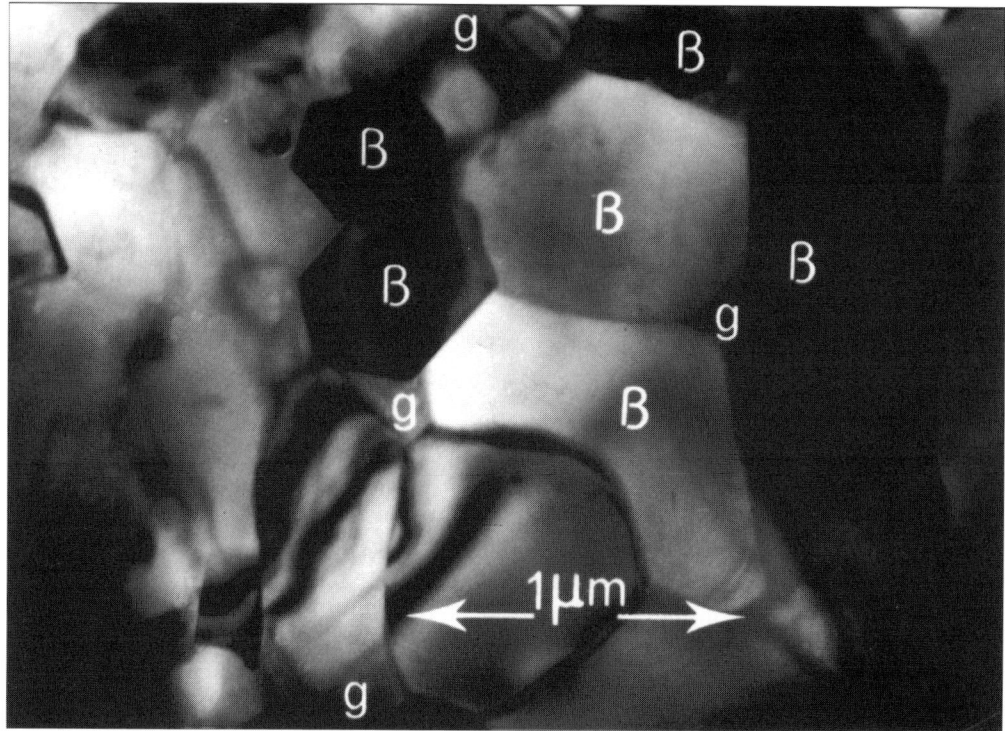

Fig. 1 Microstructure of an early hot-pressed Si_3N_4-based ceramic, illustrating the hexagonal prism $\beta'Si_3N_4$ morphology and residual intercrystalline glass (G).

chemistry and β–Si_3N_4 solid solutions to achieve microstructures with minimal grain boundary glass residue and the subsequent attainment of pressureless – sintering to theoretical density. These developments focus on the more complex Si–Al–O–N compositions, known generically as 'sialon' ceramics.

GRAIN BOUNDARY CONTROL IN SIALONS

The initial demonstration of a substitution of Al for Si in the βSi_3N_4 structure was made independently by Jack and Wilson[7] and Oyama and Kamigaito[8] from studies of reactive sintering of Al_2O_3 with Si_3N_4. This led to a first description of a β' solid solution based on a simple mixture of the stoichiometric compounds Al_2O_3/Si_3N_4 which implies a defect structure with Si vacancies. Although a vacancy-containing structure may have an attraction of enhanced diffusional sintering it was later demonstrated that simple Al_2O_3/Si_3N_4 mixtures were diphasic.[9] By reducing the O/N ratio it was concluded from X-ray diffraction and electron microscopy[10, 11] that the β' solid solution was derived by equivalent substitution of O for N and Al for Si, thus maintaining charge balance in the compound $Si_{3-x}Al_xO_xN_{4-x}$, without vacancies or interstitials, up to a limiting value $x \approx 2$.

Whereas it is possible to effect liquid phase sintering with the 'pure' quaternary Si–Al–O–N system, improved sinterability is obtained with a further oxide additive. Divalent (MO) or trivalent (M_2O_3) metal oxides are normally used with a condition of stability in oxynitride melts such that reactions of the type $2M+O_2 \rightarrow 2MO$ have larger (-ve) free energies than for oxidation of silicon nitride; $1/3Si_3N_4 + O_2 \rightarrow SiO_2 + 2/3N_2$. Thus Mg, Al, Ba, Zr, La, Nd, Ca, Be, Sc, Y, Gd, Yb have all been used as additive oxides.

A recognition of the importance of O/N ratio in achieving a β' solid solution resulted in the formation of monophase microstructures from sintering mixtures of either Si_3N_4–Al_2O_3–AlN or Si_3N_4–SiO_2–Al_2O_3–AlN. Optimum sinterability and microstructure were obtained with small (~1–2%) MgO additions which formed part of a 'transient' Mg–Si–Al–O–N sintering liquid[11] (the presence of Al and O in β' is believed to promote a small solid solubility for Mg). The liquid volume is such that pressurised sintering is necessary for full densification.

Microstructures are distinguished by their equiaxed grain structure, in the absence of β' crystal faceting or detection of intergranular glass (Fig. 2a, b). However, intergranular segregation of residual liquid elements on a scale of the β' lattice spacing is detectable by Auger electron spectroscopy of high temperature fracture surfaces (Fig. 2c).

Diphasic ceramics resulted from an increase in the sintering liquid volume which enabled pressureless sintering to the theoretical density. This has been a key step in relation to commercial application, enabling complex component shaping without expensive encapsulation of injection moulded preforms. An additional factor has been the improvement in fracture toughness and strength resulting from enhanced anisotropy in growth morphology of β' crystals. Typical microstructures (Fig. 3a) contain between 5 and 15 volume % of the liquid residue, initially in the glassy state, based on MO–Al_2O_3–SiO_2 eutectics with dissolved nitrogen. Optimum sintering and properties are obtained with M_2O_3 additives (M = Y, Nd, La etc.) and provides for greater flexibility in crystallisation of the glass phase during post sintering heat treatment for enhanced high temperature properties.

The matrix crystallisation product may be an oxide, such as YAG ($3Y_2O_3 \cdot 5Al_2O_3$) containing very small Si and N substitution,[12] or may be an intermediate M–Si–Al–O–N phase[13] such as $Nd_3Si_3Al_3O_{12}N_2$ which may crystallise from a matrix glass of similar composition. In either case the variable β' composition has the flexibility to accommodate non-stoichiometric species (eg excess Si and N) within the solid solution and hence minimise residual glass. However, it is difficult to totally crystallise β'–oxide microstructures because the glass-stabilising cations (M^{3+}) have negligible solid-solubility in β'. A transformation in near-surface microstructure (Fig. 3b) demonstrates one solution to this problem due to out-diffusion of these cations to the SiO_2 oxidation layer. The resulting isolated, particulate, morphology for YAG results from equilibration of β'–YAG and β'–β' interfacial energies. To achieve this complete state of crystallisation in the bulk requires the introduction of the α' sialon

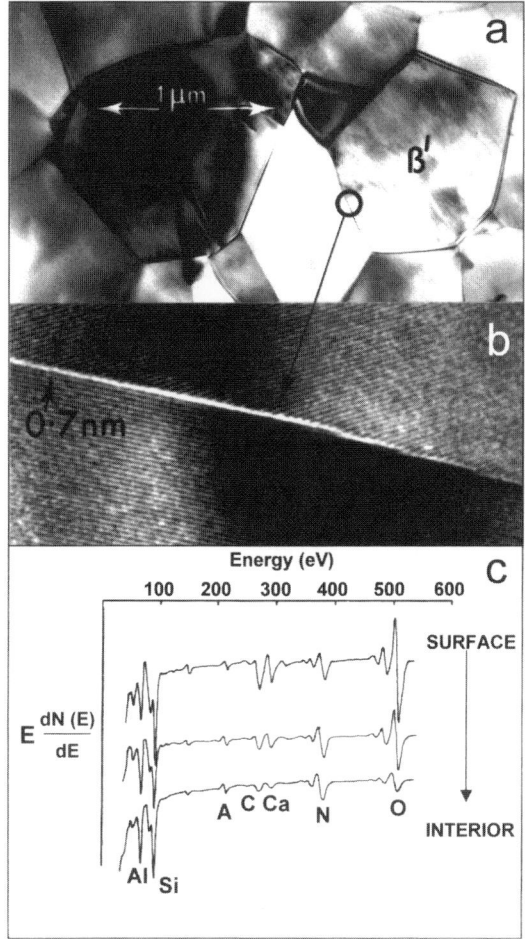

Fig. 2 Transient-liquid-sintered monophase β' sialon with lattice image recorded from the grain boundary region (b) and Auger electron spectra (c) from a progressively-sputtered intergranular fracture surface.

phase, which has an appreciable cation solubility and hence acts as a more efficient reservoir for non-stoichiometric species. The resulting post sintering crystallisation treatments are conducted between the glass-transition temperature and the liquidus (1000–1400°C). The resulting solid solutions have a planar composition range defined by $M_{m/3}Si_{12-(m+n)}Al_{(m+n)}O_nN_{(16-n)}$ with m in the range 1–2 and $n < 2$.

Sialon ceramics containing α' may be obtained by increasing the N/O ratio in the sintering mixture, moving the mean composition from the β'/liquid tie-line towards the α' plane. [14] These α'/β' ceramics have distinctive microstructures in which the residual oxide (eg YAG, Yb-garnet or GaAlO$_3$) has an isolated, non faceted, morphol-

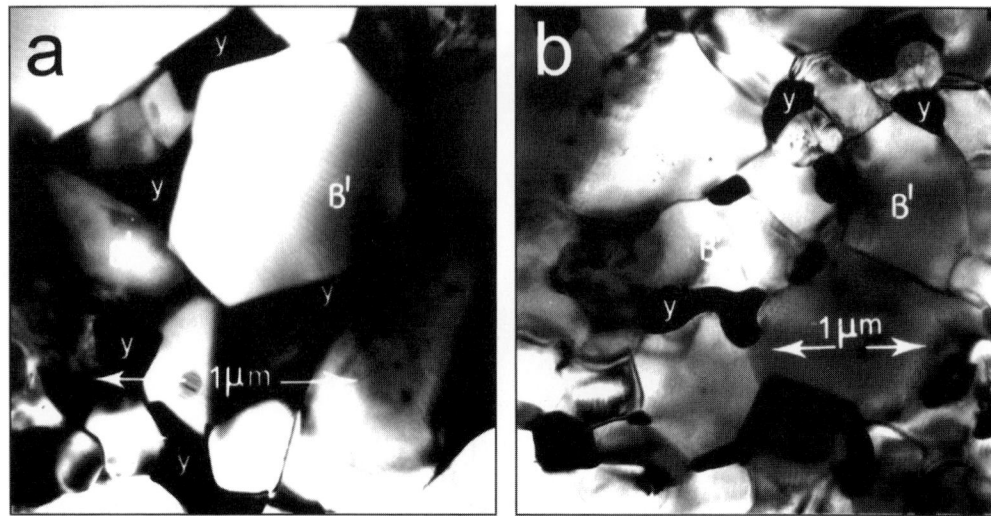

Fig. 3 Diphasic sialon microstructures (TEM images) with, (a) prismatic β′ in a semi-continuous matrix, (b) β′ with isolated granular oxide (garnet-YAG), following high temperature oxidising heat-treatment.

ogy of equilibrated grain junctions indicative of a pure solid/solid contact without residual glass(Fig. 4).

The question of 'purity' and freedom from residual non-crystalline layers at β′/α′/oxide interfaces remains a matter for debate. There is no doubt that both in $\beta - Si_3N_4$ and β′-sialons, glassy films of 1–2 nm thickness are frequently observed, especially on 'special' boundaries parallel to at least one faceted β′ hexagonal prism plane. An example is shown in Fig. 5, with an associated nanoprobe EDAX spectra from a Nd-doped β′-sialon. The phase morphology and lack of interface debonding (and hence greater incidence of transgranular fracture) indicates that α′/α′ and α′/β′ interfaces may approach pure crystalline contacts. There is a need for further high resolution analytical electron microscopy in such 'transient' liquid sintered systems.

HIGH-TEMPERATURE PERFORMANCE; THE CRITICAL INFLUENCE OF GRAIN BOUNDARIES

The most significant high temperature problem in early Si_3N_4 ceramics was their limited stress-rupture performance, due to grain-boundary creep cavitation initiated mainly within triple-junction glassy residues. This has been suppressed by modifying the amount and composition of sintering additives such that an improvement of five orders of magnitude in failure lifetime has been achieved in commercial silicon nitrides (Fig. 6). However, the nucleation of bi-grain junction cavities in the latest

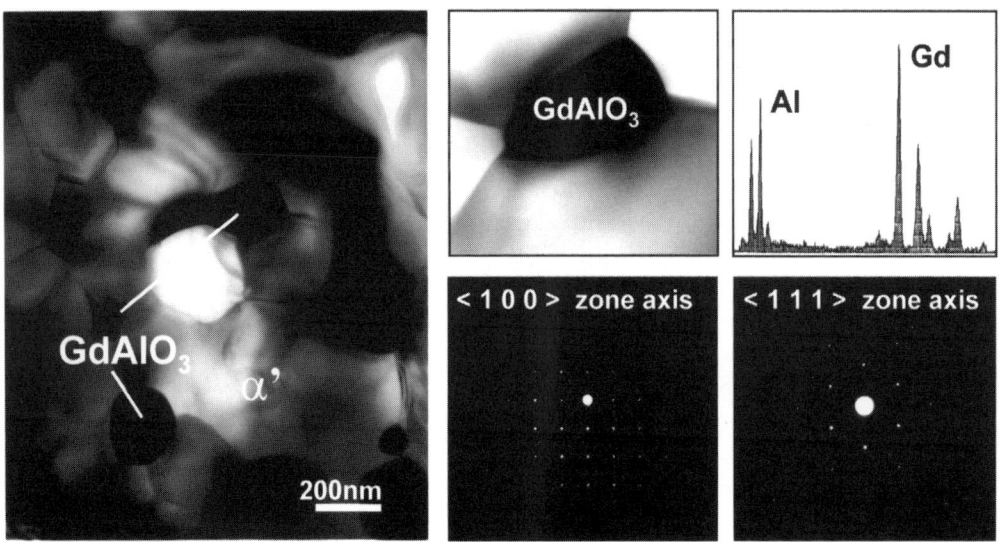

Fig. 4 Microstructure of α'/β' sialon containing isolated particulate $GdAlO_3$ following transient liquid sintering.

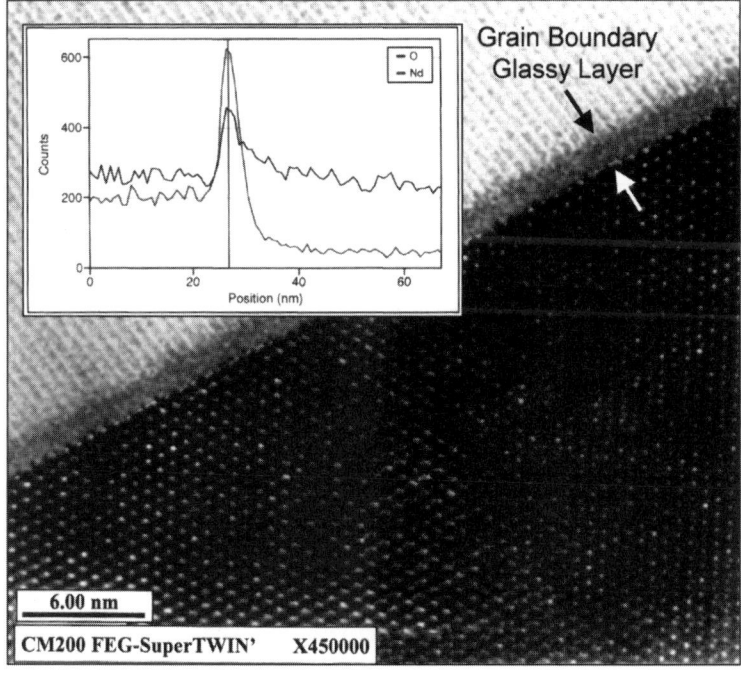

Fig. 5 High resolution TEM image of a glassy layer in a Nd-doped sialon, with nanoprobe EDAX spectra from the glassy phase.

compositions remains the origin of failure,[15] which may be modelled via a Monkman–Grant relation (lifetime α reciprocal of creep strain rate) as in many metallic systems.

One of the early successes of the near-monophase, transient-liquid-sintered, sialons was the first demonstration of a diffusional creep (with stress exponent = 1) and the suppression of creep cavitation.[16, 17] This research also demonstrated that a transformation could be effected in cavitating sialons by removal of glass-stabilising elements via out-diffusion to a surface SiO_2 oxidation layer which acts as a 'sink' for metallic cations. These changes in high temperature behaviour are demonstrated in the bend-creep and double-torsion fracture data (Figs 7, 8). In the absence of creep cavitation a threshold stress is observed above which there is limited growth (probably by diffusion) of pre-existing flaws and below which there is localised crack-blunting by diffusional creep.

The monophase sialons, although providing excellent models for grain-boundary control of high temperature mechanical behaviour, have not become important engineering ceramics because of the need for pressurised sintering and their fracture toughness values ($K_c \approx 4$ MPa m$^{1/2}$) which are significantly lower than the pressure-less-sintered diphasic sialons. Also, the high temperature creep and fracture properties of these diphasic microstructures has been shown to approach that of the monophase β'–sialons provided that the residual intergranular phase is tailored for complete crystallisation. This is reflected in the stress-rupture behaviour (Fig. 6)

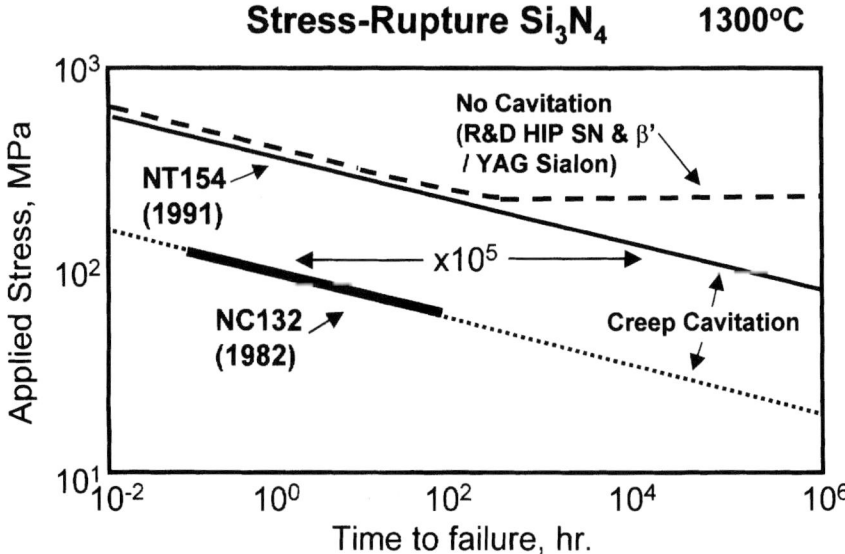

Fig. 6 Illustrating the effect of process-refinement on stress rupture properties, from an early hot-pressed Si_3N_4 (NC132) to a recent commercial ceramic (Norton NT154) and research-based sialon and silicon nitride.

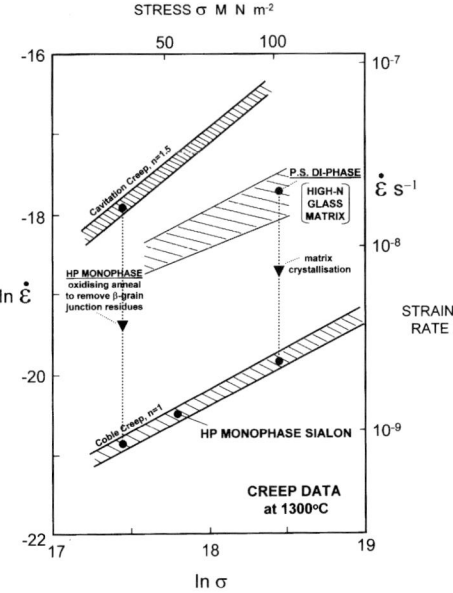

Fig. 7 Comparative creep data for monophasic and diphasic sialons (with the second phase in glassy and crystalline states).

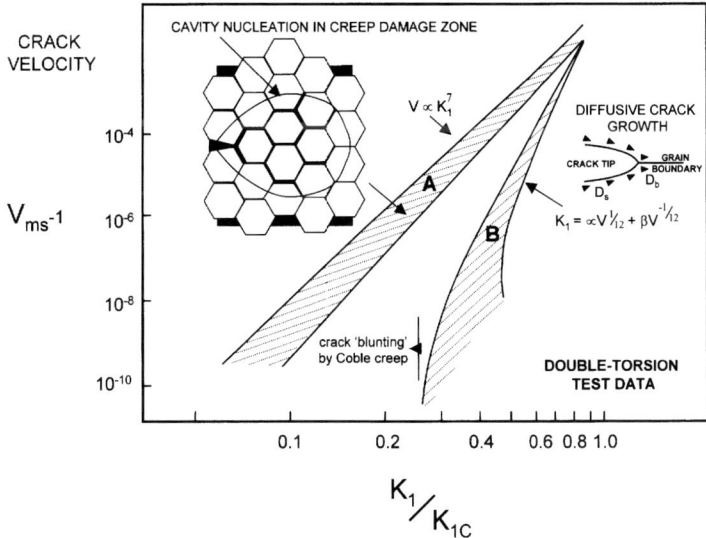

Fig. 8 Illustration of the influence of glassy grain boundary residues (A) on typical crack-velocity/stress intensity data (in double torsion) compared to monophase sialons or diphasic sialons with crystalline matrices (B).

which exhibits a threshold stress below which no failure is observed up to ~1300°C. Above this temperature the diphasic microstructures are unstable due to reversion of the intergranular phase to the liquid state (eg in $\beta' - Nd_3Si_3O_{12}N_2$ ceramics[13]) or to a reaction of the surface SiO_2 oxidation layer with the intergranular oxide to form a eutectic liquid (eg in β'–YAG ceramics[12, 18]). This is also reflected in a marked degradation in oxidation resistance, due to the rapid transport of viscosity-reducing ions to the passive oxidation layer. This is the reason underlying the superior oxidation resistance of monophase microstructures in which the more limited reservoir of cations in the grain-boundary segregated layer also have a more difficult transport path via grain boundary diffusion (Fig. 9a). Hence there is a gradation in oxidation-resistance from the 'pure' (CVD) Si_3N_4 to the diphasic β' microstructures (Fig. 9b).

The advent of α' sialons has provided for further small increments in creep and oxidation resistance due to their potential for complete solid-solubility of non-stoichiometric species (ie elements which may not be accommodated within β' or oxide phases, such as YAG). However, there is an inevitable compromise in which

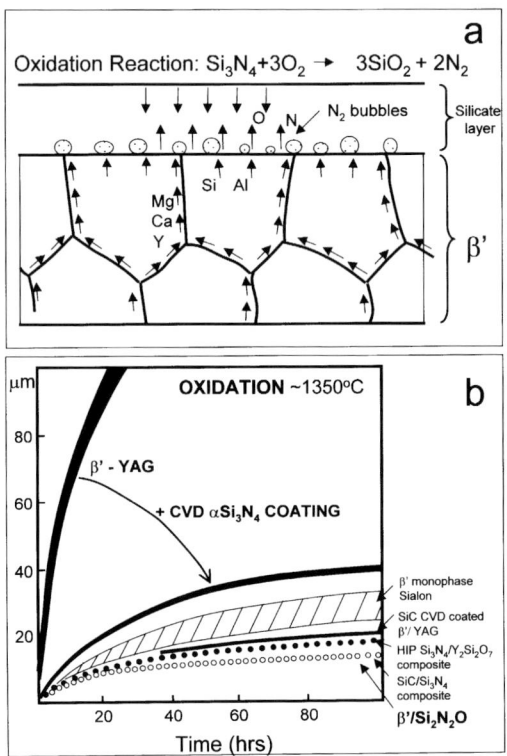

Fig. 9 Illustrating the underlying oxidation mechanism in Si_3N_4-based ceramics, with rate-control indirectly related to cation out-diffusion, and (b) typical oxidation kinetics for various microstructures (with and without CVD coatings).

the approach to a 'pure' interface between α'/β'/oxide phases is detrimental to other properties, such as toughness which requires interfacial debonding as a prerequisite for crack bridging and grain pull-out.

THE TOUGHNESS PROBLEM AND INTERPHASE COHESION

The increase in fracture-toughness (K_c), by a factor of ~2, was a major benefit in the first commercial application (eg as cutting inserts) of diphasic β'-sialons compared to the hot-pressed monophase microstructures (Fig. 10). This was originally attributed to a crack deflection mechanism at β'–glass or β'–YAG interfaces associated with the elongated prismatic morphology of β' grains which had grown with less constraint in the larger volume of sintering liquid.[19] It is now recognised that the major contribution to K_c is that of crack bridging and subsequent grain pull-out (Fig. 11). This has resulted from theoretical modelling and comparison with a range of experimental data[20] for Si_3N_4 ceramics with differing microstructural parameters obtainable with variable sintering cycle and constitution (Fig. 11). The use of a gas overpressure (GPS – at 10 MPa) suppresses Si_3N_4 dissociation and enables higher temperature sintering (1800–1900°C compared to 1700–1750°C for 'pressureless' sintering in 1 atmosphere of N_2). This encourages secondary grain growth of β or β', producing a high volume fraction (V_β) of large diameter (w) prismatic grains which are the major contributors

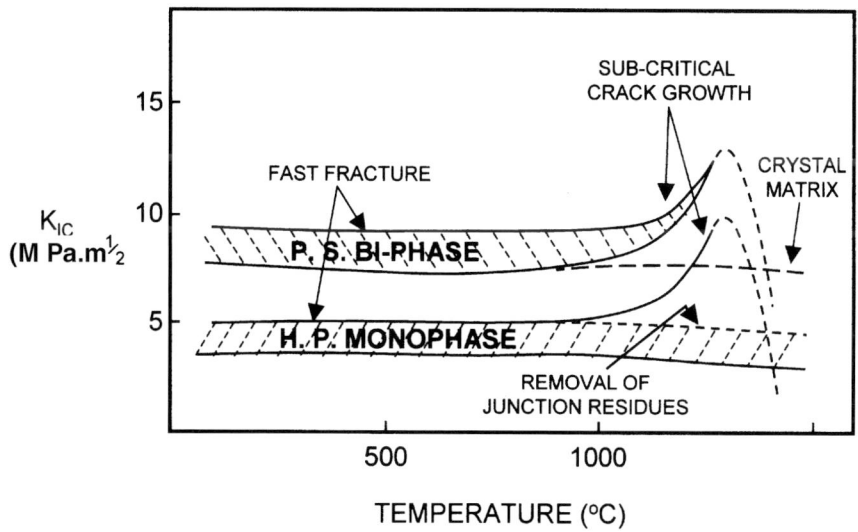

Fig. 10 Comparison of fracture toughness (K_c) for monophasic and diphasic sialons, showing the improvement for microstructures with large prismatic β' grains (the high-temperature peak in ceramics with grain-boundary residues is associated with viscous flow of the glass in the region of sub-critical crack growth).

247

to K_c (Fig. 11). Since β′ interface debonding is a prerequisite for the bridging/pull-out mechanism, interface constitution (possibly a glassy layer) is important. For high-cohesion interfaces ($G_i \rightarrow G_\beta$) transgranular fracture is more prominent, with consequent absence of the modelled relationship between K_c and $\sqrt{w}$.

Selective grain growth in β or β′ ceramics is influenced by β Si_3N_4, normally present in the α Si_3N_4 starting powders, which acts as 'seed' crystals for the α→β transformation occurring by the established solution-reprecipitation mechanism. The variable β content in the early pressureless-sintered sialons, in retrospect, was responsible for the scatter in K_c level observed (eg in Fig. 10, for notched beam tests[19]), but in some batches up to 10 MPa m$^{1/2}$. Controlled seeding, using GPS conditions, has produced isotropic values for K_c which are consistantly in excess of 10 MPa m$^{1/2}$.[21] Recent research using tape casting of preforms to align the prismatic β seed crystals in the casting direction has enabled a significant 'texture' in β grain orientation after sintering, with anisotropic values of K_c over 12 MPa m$^{1/2}$.[22]

Large prismatic grains of β, β′ or α′ major phases which are favoured for high toughness are not conducive to high fracture stress(σ). This is based on the relation between critical flaw size, c, and grain size, such that c is normally calculated to be a multiple of grain size. The origin of this relationshiip is not well-defined but may be

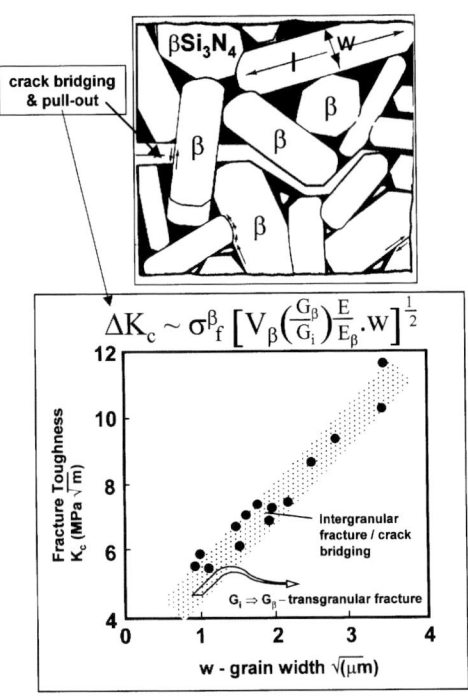

Fig. 11 Fracture Toughness K_c with varying β-Si_3N_4 grain diameter, w, plotted according to a crack-bridging mechanism $w^{1/2}$, illustrated above.

related to the penetration depth of typical surface-machining microcracks. Thus it appears that in the Griffith/Orowan/Irwin relation ($\sigma_f \alpha \, K_c / \sqrt{c}$), the ceramic flaw size is dominant over the K_c term and that to achieve very high fracture stresses small equiaxed grains are necessary. Values of σ up to 1. 5 GPa are possible using hot-isostatic processing (HIP, at ~100 MPa) of uniform α powders.

MICROSTRUCTURAL CONTROL AND PROPERTY LIMITS

The high level of understanding in the control of microstructure-property relationships for Si_3N_4-based ceramics is summarised in Fig. 12. The use of GPS or HIP in controlling grain size and distribution (via sintering temperature, time and crystal seeding) for toughness and strength variation has been described above. β grain morphology may be influenced particularly by sintering liquid volume and chemistry, with an extreme example in the sialons which may utilise a large (~10%) liquid volume to enhance prismatic β' growth or a small volume of transient liquid to produce a more equiaxed monophase microstructure.

The compromise between high toughness and high strength is a remaining prob-

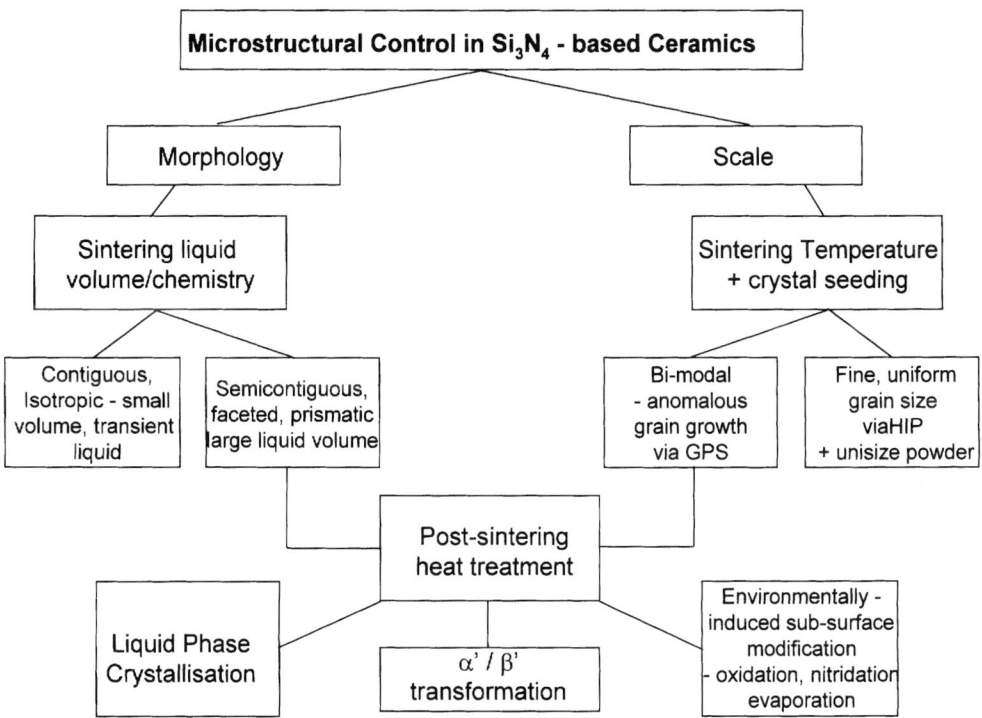

Fig. 12 A summary of strategies which have evolved for the control of microstructure in silicon nitride and sialon ceramics.

lem (Fig. 13). This has been partially resolved within the textured (seeded) ceramics but the properties are highly anisotropic.

The principles of post-sintering heat-treatment to crystallise sintering liquid residues, to enhance high temperature properties, are well-established in both Si_3N_4 and sialons. Optimum properties of creep resistance, stress-rupture and oxidation resistance favour the absence of glassy grain boundary or interphase residues and the best microstructures are monophase β′ or α′. This is normally contrary to high toughness, in relation to the development of large prismatic grains and the necessity for interfacial debonding in the path of a transverse crack.

The preference for net-shaped components raised important questions about surface modification induced by sintering gases or encapsulants; this may be beneficial- as in the case of 'out-diffusion' of components of the liquid residue, or detrimental – as in the formation of undesirable oxynitride phases in a nitrogen overpressure. A recent development involves a reversible α′/β′ transformation in sialons which occurs in the interval between ~1400°C and 1800°C.[23] This may provide an additional means of controlling relative hardness, toughness and associated tribological properties with variations in α′/β′ ratio, crystal size and morphology.

In conclusion, grain-boundary and interphase constitution has been demonstrated as a critical feature in the development of high performance engineering ceramics. It is possible that further control of grain boundary purity will result in applications which combine mechanical and, for example, optical properties. These are recent examples of α′ sialons, which have been 'transient-liquid-sintered' with fine α′ grain size and negligible

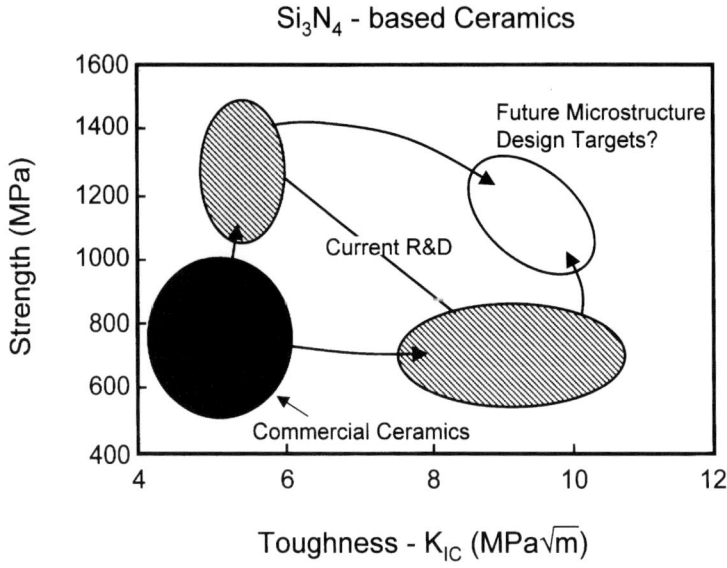

Fig. 13 Illustrating the 'compromise' in development of high strength and high toughness Si_3N_4 ceramics by grain size/morphology control.

porosity which exhibit a significant transparency to optical wavelengths. These ceramics may also be selectively coloured by controlling the valence state of the α'-stabilising rare-earth cation, e g Yb.[24] Possible applications are in abrasion or impact-resistant windows.

ACKNOWLEDGEMENT

The author wishes to recognise the inspiration provided by some of the earliest UK research on defect structure and mechanical behaviour of ceramic oxides, initiated by Professor Ray Smallman in the 1960s.

REFERENCES

1. G. G. Deeley, J. M. Herbert and N. C. Moore, *Powder Metallurgy,* 1961, **8**, 145.
2. S. Wild, P. Grieveson, K. H. Jack and M. J. Latimer, *Special Ceramics 5,* 385, P. Popper ed., Brit. Ceram. Res. Assoc., 1972.
3. P. Drew and M. H. Lewis, *J. Mat. Sci.,* 1974, **9**, 261.
4. B. D. Powell and P. Drew, *J. Mat. Sci.,* 1974, **9**, 1867.
5. R. Kossowsky, D. G. Miller and E. S. Diaz, *J. Mat. Sci.,* 1975, **10**, 983.
6. G. E. Gazza, *J. Am. Ceram. Soc.,* 1973, **56**, 663.
7. K. H. Jack and W. I. Wilson, *Nature,* 1972, **238**, 80.
8. Y. Oyama and O. Kamigaito, *Jap. J. Appl. Phys.,* 1971, **10**, 1637.
9. P. Drew and M. H. Lewis, *J. Mat. Sci.,* **9**, 1974, 1833.
10. R. J. Lumby, B. North and A. J. Taylor, *Special Ceramics 6*, 321, P. Popper ed., Brit. Ceram. Res. Assoc., 1974.
11. M. H. Lewis, B. D. Powell, P. Drew, R. J. Lumby, B. North and A. J. Taylor, *J. Mat. Sci.,* 1977, **12**, 61.
12. M. H. Lewis, A. R. Bhatti, R. J. Lumby and B. North, *J. Mat. Sci.,* 1980, **15**, 103.
13. M. H. Lewis, G. Leng-Ward and C. Jasper, *Ceramic Transactions 1: Ceramic Powder Science,* G. L. Messing, E. R. Fuller and H. Hausner eds, Am. Cer. Soc., 1988, 1019.
14. C. A. Jasper and M. H. Lewis, *Ceramic Materials and Components for Engines,* R. Carlsson, T. Johansson and L. Kahlman eds, Elsevier, 1992, 424.
15. S. D. Wiederhorn, W. E. Luecke, B. J. Hockey and G. G. Long, *Tailoring of Mechanical Properties of Si_3N_4 Ceramics,* M. J. Hoffman and G. Petzow eds, Kluwer, 1994, 305.
16. B. S. B. Karunaratne and M. H. Lewis, *J. Mat. Sci.,* 1980, **15**, 449.
17. B. S. B. Karuanaratne and M. H. Lewis, *J. Mat. Sci.,* 1980, **15**, 1781.
18. M. H. Lewis and B. S. B. Karunaratne, J. Meredith and C. Pickering, *Creep and Fracture of Engineering Materials and Structures,* B. Wilshire and D. R. Owen eds, Pineridge Press, 1981, 365.
19. M. H. Lewis, in *Proc. 1st Parsons Int. Turbine Conf.,* Inst. of Mech. Engineers, 1984, 181.
20. P. F. Becher, S. L. Hwang, H. T. Lin and T. N. Tiegs in *Tailoring of Mechanical Properties of Si_3N_4 Ceramics, 87,* M. J. Hoffmann and G. Petzow eds, Kluwer, 1994.
21. K. Hirao, T. Nagaoka, M. E. Brito and S. Kanzaki, *J. Am. Ceram. Soc.,* 1994, **77**, 1857.
22. K. Hirao, M. Ohashi, M. E. Brito and S. Kanzaki, *J. Am. Ceram. Soc.,* 1995, **78**, 1687.
23. H. Mandal, N. Camuscu and D. P. Thompson, *J. Eur. Ceram. Soc.,* 1997, **17**, 599.
24. B. S. B. Karunaratne, R. J. Lumby and M. H. Lewis, *J. Mater. Res.,* 1996, **11**, 2790.

The Role of Interfaces in the Relaxation of Strained Layers

PETER J. GOODHEW

Materials Science & Engineering, Department of Engineering, University of Liverpool, Liverpool L69 3GH, UK

ABSTRACT

In epitaxially grown layers, whether of metal or semiconductor, there is very rarely an excellent match between the lattice parameters of the substrate and the layer. This means that most epitaxial layers are initially in a strained state. Depending on the intended application it might be desirable to maintain the strained state (for example to benefit from an unusual lattice spacing or a reduction of symmetry) or alternatively it might be better to encourage relaxation to the natural lattice spacing (for instance to achieve long term stability). Most relaxation mechanisms involve dislocations and it is therefore important to understand the generation and motion of dislocations in and around the interface between a strained layer and its substrate. In this paper some recent results will be presented which show that, at least in semiconductor systems with low initial dislocation densities, the distribution of dislocation sources is not important in determining the relaxation behaviour. However interactions between the dislocations within the interface are significant in determining the rate and extent of relaxation.

INTRODUCTION

Device structures such as lasers and HEMTs now routinely use misfitting epitaxial layers. III–V compound semiconductors almost universally adopt the cubic sphalerite structure but their natural lattice parameters at room temperature range from 0.54 nm for GaP to 0.65 nm for InSb and their band gaps are between 0.18 eV and 2.45 eV. The misfit in lattice parameter in the interface plane between materials with significantly different band gaps or dielectric constants can therefore be as high as 18% (between GaP and InSb) and 7% (between GaAs and InAs) or as low as 0.05% (between $Al_{0.3}Ga_{0.7}As$ and GaAs). If an epitaxial layer is grown with a natural lattice parameter different from that of its substrate then the resultant thin layer must (at least initially) be strained.

A wafer containing one or more strained layers with stored elastic energy is intrinsically metastable. For some device purposes it may be desirable for the layer to maintain all or most of its strain and not to relax. For other purposes it may be necessary for the strain to be relieved and for the layer to relax to its natural (bulk) lattice parameter. Devices containing strained layers may rely on the breaking of cubic symmetry for their efficient operation (eg Dunstan and Adams[1]). In these cases our interest is in the long-term stability of the metastable strained system. The requirement is that such layers should not relax at the designed temperature of use over the lifetime of the device. On the other hand, relaxation is highly desirable if a layer needs to adopt a lattice parameter different from that of the (often cheaper or less defective) substrate. In these cases the device engineer needs to be able to predict and control the extent of strain relaxation and the final lattice parameter.

CRITICAL THICKNESS

The layer thickness above which it is energetically favourable for an array of misfit dislocations to exist, in order to reduce the stored elastic strain energy of the layer, is known as the critical thickness, h_c, and many analytical expressions have been derived to calculate its value. Once an epitaxial layer is thicker than h_c, relaxation is in principle possible because there is a driving force. In practice, however, kinetic factors usually ensure that a layer does not begin to relax until long after the critical thickness has been exceeded, and even then relaxation occurs only gradually. In III–V systems critical thicknesses are often only tens of nm but relaxation is unlikely to be complete until the layer is several μm thick.

STRAIN RELAXATION

If its substrate is thick enough to be considered essentially semi-infinite then a flat epitaxial layer cannot relax any elastic strain without the introduction of defects at the interface which have a component of their Burgers vector in the plane of the interface. This usually means dislocations with at least part edge character and therefore this paper is concerned with relaxation involving the generation and movement of dislocations and/or substantial movement of point defects by diffusion.

Relaxation normally occurs in several stages, and may never reach 'completion'. Full relaxation, in which the layer reaches its bulk lattice parameter and is totally free from elastic strain can never occur while the layer remains bonded to its substrate, as has been pointed out many times.

The key stages of relaxation for layers in compression are usually:

- No relaxation until the thickness of the layer is at least h_c.
- Glide of any dislocations, originating in the substrate, which thread the layer (TDs). As these dislocations move sideways they leave behind lengths of misfit dislocation (MD in Fig. 1). In the most favourable circumstances these dislocations can reach the edge of the wafer and their threading segment can be eliminated. In high-quality substrates of silicon, which may be essentially dislocation-free, this stage is effectively suppressed.
- Nucleation of new dislocations by the operation of dislocation sources, and/or
- Increase of dislocation line length by the operation of multiplication mechanisms
- Interactions between dislocations on different slip planes leading to 'work hardening'.

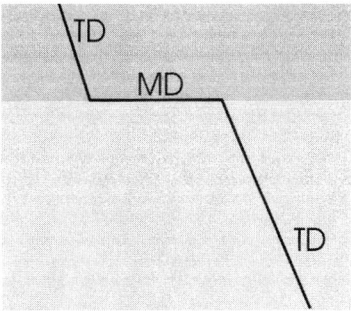

Fig. 1 A single dislocation which crosses an epitaxial layer, showing the threading (TD) and misfit (MD) segments.

Unless the substrate contains a very high density of dislocations the creation of MD segments from pre-existing TDs can only account for the relief of a small amount of strain, typically 0.05% for a substrate with 10^4 lines cm^{-2}, for instance.[2]

For III–V epilayers grown on (001) substrates TDs can glide on one of two glide planes in each of two orthogonal <110> directions. Each gliding TD therefore encounters previously-laid-down MDs orthogonal to its path, which may have Burgers vectors parallel or non-parallel to it. Freund[3] has analysed many of the possible interactions and has demonstrated that in general the gliding dislocation will be resisted by orthogonal MDs and that blocking of its passage may occur. Let us call the interaction between a gliding TD and one orthogonal MD a single barrier. Closely spaced MDs of the same Burgers vector should act as stronger blocks and we will call these double barriers. There is plenty of evidence, from TEM and surface striations (cross-hatch) that blocking does occur (eg MacPherson & Goodhew;[4] Samavedam & Fitzgerald[5]) although it is difficult to estimate the relative importance of single and double barriers. Both types of blocking must however reduce misfit relief by limiting the length of the MDs which can be laid down by gliding TDs.

In this paper an assessment is made of the effect on blocking of the initial distribution of dislocations in the substrate. A computer model is presented to substantiate the arguments.

A BLOCKING SIMULATION

We have developed a computer simulation of the blocking phenomenon which enables our analysis to be tested, and to be extended to the analytically more difficult situations in which not all barriers are effective, the dislocation population is constantly increasing and the original distribution of TDs may not be random. The basis for this simulation is as follows:

1. Threading dislocations are assumed to exist at a fixed density per unit area and to begin to glide in random sequence. This models the situation where a substrate has a significant density (typically between 10^4 and 10^5 cm^{-2}) of TDs which begin to move as the layer thickness increases beyond the critical thickness.

2. Each TD moves from its initial position in one of four directions corresponding to the two glide planes available in each of the <110>. This is shown in Fig. 2. The direction for each dislocation is selected at random to simulate the choices available to the TDs, which will actually have four different Burgers vectors.

3. As the moving TD reaches each previously laid-down orthogonal it has a probability B_1 of being blocked by the single dislocation and a different probability B_2 of being blocked by the next two MDs if they are spaced closer than a selected distance x. This enables us to model both single and double blocking. If it is blocked the TD moves no further, but if it not blocked its next intersection with an orthogonal MD is examined and so on until the edge of the defined area is reached.

4. When all TDs have been allowed to glide, the number of blocks of each type and the average MD segment length are calculated.

The choice of values for B_1 and B_2 turns out not to be very significant, but it is sensible to choose values which are likely to be realistic. Clearly $B = 0$ represents no blocking, while $B = 1$ ensures total blocking by every orthogonal dislocation. Each gliding TD might meet (in the general case) orthogonal MDs with four different Burgers vectors (two on each of two slip planes). One particular Burgers vector is likely to lead to much stronger blocking than the others and therefore a reasonable value of B should be 0.25 or less. To recognise the fact that, even with an appropri-

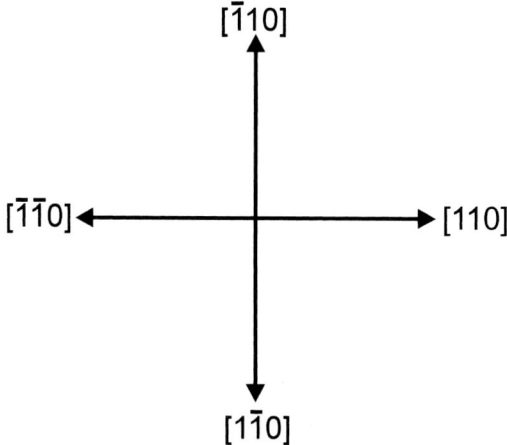

Fig. 2 The four possible glide directions in an (001) layer.

ate combination of Burgers vectors, blocking is unlikely to be perfectly efficient, a value of 0.2 has been used in most simulations when non-zero values of B_1 and/or B_2 are needed. The conclusions are not sensitive to this choice.

RESULTS OF THE SIMULATION

The output from the simulation was first checked for those simple cases where the analytical solutions described above should apply. A typical simulation is shown in Fig. 3. Such simulations execute in a second or two and all the data presented here result from averaging at least five runs with identical data. The format of the results is illustrated by Fig. 4, which shows the effect of single blocking ($B_1 = 0.2$, $B_2 = 0$) as the number of gliding dislocations increases. The starting points for the dislocations were randomly distributed. This represents the situation as the early stage of relaxation proceeds and more and more TDs start to glide.

Figure 4 shows the percentage of dislocations which are blocked (diamond symbols), and the average MD segment length (square symbols) as a function of the number of dislocations which have glided. We cannot attribute any great significance to the absolute values, since they will depend on the size of the simulation area and/or the absolute threading dislocation density. However the trends are exactly what we would expect: the percentage of dislocations which are blocked rises as the number of dislocations increases, and the corresponding average segment length decreases steadily. The same behaviour holds for double blocking ($B_1 = 0$, $B_2 = 0.2$, $x = 2$).

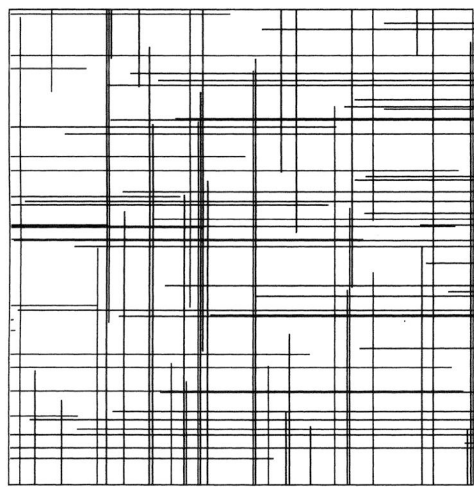

Fig. 3 A typical simulation, with no blocking and a random distribution of threading dislocation start locations.

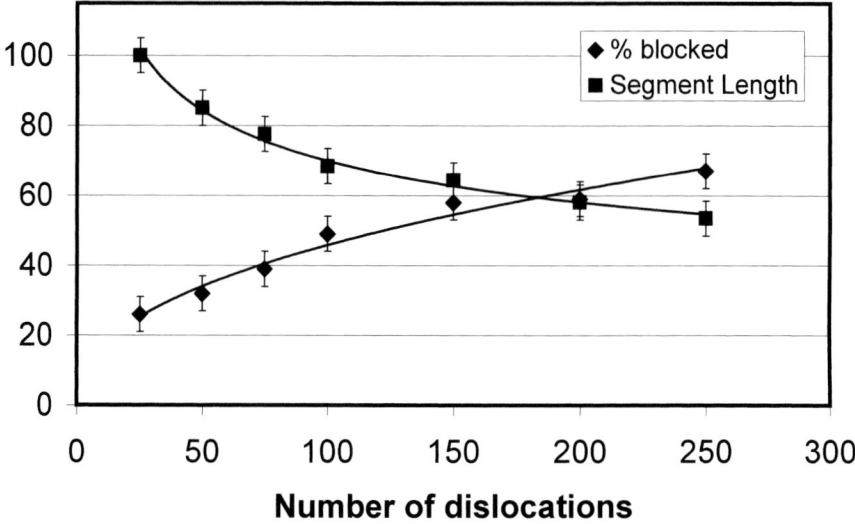

Fig. 4 The change in blocking as the number of gliding dislocations increases.

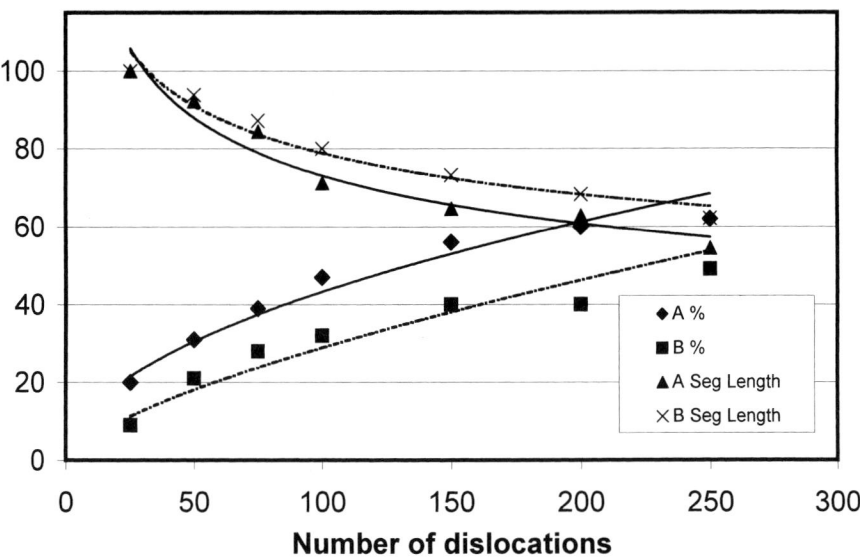

Fig. 5 The extent of double blocking as the number of gliding dislocations increases. The full lines show the effect of a 20% blocking probability while the dotted lines correspond to 10%.

Having established the validity of the model, it was used to predict behaviour in situations less amenable to exact analytical solutions. Figure 5 shows the results of simulation experiments in which both types of blocking are allowed to occur, and also shows the effect of the original distribution of dislocation sources. The full lines show the situation where single and double blocking are given equal weight (B_1 = 0.2, B_2 = 0.2, x = 2), while the dashed lines refer to the case where single blocking is half as likely as double blocking (B_1 = 0.1, B_2 = 0.2, x = 2). In each case the simulation was run with a random distribution of dislocations (as shown in Fig. 3) and with a strongly clustered distribution (shown in Fig. 6). In each case the data from the clustered dislocations fell exactly on the data for the random distribution, indicating that the conclusions are not affected by the starting configuration of TDs.

DISCUSSION AND CONCLUSIONS

The simulation results presented in this paper make it clear that we should not expect the extent of dislocation blocking to depend on the initial dislocation distribution.

While there can be little doubt that dislocation blocking plays a key role in determining the kinetics and extent of relaxation it is clear from this work that it is not by itself responsible for differences in behaviour between layers grown on different substrates. The initial distribution of threading dislocations makes almost no difference to the subsequent blocking behaviour. However the absolute value of the density of threading dislocations will determine the extent of the first stage of relaxation discussed here, and the layer thickness at which relaxation starts.

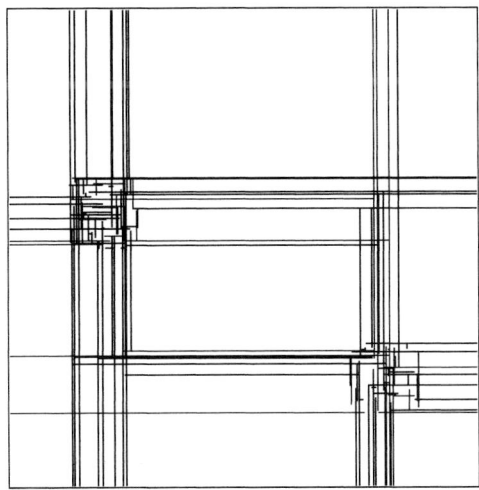

Fig. 6 The simulated distribution from strongly clustered starting points and with single and double blocking probabilities of 0.2.

REFERENCES

1. D. J. Dunstan and A. R. Adams, Analysis and design of low-dimensional structures and devices using strain: II strained layer systems. *Semicond. Sci. Technol.*, 1990, **5,** 1202–1208.
2. R. Beanland, D. J. Dunstan and P. J. Goodhew, 'Predicting relaxation in strained epitaxial layers', *Scanning Microscopy*, 1994, **8**, 859–868.
3. L. B. Freund, *J. Appl. Phys.*, 1990, **68**, 2073–2080.
4. G. Macpherson and P. J. Goodhew, *Appl. Phys. Lett.*, 1997, **70**, 2873–2875.
5. S. B. Samavedam and E. A. Fitzgerald, *J. Appl. Phys.*, 1997, **81**, 3108.

Computer Modelling of Grain Boundaries in Oxides

P. D. BRISTOWE

Department of Materials Science and Metallurgy, University of Cambridge, Pembroke Street, Cambridge CB2 3QZ, UK

ABSTRACT

The current status of grain boundary modelling in oxide materials is reviewed. The results of various studies using both classical atomistic and first principles electronic structure methodologies are summarised. Grain boundaries in the following oxides have been considered: NiO, MgO, TiO_2, $SrTiO_3$, Al_2O_3 and ZnO. Most of the studies have focussed on structural properties but some have investigated grain boundary diffusion and segregation. Comparisons with experimental data indicate that most of the classical simulations can predict the main structural features of the grain boundary but that certain characteristics, such as volume change, are more difficult to reproduce. Many of the computed grain boundary structures exhibit reduced ion density, under-coordinated atoms and reduced bond lengths. However, the first principles methods show that these imperfections in themselves do not create new electronic states deep in the band gap. Impurities seem to be the likely origin of the observed electrical properties of polycrystalline oxide ceramics.

INTRODUCTION

In 1981 when the present author first reviewed the properties of grain boundaries in ceramic oxides[1] there were only two computer modelling studies available. Both of these studies used a classical atomistic approach to simulate the structure and energy of a group of high-angle twist boundaries in MgO and NiO.[2,3] Although state-of-the-art at the time, both investigations were computationally flawed and yielded results that did not agree well with experimental observations. For example, it was predicted that (001)-type twist boundaries should be weakly bound configurations when in fact there were electron microscope studies that indicated to the contrary. It was concluded that the computational procedures employed had not incorporated all possible modes of relaxation particularly those that were necessary to relieve charge mismatch across the boundary. Three relaxation mechanisms were identified as being important for the proper modelling of grain boundaries in materials which exhibit significant ionic bonding:

- ionic reshuffling (which could be achieved by bond bending, rigid-body translations, or boundary faceting)
- creation of point defects of either polarity

• polarisation, reduction in ionicity, or change in the nature of the bonding.

Depending on the boundary geometry and the type of oxide under consideration, two or more these mechanisms could be combined. Today, computer simulations of oxide grain boundaries routinely include most of these relaxation mechanisms automatically (Refs 4 and 5 contain reviews of interface modelling techniques). Some of the mechanisms, such as point defect creation, are introduced after initial relaxation has occurred to test the stability of the boundary. They are also introduced to examine the interaction of extrinsic point defects with the boundary. The result is that most of the calculations produce physically reasonable information and some even yield atomic structures that match with detailed observations to a remarkable degree. In addition, the calculations have now been extended to grain boundaries in non-cubic oxides and to the implementation of first principles methodologies. The latter development is clearly important for determining the most accurate information possible and for examining grain boundaries in electronic ceramics. Theoretically, consideration has also been given to the simulation of polar grain boundaries and methods for quenching the electrostatic dipole that forms along the polar direction.[6]

The present paper reviews current knowledge of grain boundary properties in oxide materials obtained by computer modelling. Studies of grain boundaries in six ceramic oxides are reviewed: NiO, MgO, Al_2O_3, TiO_2, $SrTiO_3$ and ZnO. The computational methodologies employed range from classical shell models to ab initio density functional calculations. The motivation for these studies is clear. Ceramic oxides, whether used for structural or electronic applications, are invariably prepared in poly-crystalline form and therefore contain many grain boundaries. In addition, they are often doped with other oxides that can lead to the formation of second phases at the boundary either amorphous or crystalline. They are therefore complex materials whose physical properties can be significantly affected by their microstructure and chemistry. For example, doping polycrystalline alumina can yield decreased creep rates and higher toughness and although the microscopic mechanisms involved are unclear, grain boundaries play a crucial role. In the case of electronic ceramics, grain boundaries often control the functionality of a device fabricated using these materials. For example, polycrystalline strontium titanate is used to produce boundary layer capacitors with the grain boundaries acting as regions of high resistance. Determining the microscopic origin of the potential barriers that form at the boundaries is therefore of prime interest. Thus the goal of computer modelling is to provide fundamental atomic level information about the grain boundaries in these materials and how they interact with other defects and dopants. If the grain boundaries can be systematically and predictably engineered using this information then this could lead to an improvement in materials performance in whatever application the material is used.

COMPUTATIONAL METHODOLOGIES

The general approach to modelling a grain boundary at the atomic level in any material is to construct a computational cell containing an initial guess for the boundary structure and performing a systematic optimisation of this structure with respect to its energy. The optimisation process should incorporate in some way the relaxation mechanisms described above. Standard optimisation schemes include molecular statics, molecular dynamics and Monte Carlo, the latter two introducing temperature dependent effects by integrating the classical equations of motion or applying stochastic dynamics. The majority of grain boundary simulations in oxides have employed molecular statics which involves a straightforward minimisation of the potential energy of the cell using, for example, a conjugate gradients scheme (for reviews of computer simulation methods applied to interfaces in ionic solids, see Refs 7–9). The calculation thus proceeds at zero temperature but dynamical processes such as grain boundary diffusion can still be simulated using this method as described below for NiO. Although simple in principle, a molecular statics calculation of an oxide grain boundary involves some technical issues which require careful consideration and some of these are listed below:

- The calculation must properly account for the long-range electrostatic potential through an appropriate Ewald summation. Since grain boundaries are planar defects, most programs which use a classical approach implement a two dimensional scheme due to Parry.[10]
- The crystal structure far from the grain boundary, ie the borders of the cell, must be treated appropriately. The border conditions must be applied so as to avoid unintended constraints on the relaxed configuration. Since the grain boundary is usually assumed to have planar periodicity then periodic border conditions are applied parallel to the boundary plane. Fixed border conditions are then commonly applied normal to the boundary. If, instead, periodicity is imposed in this direction then the cell will contain two grain boundaries which may or may not be equivalent. A supercell with two boundaries is commonly used in quantum mechanical calculations which require the application of Bloch's theorem. The cell size must be chosen carefully to minimise spurious interactions caused by the border conditions. If a polar grain boundary is simulated using fixed border conditions then it may be important to quench any dipole moment that is formed along the polar direction. This is commonly achieved in classical calculations by adjusting the charges on the ions in the fixed region furthest from the grain boundary.[6]
- The effect of charged grain boundaries caused by the introduction of point defects or impurities needs to be considered. In classical calculations the resulting dielectric response of the crystal far from the boundary is simulated by applying atomic displacements evaluated using the Mott-Littleton method.[11]
- The crystallography of the grain boundary must be described properly. In order to simulate periodic grain boundaries, the cell edge vectors must be vectors of the appropriate coincidence site lattice.[12] As the boundary becomes more general, these vectors

become larger and the computational cell requires more atoms. Non-periodic boundaries can sometimes be described with near-coincidence site lattice vectors[13] but the
application of these vectors usually requires imposing a strain on the crystal structure.
- Metastable states need to be avoided when using static relaxation methods. One
  approach is to re-start the calculation after perturbing the optimised structure
  slightly.

Of course, the most important component of any atomistic simulation is the spec-
ification of the interatomic interactions since these determine the reliability of the
results. Until very recently almost all simulations employed a semi-empirical classical
approach in which the interactions are described by simple pair potentials.[7] In this
model of an ionic crystal, the total potential energy is composed of four terms: (i) the
long-range Coulomb potential, (ii) a short-range repulsive potential to model charge
overlap, (iii) a short-range attractive potential to model dispersion and (iv) an harmonic potential to model the electronic polarisation of the ions. Each term has a
specific functional form (eg exponential for the Pauli repulsion) and contains
adjustable parameters which are fitted to various bulk properties of the material
including the lattice, elastic and dielectric constants. The electronic polarisation
potential is usually simulated by a mechanical model in which the ions are linked to
massless shells by harmonic springs. The shells in some way represent the valence
electrons and their displacement the polarisation. The ionic shell model has been
quite successful in describing the energetics of bulk ionic solids and also simple
defects.[7] Computer programs that have implemented the above classical approach for
simulating interfaces and grain boundaries include MIDAS,[14] MARVIN[15] and GULP.[16]

Over the past decade, developments in solid state theory and computer technology
has lead to the possibility of modelling oxide grain boundaries from first principles.
Although the model sizes are still limited to around a hundred atoms, it is now possible to perform a complete atomic and electronic optimisation of a short period grain
boundary. Most of the calculations that have been performed are based on the well-
established methodology of density-functional theory (DFT) in the local density
approximation (LDA). A major advance in the application of this theory was made
by Car and Parrinello[17] and subsequently by Payne *et al.*[18] which permitted the self-
consistent ground state energy of a system to be calculated efficiently for any set of
atomic positions. Combined with a plane-wave basis set and optimised pseudopotentials for the electron-ion interactions this method has been used successfully to
investigate boundaries in TiO_2 and MgO (see below). Quantities that are readily
obtained from this approach include the grain boundary energy, relaxed atomic structure, the valence electron distribution, the density of states (full and partial), the bond
populations, the Mulliken charges, and the valence state of any impurity. For oxides
which have electronic applications, these quantities are of particular importance in
understanding the formation of potential barriers and interface states in the band gap.
Computer programs which implement this type of first principles methodology
include CASTEP[19] and CETEP.[20] Other density functional approaches have also been

applied to oxide grain boundaries including the orthogonalised linear combination of atomic orbitals method (OLCAO) and the embedded-cluster discrete-variational (DV)-Xα method. Information concerning the density of states, local chemical bonding and Mulliken charges has been obtained for grain boundaries in Al_2O_3 and $SrTiO_3$ (see below) but so far the calculations have suffered from one or more of the following deficiencies: fixed atomic geometries based on classical simulations, lack of self-consistency, lack of a parameter-free formalism, and inappropriate use of clusters to model a periodic defect. However, the investigations have revealed some important trends in the electronic structure of ordered grain boundaries in oxides as described below.

RESULTS

NICKEL OXIDE AND MAGNESIUM OXIDE

Duffy and Tasker[21-25] pioneered the simulation of grain boundaries in NiO using the ionic shell model and investigated the structure and energy of several boundary geometries including symmetric [001] and [011] twist and tilt boundaries. Their initial studies on [001] twist boundaries showed that the stability of these structures could be increased by introducing Schottky defects into the boundary core.[22] Previous investigations[26] had not considered this mode of relaxation and the result provided a better explanation for the observed occurrence of these boundaries in MgO as mentioned in the Introduction. The reduced density of ions creates a restructured boundary, as shown in Fig. 1 for the Σ5 36.9° configuration, but the periodicity remains unchanged. Using these restructured configurations, a grain boundary dislocation/

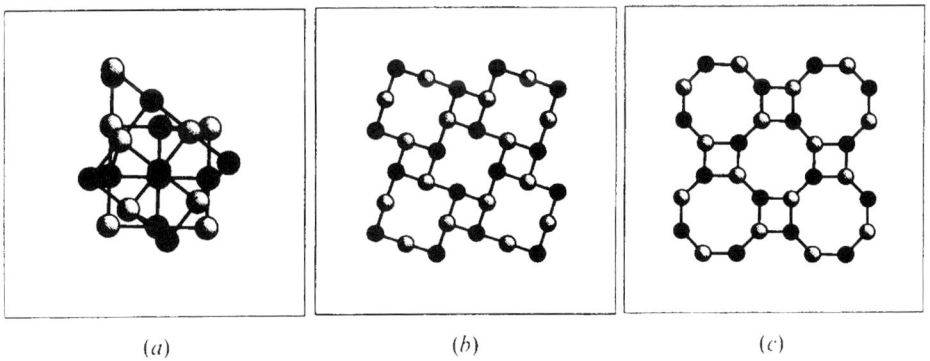

(a) (b) (c)

Fig. 1 Structure of the Σ5 36.9° [001] twist boundary in NiO and MgO viewed along the [001] twist axis. (a) perfect boundary with a full density of ions; (b) unrelaxed boundary structure obtained by creating a Schottky defect; (c) relaxed boundary structure with Schottky defect. Reproduced from Ref. 35.

structural unit model for [001] twist boundaries was proposed that was consistent with the experimental observations.[25]

Similar shell model calculations on symmetric [001] and [011] tilt boundaries in NiO yielded stable structures without the introduction of point defects. The major mode of relaxation in this case was the in-plane translation of the grains. The relaxed structures could be interpreted in terms of arrays of grain boundary dislocations and the high-angle geometries were characterised by 'pipes' or open channels oriented along the tilt axis[23, 24] as shown in Fig. 2. It was suggested that the activation energy for diffusion of vacancies or interstitials along these channels should be lower than for bulk diffusion. This was confirmed in later calculations[27, 28] which showed, for example, that the ratio of the boundary to bulk activation energy could range from 0.56 to 0.77, in general agreement with experimental measurements. Additional calculations on the interaction of various isovalent and aliovalent impurities with some of the tilt boundaries showed that in every case studied there were boundary sites that favoured segregation.[29] The magnitude of the segregation energy was found to depend on both the charge on the impurity and its atomic size. This result is, once again, in qualitative agreement with experimental observation and theoretical expectation.

The atomic structure of one of the simulated grain boundaries, the $\Sigma5$ (310) 36.9° tilt boundary, has been determined by high resolution electron microscopy.[30] It differs from the one predicted by Duffy and Tasker[23] in that it has a less open structure and smaller volume expansion, see Fig. 3. Harding *et al.*[31] have investigated the possible reasons for this discrepancy and speculate that the observed structure may contain a reduced ion density (ie vacant sites) caused by the sample preparation process which involved a quench from high temperature. Electron microscope observations on other tilt boundaries in NiO[32] also indicate the possibility of partially occupied atomic columns adjacent the boundary core. To determine the high temperature structure of the $\Sigma5$ tilt boundary, a molecular dynamics study was performed[33] which found that diffusion occurred adjacent to the boundary leaving some atomic columns with vacancies and creating interstitials in the open channels. The resulting structure was a closer match to the experimental observation providing support for the suggestion that the high temperature form of the structure, which was frozen-in by the quench, contains point defects and partially occupied atomic columns.

There have been no first principles calculations of grain boundaries in NiO so comparisons with either the classical simulations or experimental observations are unfortunately not possible. However, the [001] twist boundary in MgO has been investigated using both the shell model and DFT. The shell model calculations[34] produce results similar to that of NiO, ie the preferred structures contain a reduced ion density in the grain boundary plane. Ab initio DFT pseudopotential calculations[35] of the $\Sigma5$ twist boundary yield a relaxed atomic structure in excellent agreement with the classical calculation. Interestingly, however, analysis of the electronic density of states of this boundary reveals no new states in the band gap suggesting that the

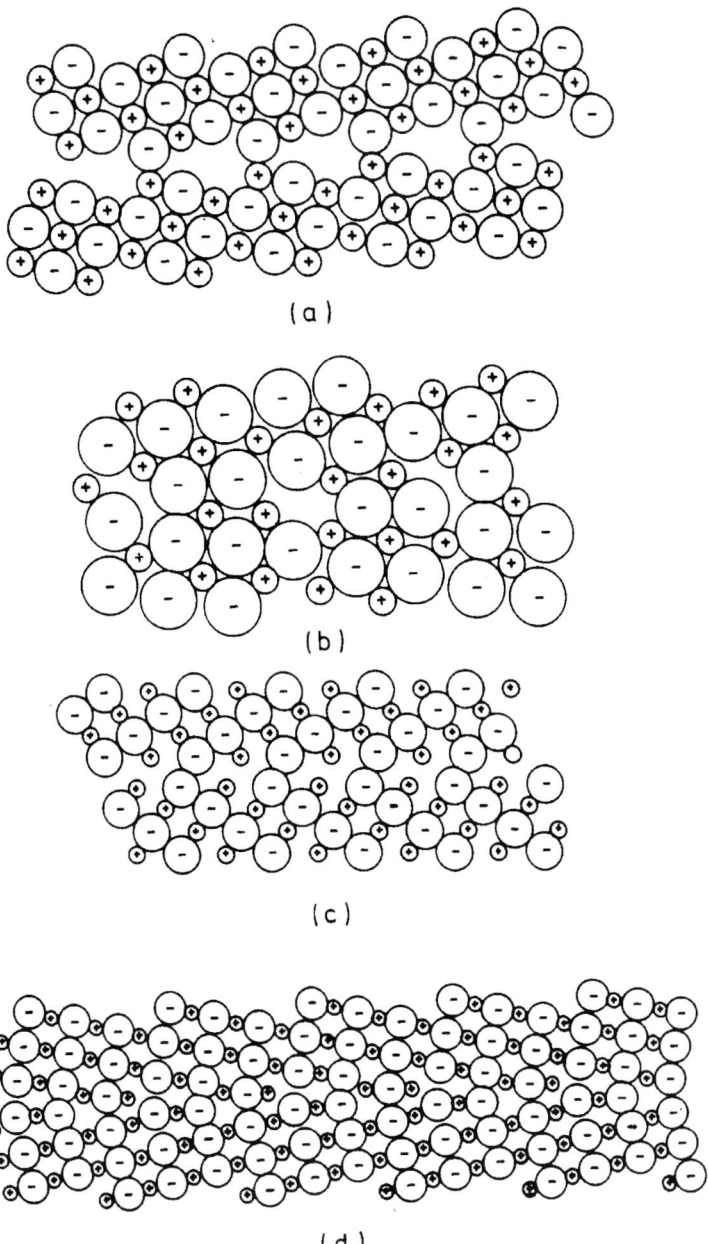

Fig. 2 Relaxed structures of four symmetric tilt boundaries in NiO. (a) $\Sigma 5$ 36.9° [001]; (b) $\Sigma 25$ 73.7° [001] ; (c) $\Sigma 11$ 129.5° [011]; (d) $\Sigma 19$ 26.5° [011]. The boundary plane passes horizontally through each figure and the circles represent the Pauling radii of the ions. Reproduced from Ref. 21.

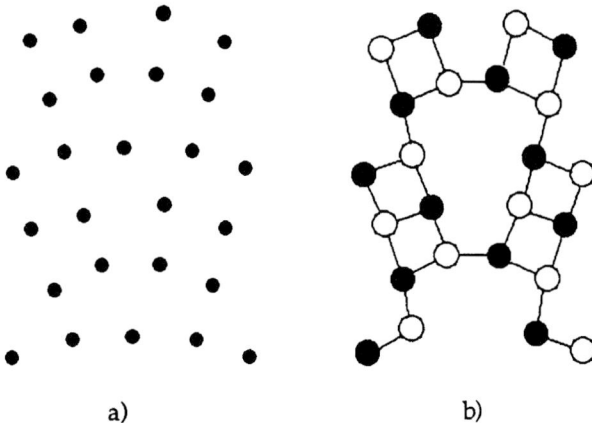

a)   b)

Fig. 3 Atomic models for the Σ5 36.9° (310) [001] tilt boundary in NiO viewed along the [001] tilt axis. (a) derived from high-resolution electron microscope studies; (b) computed structure. Reproduced from Ref. 33.

Schottky defects introduced into the boundary in order to stabilise it are an intrinsic part of the structure and do not produce an electrically active grain boundary.

Recently, a modified molecular dynamics approach[36] has been used to study the effect of pressure on diffusivity near the Σ17 (410) [001] tilt boundary in MgO. Previous diffusion simulations[28] in oxides have used quasi-static methods because the computer time required to observe many atomic jumps in a fully dynamic simulation would be prohibitively long. In the modified approach, an ion adjacent to a vacancy in the boundary core is given a force allowing it to diffuse into the vacant site without constraining the pathway. Using this method, it is found that under no applied pressure the preferred diffusion path for magnesium vacancies is parallel to the tilt axis, ie down the open channel in the boundary structure. This result is analogous to that found previously for [001] tilt boundaries in NiO.[28] Under an applied pressure of 40 GPa, it is found that the structure collapses and becomes more dense. However, the preferred diffusion path is still along the tilt axis although the activation energy for diffusion is now found to be 50% higher.

RUTILE AND STRONTIUM TITANATE

Both rutile (TiO_2) and strontium titanate ($SrTiO_3$) are important electroceramics in which grain boundaries can control the electronic properties of the material. In rutile, for example, grain boundaries are principally responsible for the formation of interfacial electronic states and Schottky barriers which lead to the non-linear conduction behaviour that is exploited in surge protectors (varistors). Potential barriers also form at grain boundaries in $SrTiO_3$ and these are used to fabricate internal boundary layer capacitors. In addition to the presence of grain boundaries, other factors such as the local impurity distribution and oxygen stoichiometry may also play a major role. Despite the importance of grain boundaries in these oxides there have been few computer simulations of their structure and chemistry. This is probably due to the crystallography of these structures which are less symmetric than rocksalt and also to the availability of accurate interatomic potential descriptions.

In rutile, an early classical shell model investigation[37] studied the structure and energy of eight [001] symmetric tilt boundaries using the molecular static approach. The calculations were possible using periodic border conditions since rotations about the c-axis in rutile, which is tetragonal, produce exact coincidence site lattice structures. Rotations about any other axis would generate non-periodic grain boundaries or, at best, near-coincidence site lattice structures. Relaxations were performed with respect to in-plane translations and also point defect creation in the boundary. Most of the relaxed structures were characterised by in-plane translations that removed the mirror symmetry of the titanium and oxygen sublattices. The calculated energy-misorientation curve showed 'cusps' at the low Σ orientations but the magnitude of the energies was rather high (~4 to 20 J m^{-2}). Unfortunately none of the relaxed structures could be compared to experimental observations.

More recently, two further classical simulation studies have been reported in which calculated tilt boundary structures in rutile have been compared to high-resolution electron microscope observations. The first study[38] involved the (101) and (301) [010] twin boundaries where excellent agreement was found between calculations and observations. The (101) twin was characterised by an in-plane translation which conserved the mirror symmetry of the metal sublattice but which imposed a displacement on the oxygen sublattice, as shown in Fig. 4. The (310) twin boundary structure

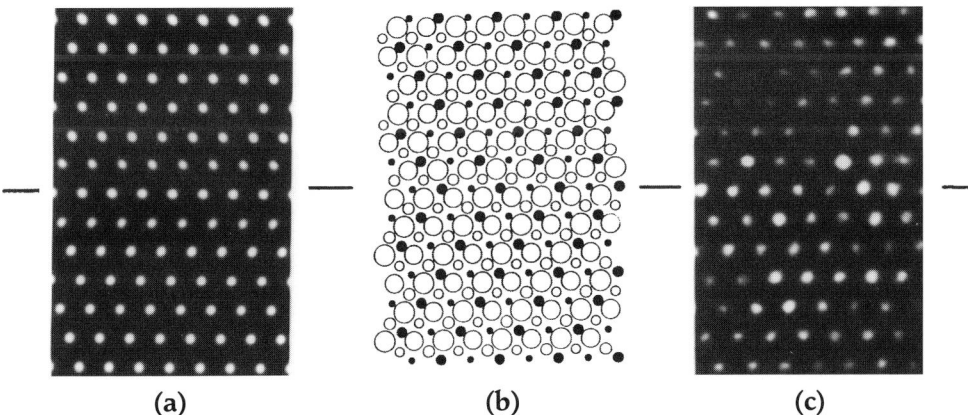

      (a)   **(b)**   **(c)**

Fig. 4 (a) high-resolution electron microscope image of the (101) [010] twin boundary in TiO$_2$; (b) computed twin boundary structure with full and open circles representing metal and oxygen ions respectively; (c) electron microscope image simulation of the computed structure. Metal ion columns are imaged as white dots. Reproduced from Ref. 38.

involved no in-plane translation and was characterised by mirror symmetry of both sublattices. The computed volume expansions, which were small (~0.01 nm), also compared well with the measurements. In both cases it was not necessary to introduce point defects to stabilise the boundaries or achieve good matching with the electron microscope images. However, the boundary structures were highly symmetric, short period and well coordinated. The second study[39] focussed on the Σ5 36. 9°(210) [001] tilt boundary which has a less ordered structure. After relaxation, this grain boundary was also characterised by an in-plane translation (~1/5[120]) which agreed reasonably well with the experimental electron microscope images.[39, 40] The resulting structure conserved the mirror symmetry of the oxygen sublattice but not the metal sublattice and is shown in Fig. 5. In addition, the Ti-O bond lengths in the boundary core were distorted from their bulk values and some of the titanium and oxygen atoms were under-coordinated. An experimentally measured contraction at the boundary was not found in the calculations possibly due to a local loss of oxygen. However, these oxygen vacancies, if present, do not apparently affect the translation state of the boundary.

In order to determine the effect of under-coordination and distorted bond lengths on the electronic structure of a grain boundary, a first principles DFT-LDA calculation using optimised pseudopotentials has been performed on the Σ5 36.9°(210) [001] tilt boundary.[41] Using the classical structure as a starting configuration, full atomic and electronic relaxation was performed. It was found that the displacements away from the starting configuration were small lending support to the ionic shell model and

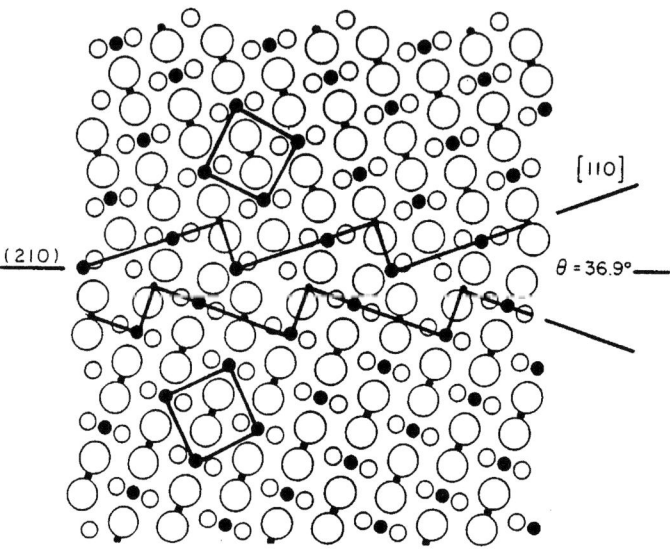

Fig. 5 Computed structure of the Σ5 36.9° (210) [001] tilt boundary in TiO$_2$ with closed and open circles representing metal and oxygen ions respectively. The circle size indicates depth along [001]. Reproduced from Ref. 39.

indicating that the principal interactions influencing the boundary structure are Coulombic. The calculations were performed using a supercell model and therefore two symmetry equivalent boundaries were present in the computational cell. An analysis was made of the valence charge density distribution, total density of states, Mulliken charges and bond populations. The valence charge density distribution indicated the presence of small flares of charge density located along the [110] (Ti-O) bond directions in the boundary core suggesting an increase in the degree of covalency in this region. However, the total density of states did not reveal any deep levels in the band gap despite the presence of bond distortions and under-coordination, see Fig. 6. Only a few shallow localised states near the valence band edge were found. It

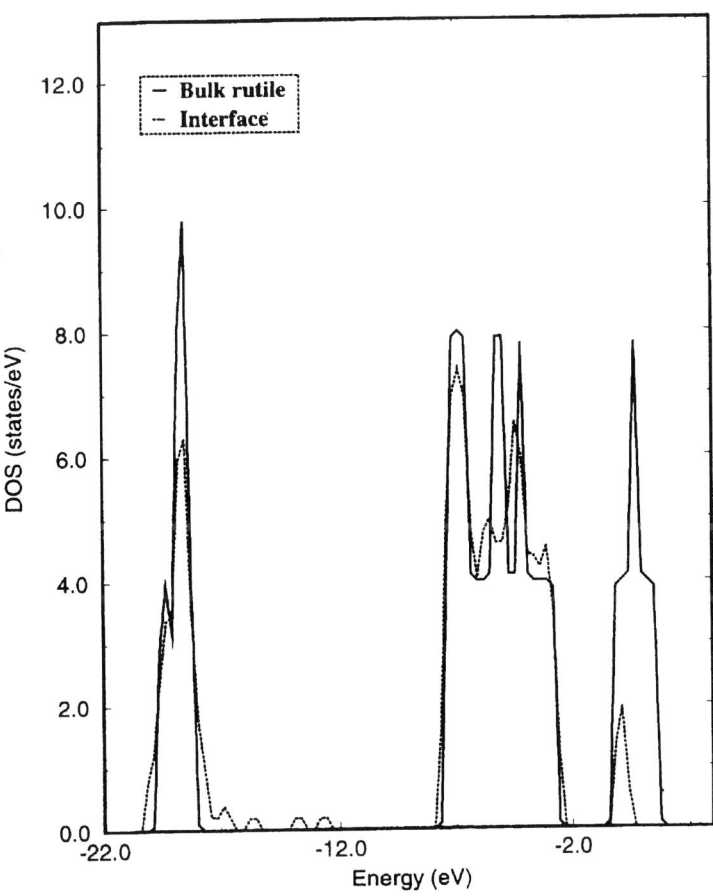

Fig. 6 Computed density of states of the $\Sigma 5$ 36.9° (210) [001] tilt boundary in TiO_2 compared to the bulk. The band gap is from -2.9 eV to -0.3 eV. Note the absence of states deep in the band gap and the increased bandwidth for the grain boundary. The latter is consistent with the overall trend towards bond length reduction in the boundary core. Reproduced from Ref. 41.

was concluded that the 'clean' defect-free boundary would not exhibit intrinsic electrical activity. The Mulliken charges and bond populations reflected the nature of the local bonding and coordination in the boundary relative to the bulk. Near under-coordinated atoms, for example, charge transfer occurred resulting in some bonds with an increased bond population (more covalent) and others with a decreased bond population (more ionic).

The effect of doping the Σ 5 (210) [001] tilt boundary with niobium has also been investigated using the DFT pseudopotential approach.[42] The goal was to determine whether impurities could create an electrically active boundary. Two sites were chosen for Nb substitution: one in the core region adjacent to an under coordinated oxygen atom and one in the bulk-like region between the two grain boundaries in the model. The calculations included spin polarisation and were relaxed with respect to the ionic forces. No significant structural relaxation was found to occur. However, examination of the electron eigenvalues for both Nb sites showed the presence of an impurity level in the gap that had shifted upwards towards the conduction band (CB) edge relative to the same level in the bulk. The Nb atom in the boundary core exhibited the largest shift with the resulting level located 0.15 eV below the CB edge. This value is comparable to the Nb impurity ionisation energy of 0.2 eV indicated in some experimental studies. There is thus a tendency for the impurity to become more donor-like on segregation and at sufficiently high temperatures could contribute to electronic conduction. Furthermore, the total energies indicate a significant driving force for segregation with a calculated segregation energy of 0.34 eV for this site.

There have been a number computer modelling studies, both quantum mechanical and classical, on grain boundaries in strontium titanate motivated, in part, by a series of detailed experimental observations on bicrystals using high resolution electron imaging and electron energy loss spectroscopy.[43–46] The simplest boundary studied to date is the Σ3 70.5°(111) [110] twin boundary using the classical shell model approach.[47] The relaxed structure is characterised by a boundary plane containing Sr–O and O columns rather than Ti columns, a small expansion (0.03 nm) normal to the boundary, and mirror symmetry across the boundary plane. This structure was found to be stable without the introduction of point defects into the core region. These general features compared well with the experimental images. However, in detail there were some discrepancies, namely in the volume expansion, which was found to be 0.06 nm experimentally, and in the position of some of the Sr–O columns adjacent to the boundary, see Fig. 7. It was concluded that the shell model had not properly accounted for charge transfer at the boundary, ie changes in ionicity and nature of the local bonding. A second classical simulation study has been performed on a more complex boundary geometry, ie the Σ5 36.9°(310) [001] tilt boundary.[48] The relaxed structure is characterised by an in-plane translation of ~0.06 nm, no appreciable volume change and some partially occupied atomic columns in the boundary core, see Fig. 8a. It also contains under-coordinated atoms and reduced bond lengths. These characteristics compare reasonably well with the experimentally deduced structure[45] particularly with regard to the translation state and the presence

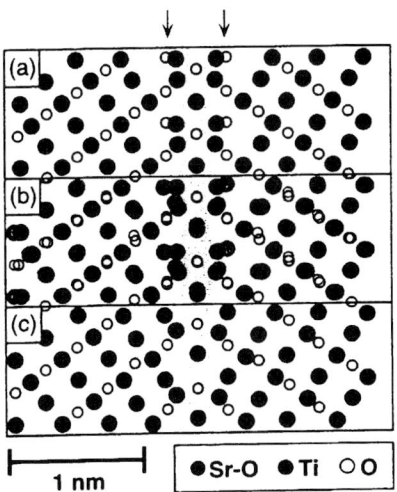

Fig. 7 (a) Computed structure of the Σ3 70.5° (111) [110] twin boundary in SrTiO₃ viewed along [110]; (b) comparison between the computed structure and the observed model structure; (c) model structure of the boundary derived from high-resolution electron microscope images. Reproduced from Ref. 47.

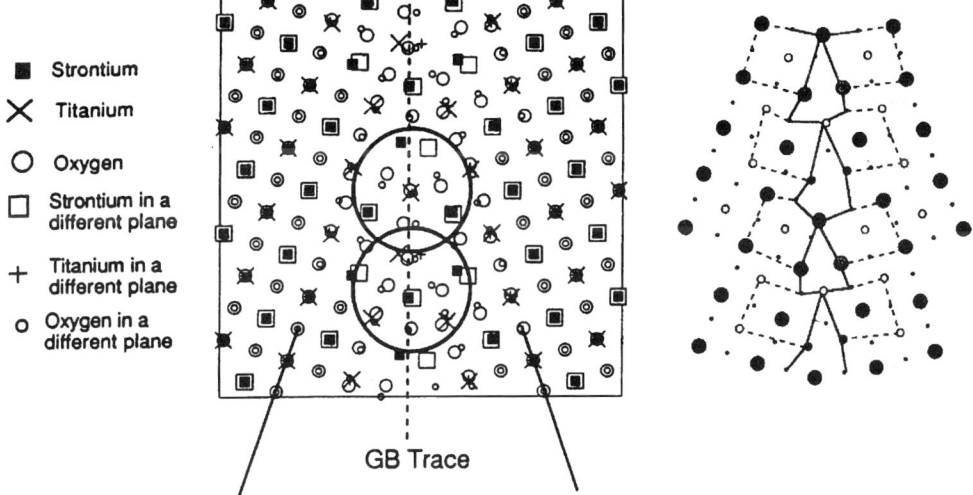

Fig. 8 (a) Computed structure of the Σ5 36.9° (310) [001] tilt boundary in SrTiO₃ viewed along the [001] tilt axis. Circles indicate regions where cluster calculations were performed. Reproduced from Ref. 48. (b) Model structure of the Σ5 36.9° (310) [001] tilt boundary in SrTiO₃ derived from high-resolution electron microscope images. Some columns of atoms in the boundary core are found to be partially occupied. Reproduced from Ref. 45.

of partially occupied atomic columns, see Fig. 8b. It is clear, however, that there are discrepancies between the observations and simulations with regard to the detailed atomic positions in the boundary core. Despite these discrepancies, further calculations have been performed to determine the electronic structure of this boundary geometry using a DFT embedded cluster discrete-variational method.[48, 49] In this method, a molecular orbital calculation is performed on small clusters of fixed atoms within the grain boundary core and can be centered, in this case, on either Sr or Ti atoms. Circles in Fig. 8a illustrate two such clusters. Information that can be obtained from this calculation include the valence charge density distribution, the partial density of states and the Mulliken charges. For the clean, dopant-free, $\Sigma 5$ boundary it is found that the Ti–O bonds in the cluster become more covalent compared to the perfect crystal while the Sr–O bonds retain their ionic character. Charge transfer occurs from the O sites to the Ti sites suggesting that Ti is more electrically active than Sr in the boundary core. The effect of doping the boundary with an impurity was also studied by replacing the electrically active Ti atom with Mn.[49] It is found that Mn prefers to occupy this site with a valence charge of about +2 which is consistent with electron holography experiments on the same doped boundary that indicate that the boundary core is negatively charged and there is a corresponding positive space charge. A molecular orbital cluster approach has also been used to determine the effect of doping the $\Sigma 5$ boundary with Fe.[50] By examining the computed density of states and molecular orbitals and comparing their results with electron energy loss spectra it is concluded that the boundary is deficient in Sr and that there is a change in the connectivity of the TiO_6 octahedra in the boundary core relative to the bulk.

ALUMINA AND ZINC OXIDE

The number of grain boundary simulations in both Al_2O_3 and ZnO is extremely limited due the hexagonal nature of their crystal structures and the lack of a centre of symmetry. In Al_2O_3, Kenway[51] has performed classical shell model calculations of stacking faults and twin boundaries on basal and non-basal planes and predicted the lowest energy configurations. For the {0001} twin boundaries, several candidate structures were considered and it was found that their relative stability depended on the difference in the coordinations of the aluminium atoms at the interface relative to the bulk. For twins on {$10\bar{1}4$}, two different configurations were simulated, both with mirror glide symmetry. The one with lowest energy was characterised by aluminium atoms at the boundary and an in-plane translation of the crystals. In neither case were point defects introduced into the boundaries to stabilise them. Kenway[51] also simulated the structure of a $\Sigma 11$ 35.2° {$10\bar{1}1$} <$1\bar{2}10$> tilt boundary since high-resolution electron microscope images of this boundary are available for direct comparisons. In determining the lowest energy structure for this boundary both in-plane translations and the creation of Schottky defects were considered. Schottky defects were not found to lower the boundary energy but the in-plane translation yielding the most energetically favourable structure was found to compare well with the experimental observation, see Fig. 9. This structure was characterised by shortened bond lengths and a significant number of under-coordinated atoms in the boundary core.

The electronic structure of the Σ11 tilt boundary in Al_2O_3 has been investigated to determine whether the distortion in the grain boundary produces defect states in the band gap.[52, 53] Employing the OLCAO method and using the Kenway model for the fixed atomic geometry, it is found that the grain boundary does not introduce deep states in band gap but only localised states near the valence and conduction band edges. The main effect of the under-coordinated atoms and shortened bond lengths is an increase in charge transfer from aluminium to oxygen with a corresponding increase in ionicity at the boundary. This boundary therefore, which is dopant free, is not expected to be electrically active.

In ZnO, there have been two recent classical simulations of grain boundary structure and segregation. Using shell model potentials and a molecular dynamics approach, the structure of a series of [10$\bar{1}$0] symmetical tilt boundaries has been investigated.[54] These are near-coincidence geometries and therefore formally non-periodic. After equilibration at 300 K, it was found that three of the boundaries (Σ35 38.7°, Σ15 85.6°, Σ29 101.6°) exhibited periodic structures containing structural units characterised by rings of 4, 6, 8 and 10 atoms. Rings with 8 and 10 atoms are larger than found in the bulk and represent large open channels in the boundary core. It is postulated that these boundaries could be electrically active. Unfortunately the boundary energies, volume expansions and many of the computational details such

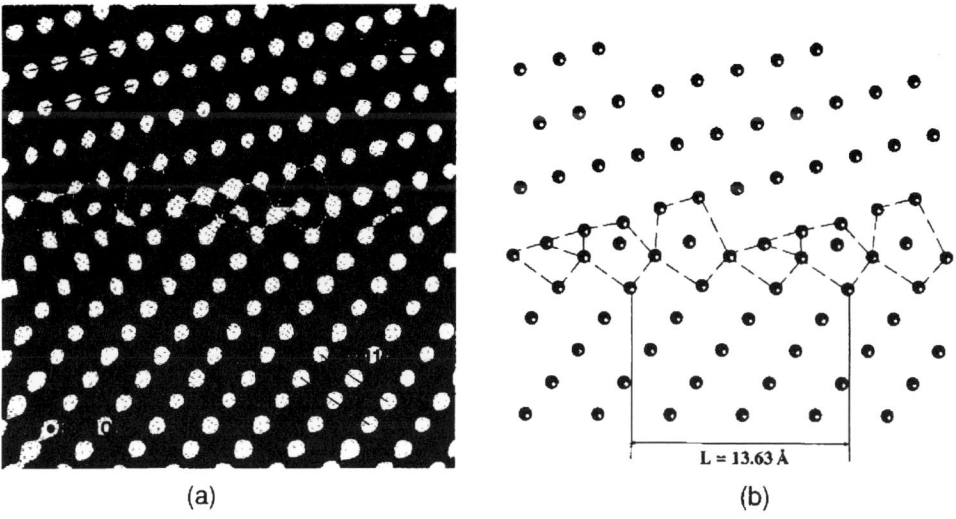

(a)   (b)

Fig. 9 (a) High-resolution electron microscope image of the Σ11 35.2° {10$\bar{1}$1} <1$\bar{2}$10> tilt boundary in Al_2O_3; (b) computed structure of the boundary showing only the oxygen atoms. Reproduced from Ref. 52.

as the border conditions are not presented. In addition, the boundary structures could not be compared to experimental observations. In the second classical study, a series of [0001] twist boundaries has been simulated together with their interaction with impurities.[55] A molecular statics approach is used and shell model potentials are employed. A periodic supercell containing two grain boundaries is used to avoid the dipole moment associated with the polar (0001) surface. Eleven boundary geometries were considered each of which corresponded to an exact coincidence site lattice structure. Point defect formation in the boundary core was considered as a mode of relaxation but, unlike [001] twist boundaries in NiO, a Schottky defect configuration which lowered the boundary energy was not found. The energy-misorientation curve showed cusps at 0° and 60° and the relaxed structures near those misorientations exhibited arrays of screw dislocations in agreement with the grain boundary dislocation model for grain boundaries, see Fig. 10. Segregation simulations were per-

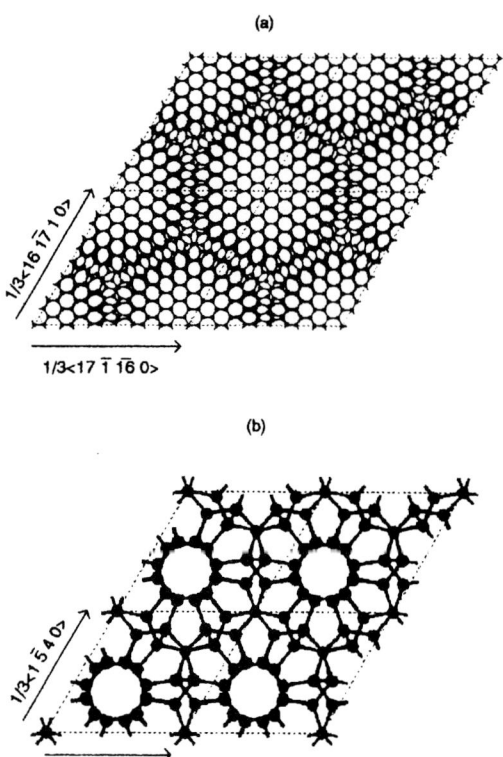

Fig. 10 Computed structures of the (a) Σ91 6.0° and (b) Σ7 38.2° [0001] twist boundaries in ZnO viewed along the [0001] twist axis. Four unit CSL cells are shown in each case together with the CSL vectors. A hexagonal array of grain boundary dislocations is seen in the low-angle boundary. Reproduced from Ref. 55.

formed by substituting either Ba or Co for zinc atoms in and around the core of the $\Sigma 7$ 38.2° boundary. The segregation characteristics could be explained on the basis of their relative atomic sizes since both impurities are isovalent with zinc. It was found that both Ba and Co have a tendency to segregate to the boundary although the driving force for Ba is much greater since it is larger than Co with non-CSL sites being the preferred locations for the impurities.

There are few electronic structure calculations on grain boundaries in ZnO. This is surprising since polycrystalline doped ZnO is commonly used in commercial varistors. The non-linear conductive behaviour of the material is directly attributable to the grain boundaries and dopants. It has one of the largest non-linearity coefficients of all semiconducting oxides. Most of the modelling studies have employed the molecular orbital cluster approach to simulate the atomic environment of a grain boundary by introducing various defects into the cluster such as oxygen and zinc vacancies, dangling bonds, adsorbed molecular oxygen and substitutional impurities. One such study[56] has used the self-consistent field X-α scattered wave cluster method to investigate the 'interface' states introduced by some of these defects. It was found that dangling bonds do not introduce energy levels into the band gap and therefore they are unlikely to contribute to the electronic properties of this oxide. However, molecular oxygen forms a deep acceptor state having π character which would represent a potential barrier to conduction in n-type ZnO. Similarly, Zn vacancies form shallow acceptors (although this is not observed experimentally) and oxygen vacancies form deep donors. A second series of calculations[57] using the same methodology have investigated the effect of substitutional Mn, Co and Cu on the band structure of the cluster. The impurities all form deep acceptor levels having 3d character. Since grain boundaries can be sources and sinks for vacancies and interstitials and also sites for the preferential adsorption of solute atoms, all of the defects considered in the cluster calculations could be present in these interfaces and therefore influence the electronic properties of the material. Further electronic structure calculations are needed, preferably using more accurate DFT methods, on specific grain boundary structures containing some of these defects to provide additional insight

CONCLUSIONS

The computer modelling of grain boundaries in oxides has progressed significantly over the past two decades. Classical codes have been developed that can relax complex structures efficiently, perform molecular dynamics and handle charged and polar interfaces. This has led to the investigation of a wide range of boundary geometries in six different oxides using computational cells containing several thousand atoms. The only major drawback to these calculations remains the reliability of the interatomic potentials. Their transferability to a crystal imperfection involving large atomic displacements and possible charge transfer effects will always be questionable. However, when it has been possible to make comparisons with experimental observations, the simulations nearly always reproduce the major characteristics of the grain boundary structure such as the in-plane translation state. Volume changes and detailed atomic positions in the boundary core are more difficult to predict correctly. The volume changes are often overestimated resulting in more open structures than are observed. The recent ability to perform first principles quantum mechanical calculations on oxide grain boundaries should remedy these deficiencies. Charge transfer effects are automatically accounted for in these calculations and provided all possible modes of relaxation are considered, including point defect formation, the results should be very accurate.

The structural characteristics of oxide grain boundaries determined thus far using computer modelling techniques may be summarised as follows:

- when the grain boundary exhibits a high degree of symmetry and particularly low energy, eg a mirror twin, point defects are not needed to stabilise the boundary or achieve a reasonable match to experimental observation.
- grain boundaries other than mirror twins invariably exhibit a relative translation of the adjoining grains and also reduced ion density resulting from point defect creation. Examples of the latter include Schottky defects in the interface plane of twist boundaries and partially occupied atomic columns parallel to the interface plane of tilt boundaries.
- tilt boundary structures are often characterised by under-coordinated atoms and reduced bond lengths. Despite these distortions their electronic structures do not exhibit interface states deep in the band gap. Calculations on tilt boundaries in germanium and silicon show a similar effect.[58, 59]
- substitutional impurities in oxide grain boundaries can introduce new states into the band gap.

There are indications from several studies that clean, dopant-free, grain boundaries in oxides are not in themselves electrically active. This suggests that impurities and/or oxygen adsorption or loss are responsible for the observed electrical properties of polycrystalline oxide ceramics. To shed further light on the microscopic origins of these properties, further ab initio calculations on grain boundary segregation effects are clearly desirable. In addition, there have been no simulations of grain boundaries containing a second phase caused by doping. This phase can be of significant thickness and either glassy or crystalline. Determining the electronic properties of the two interfaces present in such cases would be of considerable interest.

REFERENCES

1. R. W. Balluffi, P. D. Bristowe and C. P. Sun, *J. Am. Ceram. Soc.,* 1981, **64**, 23.
2. P. Chaudhari and H. Charbnau, *Surf. Sci.,* 1972, **31**, 104.
3. D. Wolf, *Proc. Int. Conf. on Defects in Ionic Crystals, Canterbury,* UK, 1979.

4. D. Wolf and S. Yip eds, *Materials Interfaces,* Chapman & Hall, 1992.
5. A. P. Sutton and R. W. Balluffi, *Interfaces in Crystalline Materials,* Oxford, 1995.
6. J. H. Harding, *Surf. Sci.,* 1999, **422**, 87.
7. J. H. Harding, *Rep. Prog. Phys.,* 1990, **53**, 1403.
8. P. J. Lawrence and S. C Parker, in *Computer Modelling of Fluids, Polymers and Solids,* C. R. A. Catlow *et al.* eds, Kluwer Academic, 1990, 219.
9. E. A. Colbourn, *Surf. Sci. Rep.,* 1992, **15**, 281.
10. D. E. Parry, *Surf. Sci.,* 1975, **49**, 433.
11. N. F. Mott and M. J. Littleton, *Trans. Faraday Soc.,* 1938, **34**, 485.
12. W. Bollmann, *Crystal Defects and Crystalline Interfaces,* Springer, 1970.
13. R. Bonnet, E. Cousineau and D. H. Warrington, *Acta Cryst.,* 1981, **A37**, 184.
14. J. H. Harding, Harwell Report No. AERE-R13127, 1988.
15. D. H. Gay and A. L. Rohl, *J. Chem. Soc. Faraday Trans.,* 1995, **91,** 925.
16. J. D. Gale, *J. Chem. Soc. Faraday Trans.,* 1997, **93**, 627.
17. R. Car and M. Parrinello, *Phys. Rev. Lett.,* 1985, **56**, 2471.
18. M. C. Payne, M. P. Teter, D. C. Teter, D. C. Allan, T. A. Arias and J. D. Joannopoulos, *Rev. Mod. Phys.,* 1992, **64**, 1045.
19. P. J. D. Lindan, *Guide to CASTEP 3.9,* Daresbury Laboratory, UK, 1999
20. L. J. Clarke, I. Stich and M. C. Payne, *Comput. Commun.,* 1992, **72**, 14.
21. D. M. Duffy and P. W. Tasker, *J. de Physique,* 1985, **46**, C4–185.
22. P. W. Tasker and D. M. Duffy, *Phil. Mag. A,* 1983, **47**, L45.
23. D. M. Duffy and P. W. Tasker, *Phil. Mag. A,* 1983, **47**, 817.
24. D. M. Duffy and P. W. Tasker, *Phil. Mag. A,* 1983, **48**, 155.
25. D. M. Duffy and P. W. Tasker, *Phil. Mag. A,* 1986, **53**, 113.
26. D. Wolf and R. Benedek, *Advances in Ceramics,* 1981, **1**, 107.
27. D. M. Duffy and P. W. Tasker, *Phil. Mag. A,* 1984, **50**,143.
28. D. M. Duffy and P. W. Tasker, *Phil. Mag. A,* 1986, **54**, 759.
29. D. M. Duffy and P. W. Tasker, *Phil. Mag. A,* 1984, **50**, 155.
30. K. L. Merkle and D. J. Smith, *Phys. Rev. Lett.,* 1987, **59**, 2887.
31. J. H. Harding, S. C. Parker and P. W. Tasker in *Non-Stoichiometric Compounds, Surfaces, Grain Boundaries and Structural Defects,* J. Nowotny and W. Weppner eds, New York Plenum, 1989, 337.
32. K. L. Merkle, J. F. Reddy, C. L. Wiley and D. J. Smith, *J. Physque,* 1988, **49**, C5–251.
33. M. Meyer and C. Waldburger, *Mat. Sci. Forum,* 1993, **126–128**, 229.
34. D. C. Sayle, T. X. T. Sayle, S. C. Parker, J. H. Harding and C. R. A. Catlow, *Surf. Sci.,* 1995, **334** 170.
35. J. H. Harding and C. Noguera, *Phil. Mag. Lett.,* 1998, **77**, 315.
36. S. C. Parker, D. J. Harris, F. M. Higgins, N. H. de Leeuw, P. M. Oliver and G. W. Watson in *Ceramic Interfaces: Properties and Applications,* R. St C. Smart and J. Nowotny eds, IOM Communications, 1998, 45.
37. F. Matsushima, H. Fukutomi and E. Iguchi, *Trans. Japan Inst. Metals,* 1987, **28**, 869.
38. W.-Y. Lee, P. D. Bristowe, Y. Gao and K. L. Merkle, *Phil. Mag. Lett.,* 1993, **68**, 309.
39. W-Y Lee, P. D. Bristowe, I. G. Solorzano and J. B. Vandersande, *Mat. Res. Soc. Proc.,* 1994, **319**, 239.
40. U. Dahmen, S. Paciornik, I. G. Solorzano and J. B. Vandersande, *Inter. Sci.,* 1994, **2**, 125.
41. I. Dawson, P. D. Bristowe, M.-H. Lee, M. C. Payne, M. D. Segall and J. A. White, *Phys. Rev B,* 1996, **54**, 13727.
42. I. Dawson, Ph. D. Thesis, University of Cambridge, 1998
43. M. M. McGibbon, N. D. Browning, M. F. Chisholm, A. J. McGibbon, S. J. Pennycook, V. Ravikumar and V. P. Dravid, *Science,* 1994, **266**, 102.

44. D. J. Wallis and N. D. Browning, *J. Am. Ceram. Soc.,* 1997, **80**, 781.
45. N. D. Browning and S. J. Pennycook, *J. Phys. D: Appl. Phys.,* 1996, **29**, 1779.
46. N. D. Browning, J. P. Buban, H. O. Moltaji, S. J. Pennycook, G. Duscher, K. D. Johnson, R. P. Rodrigues and V. P. Dravid, *Appl. Phys. Lett.,* 1999, **74**, 2638.
47. O. Kienzle, M. Exner and F. Ernst, *Phys. Stat. Sol. (a),* 1998, **166**, 57.
48. H. Chang, R. P. Rodrigues, J-H. Xu, D. E. Ellis and V. P. Dravid, *Ferroelectrics,* 1997, **194**, 249.
49. H. Chang, J. D. Lee, R. P. Rodrigues, D. E. Ellis and V. P. Dravid, *J. Mat. Synth. Proc.,* 1998, **6**, 323.
50. I. Tanaka, T. Nakajima, J. Kawai, H. Adachi, H. Gu and M. Ruhle, *Phil. Mag. Lett.,* 1997, **75**, 21.
51. P. R. Kenway, *J. Am. Ceram. Soc.,* 1994, **77**, 349.
52. S.-D. Mo, W. Y. Ching and R. H. French, *J. Am. Ceram. Soc.,* 1996, **79**, 627.
53. S.-D. Mo, W. Y. Ching and R. H. French, *J. Am. Ceram. Soc.,* 1996, **79**, 1761.
54. W. Wunderlich, *Phys. Stat. Sol. (a),* 1998, **170**, 99.
55. H. Domingos and P. D. Bristowe, *Scripta Mater.,* 1999, **41**, 1347.
56. M. H. Sukkar, K. H. Johnson and H. L. Tuller, *Mat. Sci. Eng. B,* 1990, **6**, 49.
57. Y. Yano, Y. Takai and H. Morooka, *J. Mater. Res.,* 1994, **9**, 112.
58. E. Tarnow, T. Arias, P. D. Bristowe. P. Dallot, G. P. Francis, J. D. Joannopoulos and M. C. Payne, *Mat. Res. Soc. Proc.,* 1990, **193**, 235.
59. M. Kohyama, R. Yamamoto, Y. Ebata and M. Kinoshita, *J. Phys C.,* 1988, **21**, 3205.

Grain Boundary Phenomena in Oxygen Ion Conductors

J. A. KILNER

Centre for Ion Conducting Membranes (CICM), Department of Materials,
Imperial College of Science, Technology and Medicine, London SW7 2BP, UK

INTRODUCTION

Fast oxygen ion conductors are solids that possess some unique mass transport properties. They are polycrystalline, solid materials that exhibit a high mobility of the oxygen ions at elevated temperatures, with diffusion coefficients (ionic conductivities) approaching levels normally found in the liquid state. This unusual phenomenon occurs because of the high degree of disorder on the oxygen sub-lattice in these oxides, arising from either intrinsic disordering, or from intentional aliovalent cation substitutions.

Currently, these materials are the focus of much industrial interest as they provide the active components for use in novel high temperature electrochemical devices, such as the Solid Oxide Fuel Cell (SOFC), Ceramic Oxygen Generators (COG's) and partial oxidation membranes. These new devices have the potential to deliver high economic and ecological benefits, provided they can be developed to achieve satisfactory performance. Of principal importance for all the devices mentioned above is the total oxygen ion conductivity of the active materials. This normally consists of two main components, one arising from the grain interiors and the other from the grain boundaries. A great deal of effort has gone into the understanding of the defect chemistry and defect mobility in these oxides, which gives rise to the grain interior component of the conductivity.[1,2] Much less effort has been put into an understanding of the grain boundary properties and thus current understanding of grain boundary oxygen transport is, at best, incomplete. The aim of this article is to outline the progress that has been made in the understanding of the effects of grain boundaries on the transport properties of fast ion conducting polycrystalline ceramics.

OXYGEN ION CONDUCTORS

Oxygen ion conductivity is limited to a few oxide structures where large deviations from stoichiometry can occur giving rise to high defect concentrations. In the main, this means materials with oxygen deficiency and the consequent introduction of vacancies onto the oxygen sub-lattice. Two main crystal structures show extensive non-stoichiometry and a high mobility of oxygen in the resulting defective oxygen sub-lattice. These are the oxides with the fluorite structure, AO_2, such as CeO_2 and ZrO_2, and those with the perovskite structure, ABO_3, eg $LaGaO_3$. Two important distinctions should be made at this point. The first is between materials that are effectively electronic insulators, in which charge transport takes place by the passage of oxygen ions alone, and those that display appreciable electronic and ionic conductivities. The former materials are often known as ionic conductors or solid electrolytes, and the latter are commonly known as mixed conductors. The second distinction is that the materials in their pure state do not usually achieve very high oxygen ion conductivities. Consequently they are doped with lower valent cations to promote the formation of oxygen vacancies. The following solid solutions are typical examples of this type of oxide: ionic conductors $Ce_{1-x}Gd_xO_{2-x/2}$ (CGO), $Zr_{1-x}Y_xO_{2-x/2}$ (YSZ), $La_{1-x}Sr_xGa_{1-y}Mg_yO_{3-\delta}$ (LSGM), mixed conductors $La_{1-x}Sr_xCoO_{3-\delta}$ (LSC), $La_{1-x}Sr_xCo_{1-y}Fe_yO_{3-\delta}$ (LSCF).

MEASUREMENT OF OXYGEN TRANSPORT RATES

The measurement of the total oxygen transport rates in these materials differs, depending upon whether the material is an ionic conductor or whether it is a mixed conductor. For the ionic conductors, the simple technique of impedance spectroscopy can be used to extract the grain boundary and grain interior components of the total electrical conductivity of a material. This technique is outlined below, and an example is given as to the effect of the grain boundaries on the total oxygen conductivity for a material of current interest. For the mixed conducting materials, it is more difficult to measure the oxygen transport rates, because the presence of electronic conductivity dominates the electrical properties of the material and makes it tricky to measure the ionic component directly. Because of the inherent ambiguities of the electrical methods for mixed conductors, other methods of determining the oxygen transport rates have been devised. The method explored here uses gas/solid exchange of the stable isotope ^{18}O followed by a depth profiling technique to obtain the kinetic parameters associated with the exchange. These isotopic exchange experiments for the mixed conductors and impedance spectroscopy used for ionic conductors are described in more detail below.

IMPEDANCE SPECTROSCOPY

The two components contributing to the total conductivity of an ionic conductor are difficult to extract using conventional dc conductivity techniques. This can be done by dc methods, however it involves the measurement of a large number of samples of different geometry. In contrast, impedance spectroscopy has successfully been used over the past 20 years to separate the relative contributions arising from the grain interiors and the grain boundaries by the measurement of a single sample. The method is simple; it is a two terminal measurement involving the application of a small voltage, variable frequency ac signal and measuring the complex impedance, z, of the sample at each applied frequency. The frequency range involved is large, from 10 MHz down to ~100 mHz with a number of measurements being taken in each frequency decade. Thus, although the method is simple, large amounts of data are collected for each sample. The advent of microprocessor controlled Frequency Response Analysers (FRA's) has meant that this type of analysis is now routine, whereas when the data had to be collected by hand, it was a task undertaken only when essential.

The complex impedance data is usually analysed by plotting the data on a modified Argand diagram as shown in Fig. 1. Several semicircles are seen in a typical impedance spectrum. The equivalent circuit used to analyse this data is also shown in Fig. 1. The sample is represented by three lumped RC circuits, representing the grain interior, grain boundary and electrode components. The higher frequency semicircles are due to the grain interior (bulk) and the grain boundary response of the ceramic, the low frequency semicircles are due to the response of the electrodes. The intercepts on the real axis can be used to obtain the grain interior, grain boundary and electrode (charge transfer) resistances of the sample. From the impedance spectra themselves, it is difficult to extract the trends with temperature and with changes in composition, data are thus presented as either Arrhenius or isothermal plots of the resistive components.

At this point, a practical example will help to understand the principles involved. A lot of recent work on high temperature solid-state ionic devices has been directed towards the development of devices with lower operating temperatures, in the region of 500–700°C.[3] The ionic conductors for the low temperature devices need to display high total conductivities at low temperatures. One such family of materials is the ceria-gadolinia solid solutions $Ce_{1-x}Gd_xO_{2-x/2}$ (CGO). The CGO's have some interesting properties but there has been some debate as to the optimum composition for practical applications.

Until recently there has been no systematic investigation of the contribution made to the total conductivity by the grain interior and grain boundary components. Figure 2 shows the complex impedance spectrum at low temperatures of some CGO materials prepared by Ralph *et al.*[4,5] It is clear that the resistances vary quite dramatically and for the frequency range chosen, the full impedance spectra are not visible. Only for the material with a Gd content of 0.2 is there a clearly visible grain interior semicircle plus the majority of the grain boundary component.

The main aim of Ralph's work was to follow the changes in the grain and grain boundary components of the ionic conductivity using materials prepared by standard methods, and at such concentration intervals as to determine the major changes in composition for the two components. This data is shown in Fig. 3 as an isothermal conductivity plot at 500°C. From this figure, it is interesting to note that there are two peaks in the conductivity of these materials. These occur at Gd contents of roughly

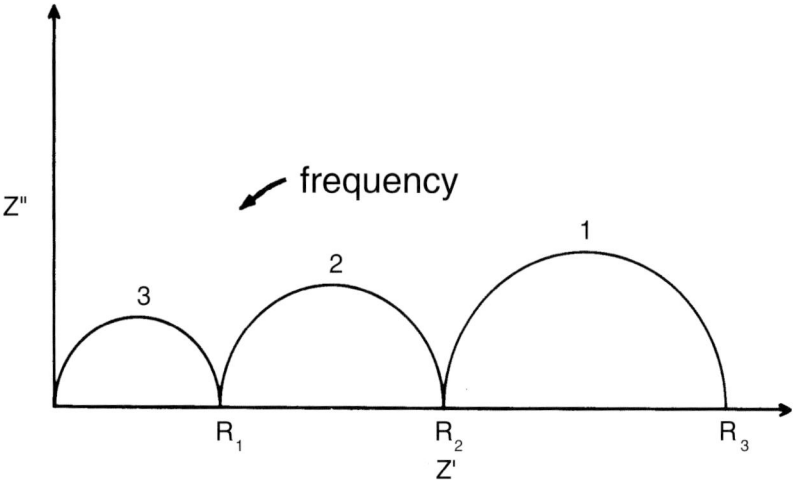

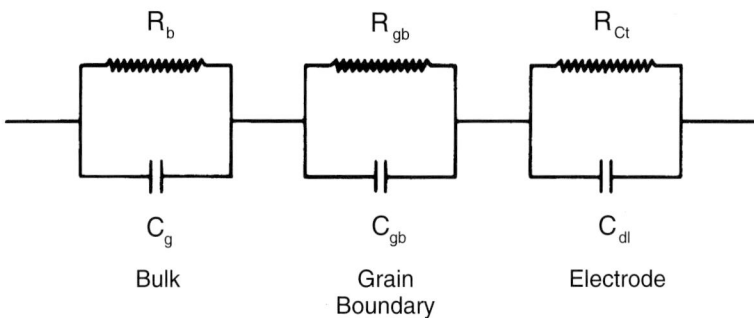

Fig. 1 Impedance spectrum and equivalent circuit for a hypothetical ionic conductor. $R_1 \equiv R_b,\ R_2 \equiv R_1 + R_{gb},\ R_3 \equiv R_2 + R_{ct}$.

0.1 for the grain interior component and 0.25 for the grain boundary component. The total conductivity is also shown in the figure and this is obtained from the sum of the two component *resistivities*. For the lower Gd contents, the total conductivity is dominated by the low grain boundary component, however this rises sharply leading to an increasing total conductivity. Past a Gd content of about 0.25, both the grain interior and the grain boundary components begin to decrease, leading to a peak in the total conductivity at around this point. This is an interesting result because it indicates that although the maximum in the grain interior (bulk) conductivity occurs at

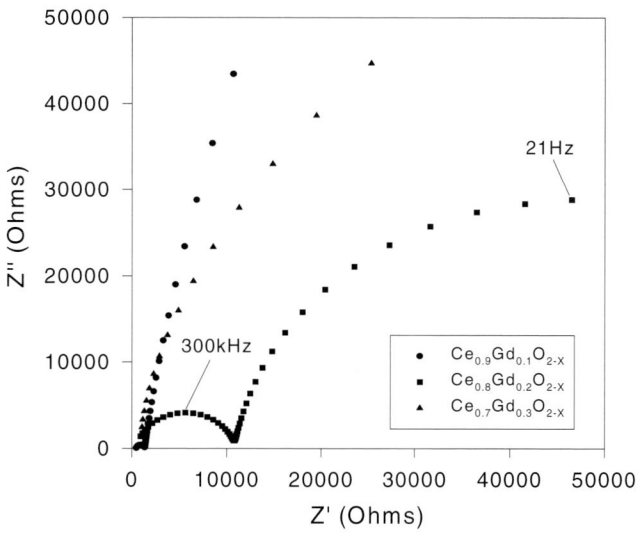

Fig. 2 Impedance spectra for several CGO samples, measured in air at 250°C.[4]

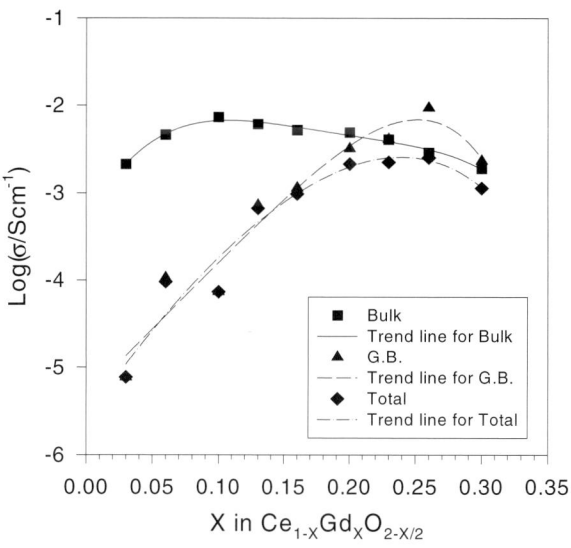

Fig. 3 Grain interior (bulk), grain boundary (GB) and total conductivity of CGO solid solutions obtained by ac impedance spectroscopy at 500°C.[4]

much lower Gd content, the more practical material in terms of the dc (total) conductivity will be of compositions close to 0.2–0.25 Gd, exactly the composition range favoured by most fuel cell constructors. Care must be taken in any comparison of the extracted grain boundary components because the resistivity value obtained from the grain boundary arc is sample dependent, and would need to be normalised to grain size to determine the specific grain boundary conductivity. However, if the grain size does not change between the various samples, as is the case here, then the comparison of grain boundary values is valid.

The cause of the maxima observed in the conductivity curves is complex and only that for the grain interior component is even partially understood. The grain interior conductivity peaks at low additions of Gd because of defect interactions between the added Gd and the resultant oxygen vacancies. As yet, the exact nature of the interaction is obscure but the outcome of the process is a minimum in the activation enthalpy for grain interior conduction causing a maximum in the conductivity.[2] The grain boundary component is even less well understood. Ralph,[5] investigated the boundary composition by TEM and surface composition by Low Energy Ion Scattering (LEIS) and found that the chemistry of the surfaces and boundaries was complex. Both his investigations, and the LEIS of Scanlon *et al.*[6] have revealed that the Gd is enriched at surfaces (and presumably to grain boundaries). Ralph and Kilner[4] speculate that the Gd enrichment of the grain boundaries has some interplay with the residual silica in the material to produce the more conductive grain boundaries found at the high Gd contents. As the silica content of each of his materials is identical, because each sample was produced by an identical route, this implies that the Gd content of the sample affects the silica distribution in the grain boundary region. Other workers have shown that by a reduction of the silica content of the initial CGO powders to very low levels, the total conductivity of the 0.1Gd content ceramic can be considerably improved. In fact this composition then shows a superior total conductivity to the 0.2Gd material, as would be expected if the bulk behaviour were to be dominant.[3]

From this work and earlier studies on zirconia-based materials, it would appear that grain boundary transport in these fluorite-based ceramics is a slower process than in the grain interior, and the boundaries can have a blocking effect to the passage of high fluxes of oxygen ions. This is most likely due to two effects. Firstly, the bulk structure of these materials is ideally suited to the migration of the anions. The high symmetry of these cubic structures provides a cationic framework that is open enough to allow the large oxygen ions to move with little hindrance. In the grain boundary regions, the situation is much less favourable. This has been confirmed by recent molecular dynamic simulations by Fisher and Matsubara,[7, 8] who have shown that oxygen ion transport in the fluorite structured yttria stabilised zirconia is faster in the bulk than in the grain boundaries. They found that for the $\Sigma5(310)/[001]$ and $\Sigma13(320)/[001]$ tilt grain boundaries the oxygen mobility was much higher in the bulk than in the grain boundary region, only in the more unusual $\Sigma5(111)60°$ twist boundary was rapid diffusion observed in the boundary region. These results essentially confirm earlier simulations by Bingham *et al.*[9]

286

The second important aspect is that of the segregation of impurities to the grain boundaries. It has been known for some time that the presence of silica in these ceramic materials has a deleterious effect on the total conductivity, owing to the tendency of the silica to segregate at the grain boundaries forming blocking phases.[10] Interestingly, recent work on nano-crystalline materials has shown that when low silica content materials are investigated the specific grain boundary contribution is much lower than that for microcrystalline materials. This is presumably resulting from the fact that the grain boundary area is significantly increased for the nano-crystalline materials and the silica is not able to spread to all the grain boundaries giving a large number of clean grain boundaries displaying intrinsic properties.[11, 12] Clearly, the grain boundary contribution can be the dominant component of the total conductivity of oxides, such as CGO, and care must be taken in the optimisation of microstructure and purity to achieve high performance materials for practical applications.

OXYGEN ISOTOPIC EXCHANGE

For electronic and mixed conducting materials, ac impedance spectroscopy is not the method of choice. The presence of the highly mobile electrons means that the electronic conductivity is orders of magnitude greater that that of the ionic. Electrical methods are thus reduced to indirect measurements of the rate of oxygen diffusion that rely upon the implied changes in electrical properties with oxygen composition. A further complication is that the range of diffusivity in these materials is enormous with some compositions clearly qualifying as fast oxygen conductors and others where the diffusivity is very low. Activation enthalpies for the oxygen transport process can also be large, meaning large changes in diffusivity with temperature.

One unambiguous method to obtain the diffusivity is to measure the kinetics of oxygen exchange with a gas containing $^{18}O_2$. This can be done by either monitoring the isotopic concentration of the exchanging gas phase, or by the depth profiling of an isotopically exchanged sample, to determine the diffusion profile of the stable isotope. This later method, sometimes called the Isotope Exchange Depth Profile (IEDP) method, has been well documented over recent years.[13, 14] One of the commonest ways of determining the isotopic diffusion profile is to use Secondary Ion Mass Spectrometry (SIMS) as the depth profiling technique. The ceramic samples can be either sputter depth profiled, or sectioned then transversely line scanned, depending upon the depth of penetration of the isotope. Both the techniques yield the same result, i.e. a depth profile in the material of interest. The method will not be described in full but some recent results pertaining to the presence of grain boundaries will be discussed.

During the ^{18}O isotope anneal, the net isotope flux crossing the gas/solid interface is directly proportional to the difference in isotope fraction between the gas and the solid. This flux is equal to the ^{18}O flux diffusing away from the surface into the solid. This leads to the following boundary condition

$$D_b^* \left. \frac{\partial C}{\partial x} \right|_{x=0} = k_s^* (C_g - C_s) \tag{1}$$

where D_b^* is the bulk oxygen tracer diffusion coefficient, k_s^* is the oxygen tracer surface exchange coefficient, and C_g and C_s refer to the ^{18}O isotope fraction of the gas and at the surface respectively. The solution for a semi-infinite medium with the above boundary condition has been given by Crank[15]

$$C'(x,t) = \frac{C(x,t) - C_{bg}}{C_g - C_{bg}} = \text{erfc} \left[\frac{x}{2\sqrt{D_b^* t}} \right] -$$

$$\left[\exp \left(\frac{k_s^* x}{D_b^*} + \frac{k_s^{*2} t}{D_b^*} \right) \times \text{erfc} \left(\frac{x}{2\sqrt{D_b^* t}} + k_s^* \sqrt{\frac{t}{D_b^*}} \right) \right] \tag{2}$$

Where $C(x,t)$ is the ^{18}O isotope fraction in the solid obtained by SIMS, C_{bg} ($= 0.2 \%$) is the natural isotopic background level of ^{18}O, C_g ($= 97\%$) is the isotope enrichment of the gas and t is the time of the isotope anneal.

Much recent work using this technique has centred on electronic/mixed conducting materials for SOFC's. In particular the perovskites with the general formula $La_{1-x}Sr_x(B)O_{3\pm\delta}$, where B is Fe, Co, Mn or Cr, have received considerable attention. For example for the high temperature zirconia based SOFC, $La_{1-x}Ca_xCrO_3$ is used as an interconnect material, and $La_{1-x}Sr_xMnO_{3\pm\delta}$ (LSM) is used as a cathode material. For lower temperature cells, operating with a CGO electrolyte, the doubly substituted $La_{1-x}Sr_xCo_{1-y}Fe_yO_{3-\delta}$ (LSCF) is used as a cathode material. These materials are not particularly fast oxygen diffusers but are closely related to materials that are. One feature of the materials is that they tend to show the effect of grain boundaries when the bulk diffusivity is low (at low temperatures) and this manifests as fast (short circuit) diffusion paths, rather than the blocking behaviour shown by the electrolyte materials discussed above.

Again, a specific material will be taken as an example. Much work has centred upon LSM because an understanding of the oxygen diffusion mechanisms in this material is important for modelling the behaviour of SOFC cathodes. There have been four recent studies of LSM based materials[16-19] using the IEDP technique. All these studies show the effect of grain boundaries on the diffusion of oxygen through this material. Figure 4 shows an oxygen-18 depth profile for $La_{0.9}MnO_{3\pm\delta}$ obtained after annealing samples at a temperature of 801°C. Included in this figure is a fitted line showing the results of a non-linear least squares fit to the experimental data using the solution to the diffusion equation appropriate for a semi-infinite medium (equation (2)) with a surface exchange process.[13] It is apparent that the experimental data deviates from that expected theoretically at large depths, in a manner usually described as a 'tail'. This phenomenon is widely ascribed to the effect of grain boundaries. Unfortunately, the analysis of such depth profiles is not simple and there

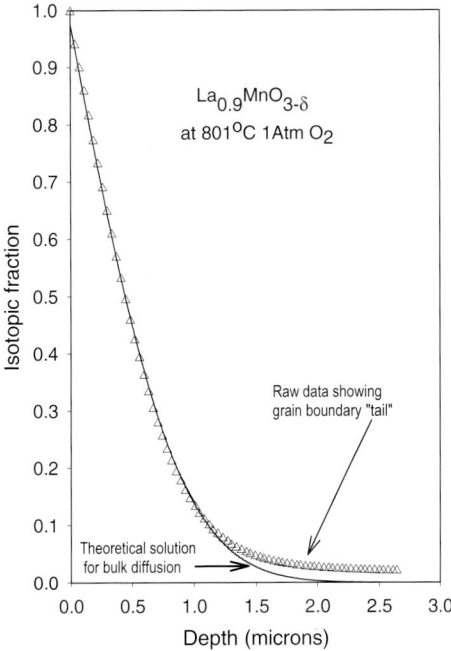

Fig. 4 Depth profile of $La_{0.9}MnO_{3\pm\delta}$ annealed at 801°C at 1 atm.[19] Solid line shows a non-linear least squares fit to the diffusion equation assuming only bulk oxygen diffusion (equation (2)).

is no entirely satisfactory method of obtaining values of the grain boundary diffusivity.

An early analysis of the problem was undertaken by Whipple.[20] He treated mathematically the case of a bi-crystal with the grain boundary represented as an isotropic slab of fast diffusing material of diffusivity D^*_{gb} and width δ normal to the diffusing surface. This gave an analytical solution to the problem, but in integral form, inapplicable to the treatment of experimental data. Le Claire[21] and Chung and Weunsch[22] have both addressed the problem of how to extract numerical data from these grain boundary affected profiles. Both have used a numerical integration to obtain relationships from which the product $D^*_{gb}\,\delta$ can be obtained.

Equation (2) then becomes

$$C'(x,t) = \frac{C(x,t) - C_{bg}}{C_g - C_{bg}} = \text{erfc}\left[\frac{x}{2\sqrt{D^*_b t}}\right] -$$

$$\left[\exp\left(\frac{k^*_s x}{D^*_b} + \frac{k^{*2}_s t}{D^*_b}\right) \times \text{erfc}\left(\frac{x}{2\sqrt{D^*_b t}} + k^*_s\sqrt{\frac{t}{D^*_b}}\right)\right] + A_{gb}\exp\left(-Z_{gb}x^{6/5}\right) \quad (3)$$

where A_{gb} and Z_{gb} are the grain boundary tailing function parameters. The grain boundary diffusion product is usually obtained from the Le Claire relation

$$D^*_{gb}\, \delta = 0.66\left(\frac{D^*_b}{t}\right)^{1/2} Z_{gb}^{-5/3} \tag{4}$$

It is clear from an inspection of these relationships that a plot of Log (C') versus $x^{6/5}$ will be a straight line in the grain boundary tail region. Figure 5 shows the data from Fig. 4 on such a plot, and the deviation of the experimental data from the ideal bulk solution is obvious. Also included in Figure 5 is the result of using equation (3) to fit to the data with the grain boundary tail. Recently, Chung and Wuensch[22] have shown that the use of the Le Claire relation in analysing SIMS data can lead to large errors in the grain boundary diffusion product obtained. Consequently, they derived a highly accurate set of relations, also based on numerical integration of Whipple's exact solution, specifically for the analysis of this type of high-resolution data. The set of relations takes the form

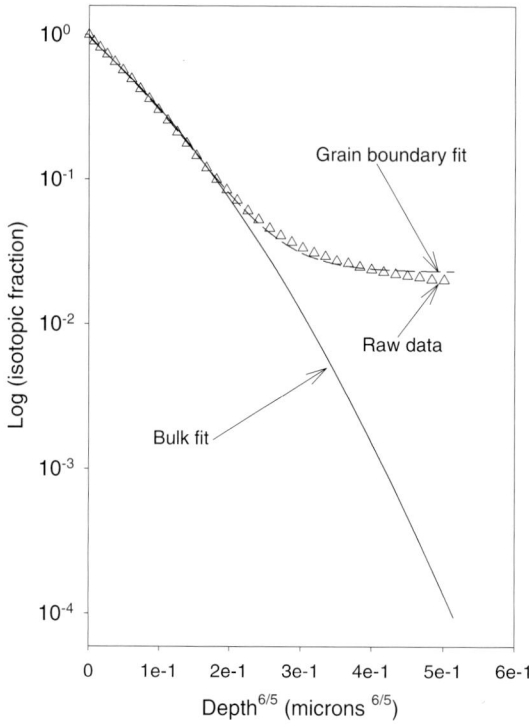

Fig. 5 Data from Fig. 4 re-plotted on a semi-log plot to enhance the tail region of the depth profile, showing the fits to equations (2) and (3).

$$D^*_{gb}\,\delta = 2D^{*\,3/2}_{b}\,t^{1/2}\left[10^A\left(-\frac{\partial\,\ln C'(x,t)}{\partial\,\eta^{6/5}}\right)^B\right] \tag{5}$$

where

$$\eta = \frac{x}{\sqrt{D^*_b t}} \tag{6}$$

and A and B are parameters whose values depend on the experimental value of the tail slope, $\partial\,\ln C'(x,t)\,/\,\partial\,\eta^{6/5}$, between $6 \leqslant \eta \leqslant 10$. Only data within the specified depth range, $6 \leqslant \eta \leqslant 10$, is used to determine the grain boundary diffusion product, $D^*_{gb}\,\delta$, from this relationship (C-W method). Note that these relations, being based on Whipple's solution, have only been established for the constant surface concentration boundary condition, but it has been assumed that they apply to the case of the surface reaction boundary condition, equation (1). Figure 6 shows a plot of this type of analysis used to obtain the grain boundary product for a $La_{0.8}Sr_{0.2}MnO_{3+\delta}$ sample at 800°C.[18]

A measure of the relative magnitudes of D^*_{gb} and D^*_b is given by the dimensionless parameter β

$$\beta = \frac{D^*_{gb}\,\delta}{D^*_b\,2\sqrt{D^*_b t}} \tag{7}$$

The physical interpretation of β is that preferential penetration down the boundary is much more evident for large β, that is large D^*_{gb} relative to D^*_b.

It must be highlighted that the solutions mentioned above are restricted in extent and have validity only under certain limited conditions. As noted by Le Claire,[21] the determination of grain boundary diffusivities is best carried out under conditions where β is large (>10), while the investigation of bulk diffusion is aided by small β. It must also be remembered that for a ceramic material the SIMS depth profiling averages over many grain boundaries and that the solutions only strictly apply to bi-crystals. However, Harrison[23] has stated that this limitation is removed if the following relationship holds

$$20\sqrt{D^*_b t} < d_{gr}$$

where d_{gr} is the average grain size.

As mentioned above, several studies have been performed on perovskites such as the LSM, $La_{1-x}Sr_xCo_{1-y}Fe_yO_{3-\delta}$ (LSCF) and the chromate materials. These materials are characterised by moderate to high activation enthalpies for the diffusion of bulk oxygen (>200 kJ mol^{-1}). At high temperatures, the bulk oxygen diffusivity is high and boundary diffusion is generally not observed. As the temperature is lowered, grain boundary effects of the type shown in Figs. 4–6 become more evident. A selection of the data available from the literature is given in Table 1 which lists the bulk diffusivity, the grain boundary product obtained using the C-W method, the value of

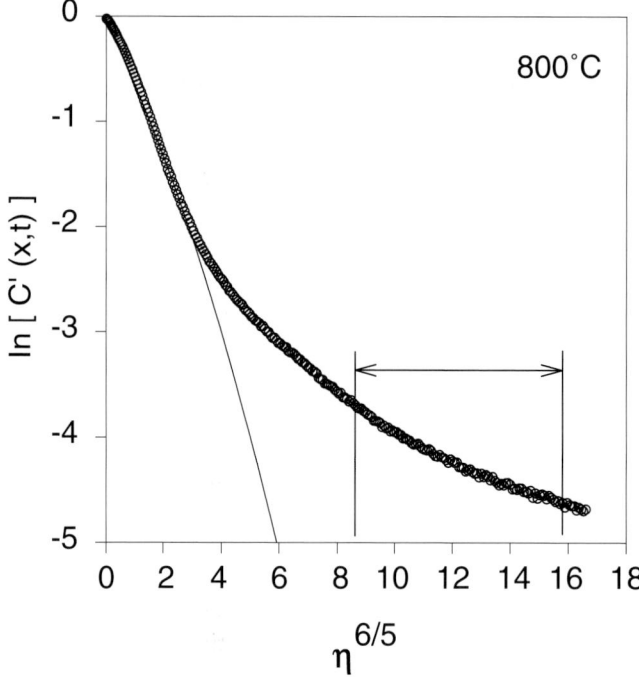

Fig. 6 ^{18}O penetration profile determined for $La_{0.8}Sr_{0.2}MnO_{3+\delta}$. The section of the curve used in the calculation of the grain boundary product using the C-W method is indicated.[18]

β and the grain boundary diffusivity, assuming a grain boundary thickness δ of 1 nm. Although extensive studies have not been performed, primarily because of the cost and difficulty of performing SIMS analyses on large numbers of samples, several general observations can be made from these studies: in the range of bulk (lattice) diffusivities $D_{b}^{*} > 10^{-11}$ cm^2 s^{-1}, grain boundary diffusion phenomena are not usually observed and in the range of bulk diffusivity $D_{b}^{*} < 10^{-11}$ cm^2 s^{-1}, grain boundary diffusion can provide short circuit diffusion paths which has an increasing effect on the diffusion mechanisms as the bulk diffusivity lowers.

This second finding may be of importance if the material is used for applications where oxygen diffusion has to be kept to a minimum, the exact reverse of the fast ion conductor. One such application is the bipolar plate in the high temperature SOFC. This is a piece of ceramic that provides electrical contact between successive cells in a stack, and therefore is exposed to the fuel and air environments on opposite faces. Any diffusion of oxygen through the bipolar plate to the fuel compartment gives rise to the unwanted oxidation of fuel, ie an effective leakage current. Measurements of the grain boundary diffusion in $La_{0.7}Ca_{0.3}CrO_3$ at 1000°C have determined an effective leakage current of 300 mA cm^{-2} owing to the presence of grain boundary short circuit diffusion paths.[26] Clearly, knowledge of these diffusional mechanisms is important for both academic and technological reasons.

One final point should be made. The solution to the diffusion equation used in all the cases mentioned above involves the product $D^*_{gb}\delta$ of the grain boundary diffusivity and the grain boundary width. For many of the materials investigated δ is unknown and a 'typical' value of 1 nm is used to extract values of D^*_{gb}. This generally leads to values of D^*_{gb} several orders of magnitude larger than the bulk components. This is usually used to justify the argument of boundary dominance in the diffusion mechanism, however it must be noted that δ may be temperature dependent leading to errors in the characterisation of the true grain boundary component.

CONCLUSIONS

Grain boundary phenomena are of considerable importance in the study of fast ionic conductors and other component materials in high temperature electrochemical devices. For the fast ionically conducting materials, used as electrolytes, ac impedance studies have shown that the grain boundaries have a blocking effect. This is thought to be due to their intrinsic nature, plus the tendency for impurities and dopants to segregate to these regions. For the mixed conducting materials, where the range of diffusivites is much larger, the main effect of the boundaries is short circuit diffusion occurring when the bulk diffusivity is low.

Table 1 Diffusion parameters for several perovskite mixed conductors

Temperature, °C	D^*_b, cm^2 s^{-1}	D^*_{gb}, cm^3 s^{-1}	D^*_{gb}, cm^2 s^{-1}	β
La$_{0.6}$Sr$_{0.4}$Fe$_{0.8}$Co$_{0.2}$O$_{3-\delta}$ (Ref. 24)				
388	7×10^{-14}	2×10^{-16}	2×10^{-9}	98
430	7×10^{-13}	4×10^{-15}	4×10^{-8}	115
502	4×10^{-12}	9×10^{-15}	9×10^{-8}	25
La$_{0.8}$Sr$_{0.2}$MnO$_{3+\delta}$ (Ref. 18)				
1000	6.5×10^{-13}	5.8×10^{-17}	5.8×10^{-10}	1.5
900	1.6×10^{-13}	6.2×10^{-17}	6.2×10^{-10}	3.7
800	$.4 \times 10^{-15}$	3.4×10^{-18}	3.4×10^{-11}	22.3
700	3.1×10^{-16}	7.8×10^{-20}	7.8×10^{-13}	24.3
La$_{0.9}$MnO$_{3+\delta}$ (Ref. 19)				
706	7.3×10^{-16}	1.1×10^{-20}	1.1×10^{-13}	1.7
801	1.3×10^{-14}	5.5×10^{-18}	5.5×10^{-11}	4.4
897	2.0×10^{-13}	2.9×10^{-17}	2.9×10^{-10}	1.1
La$_{0.7}$Ca$_{0.35}$CrO$_{3-\delta}$ (Ref. 25)†				
900	5.0×10^{-13}	3.4×10^{-15}	3.4×10^{-8}	270
1000	5.9×10^{-13}	1.4×10^{-14}	1.4×10^{-7}	–
1100	9.1×10^{-11}	6.0×10^{-14}	6.0×10^{-7}	–

† Data obtained using the LeClaire method.

REFERENCES

1. J. A. Kilner and B. C. H. Steele: *Nonstoichiometric Oxides*, O. T. Sorensen ed., Academic Press, New York, 1981, p. 233.
2. J. A. Kilner: *Solid State Ionics*, 2000, **129**, 13.
3. B. C. H. Steele: *Solid State Ionics*, 2000, **129**, 95.
4. J. M. Ralph and J. A. Kilner: Second European Solid Oxide Fuel Cell Forum, B. Thorstensen ed., 1996, p. 773.
5. J. M. Ralph: PhD, University of London, 1998.
6. P. J. Scanlon, R. A. M. Brink, F. P. F. van Berkel, G. M. Christie, L. J. van Ijzerdoorn, H. H. Brongersma and R. G. van Welznis: *Sold State Ionics*, 1998, **112**, 123.
7. C. A. J. Fisher and H. Matsubara: *Solid State Ionics*, 1998, **113–115**, 311.
8. C. A. J. Fisher and H. Matsubara: *J. Eur. Ceram. Soc.*, 1999, **19**, 703.
9. D. Bingham, P. W. Tasker and A. N. Cormack: *Phil. Mag.*, 1989, **A60**, 1.
10. J. E. Bauerle: *J. Phys. Chem. Solids*, 1969, **30**, 2657.
11. P. Mondal, A. Klein, W. Jaegermann and H. Hahn: *Solid State Ionics*, 1999, **118**, 331.
12. M. Aoki, Y.-M. Chiang, I. Kosacki, L. J.-R. Lee, H.-L. Tuller and Y. Liu: *J. Am. Ceram. Soc.*, 1996, **79**, 1169.
13. J. A. Kilner and R. A. De Souza: *Proc. 17th Int. Risø symposium on Materials Science*, F. W. Poulsen, *et al.* eds, Risø National Laboratory, Roskilde, Denmark, 1996, p. 41.
14. J. A. Kilner, B. C. H. Steele and L. Ilkov: *Solid State Ionics*, 1984, **12**, 89.
15. J. Crank: *The Mathematics of Diffusion*, Oxford University Press, Oxford, 1975.
16. I. Yasuda, K. Ogasawara, M. Hishinuma, T. Kawada, and M. Dokiya: *Solid State Ionics*, 1996, **86–88**, 1197.
17. R. A. De Souza and J. A. Kilner: *Solid State Ionics*, 1998, **106**, 175.
18. R. A. De Souza, J. A. Kilner and J. F. Walker: *Mater. Lett.*, 2000, **43**, 43.
19. A. V. Berenov, J. L. Macmanus-Driscoll and J. A. Kilner: *Solid State Ionics*, 1999, **122**, 41.
20. R. T. P. Whipple: *Philos. Mag.*, 1963, **45**, 1225.
21. A. D. Le Claire: *Br. J. Appl. Phys.*, 1963, **14**, 351.
22. C. Chung and B. J. Wuensch, *J. Appl. Phys.*, 1996, **79**, 8323.
23. L. G. Harrison: *Trans. Faraday Soc.*, 1961, **57**, 1191.
24. S. J. Benson: PhD thesis, University of London, 1999.
25. I. Yasuda and M. Hishinuma: in T. Ramananarayanan, W. L. Worrell, H. L. Tuller (Eds.), *Proc. 2nd Int. Symp. on Ionic and Mixed Conducting Ceramics*, T. Ramananarayanan *et al.* eds. Electrochemical Society, Pennington, NJ, (1994) p. 209.
26. T. Kawada, T. Horita, N. Sakai, H. Yokikawa and M. Dokiya: *Solid State Ionics*, 1995, **79**, 201.

Interfaces in PZT Thin Films

I. M. REANEY AND P. WANG

Department of Engineering Materials, Sir Robert Hadfield Building, Mappin St., University of Sheffield, Sheffield, S1 3JD, UK

ABSTRACT

This article briefly reviews some of the key features associated with the interface structure of ferroelectric PZT thin films deposited onto platinised Si/SiO_2 electrodes. High resolution and analytical electron microscopy have been used to compare the Pt/PZT interfaces in optimised films deposited using sol–gel spinning, *in situ* reactive sputtering and metallorganic chemical vapour deposition. PZT/PZT interfaces in multilayer sol–gel films have also been investigated.

INTRODUCTION

Ferroelectric materials have many applications in the electronics industry arising from various electrical properties. Their high piezoelectric and pyroelectric coefficients are used to manufacture sensors and actuators and their large dielectric constants, ε_r, are utilised to make capacitors with a high volumetric efficiency.[1] In addition, rapid switching of polarisation and the non-volatility associated with a FE make them suitable for the development of memories. The leading compound for most applications is the solid solution, $PbZrO_3$–$PbTiO_3$ (PZT).[2] For piezoelectric applications, PZT with a Zr:Ti ratio ~53:47 is often preferred. However, other applications may use different compositions within the solid solution range.

As always in the electronics industry, there is a driving force to produce smaller and cheaper devices, preferably integrated into Si technology. Ferroelectric materials are no exception and recently there has been considerable interest in the deposition of thin PZT films onto electroded Si substrates that can be micro-machined to form integrated devices, eg the PZT infra red imaging array developed by GEC Marconi and discussed by Shorrocks *et al.*[3]

FE memories generally only require films of around 200 nm whereas sensors and actuators need to be thicker (1–10 µm). There are many deposition methods that are used to generate PZT films. However, three main routes are commercially viable: sol–gel spin coating, sputtering and metallorganic chemical vapour deposition (MOCVD). The latter two are techniques in which the perovskite crystal phase is grown *in situ* on the substrate inside a vacuum chamber. Sol–gel spinning coating however, is an *ex situ* technique. Briefly, an amorphous layer is spun down on to the substrate under ambient conditions and is subsequently crystallised typically at 600°C in air.[4, 5]

The development of reproducibly high quality films using any of the techniques mentioned above requires an understanding of the relationship between deposition

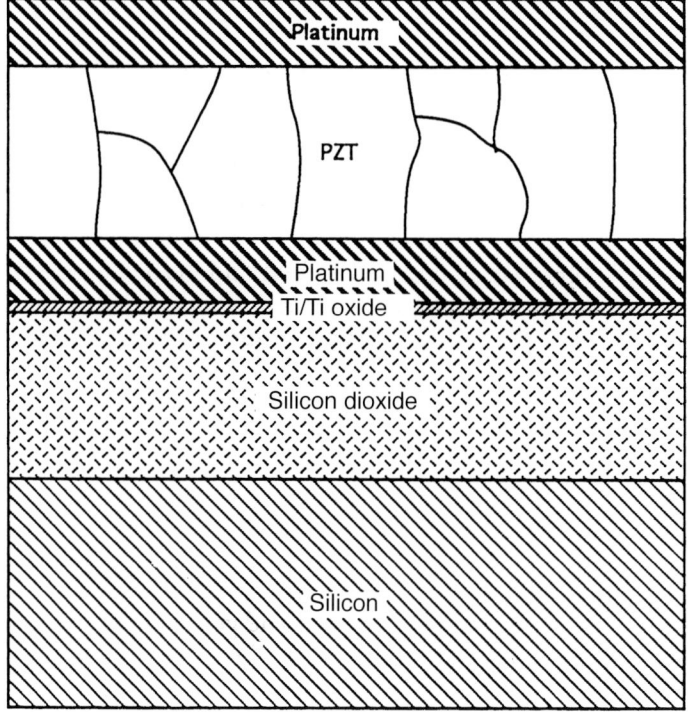

Fig. 1 Schematic showing the microstructure of a typical Si/SiO$_2$/Pt/PZT/Pt 'FEstack'.

parameters and the resulting microstructure. The term microstructure in this case refers to the 'FEstack', typically, Si/SiO$_2$/Pt/PZT/Pt, and not only the film (Fig. 1). The interfaces and film need to be of the highest quality if the required properties are to be routinely achieved.

High-resolution transmission electron microscopy (HRTEM) is an ideal tool to assess the interface structure in integrated devices and has been extensively used to study and characterise more conventional semiconductor systems. This paper will review the interface structure observed in PZT FEstacks fabricated by sol–gel spin coating, MOCVD and *in situ* reactive sputtering. Some new results from sol–gel multilayer deposition techniques will also be discussed.

EXPERIMENTAL PROCEDURE

The deposition methods associated with sol–gel spin coating,[5] *in situ* reactive sputtering[6] and MOCVD[7] have been dealt with elsewhere. Here the authors will concentrate on those experimental procedures associated with TEM.

Four pieces (3 × 3 mm) were initially fractured from the sampled and glued together using Epoxy Resin with the PZT films pointing inward. The glued sample was mounted with the film perpendicular to the surface of a Gatan Disc Grinder stub using Lakeside Resin. The sample was ground flat on course and then polished using 1200 grade SiC paper. The sample was removed from the stub by heating the lakeside resin and remounted with the freshly ground surface in contact with stub. The thickness was reduced until the sample was ~20 μm and a copper support ring glued onto the specimen. The sample was removed from the stub by heating on a hot plate, allowing the lakeside resin to soften. The sample was cleaned in acetone.

A Gatan Duo Mill operating at 6 kV, 0.1 mA and an incident angle of 15° was used to thin most samples. However, some samples were thinned using a precision ion polishing unit at an angle of ~5°. The samples were examined using either a Philips EM430, JEOL 3010 or 2010 TEM. The latter two were equipped with a link energy dispersive X-ray detectors and appropriate spectrum processing software/hardware.

RESULTS AND DISCUSSION

SOL–GEL SPIN COATING (53/47)

Figure 2 is a bright field (BF) TEM image of the general microstructure of a PZT thin film deposited by sol–gel spinning and crystallised at 600°C (Ref. 8 gives further details of the deposition process). The bottom and top electrode are marked and the PZT thin film is composed of (111) columnar grains perpendicular to the surface of the Pt, approximately 100–200 nm wide. The general crystalline quality is good and no second phase is apparent in the bulk of the film. At the top electrode/film interface however, there is some evidence of presence

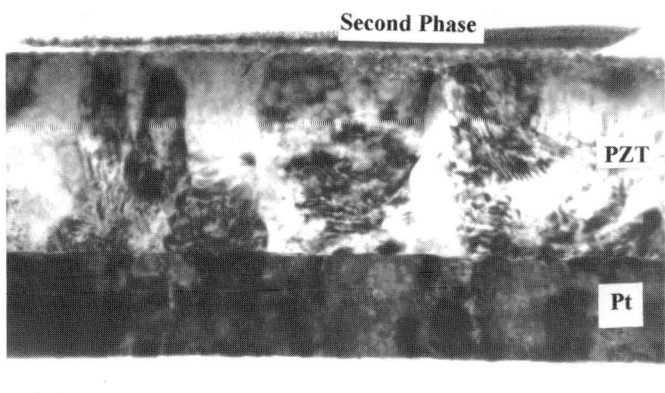

Fig. 2 Bright field (BF) TEM image of the general microstructure of a PZT thin film deposited by sol–gel spinning.

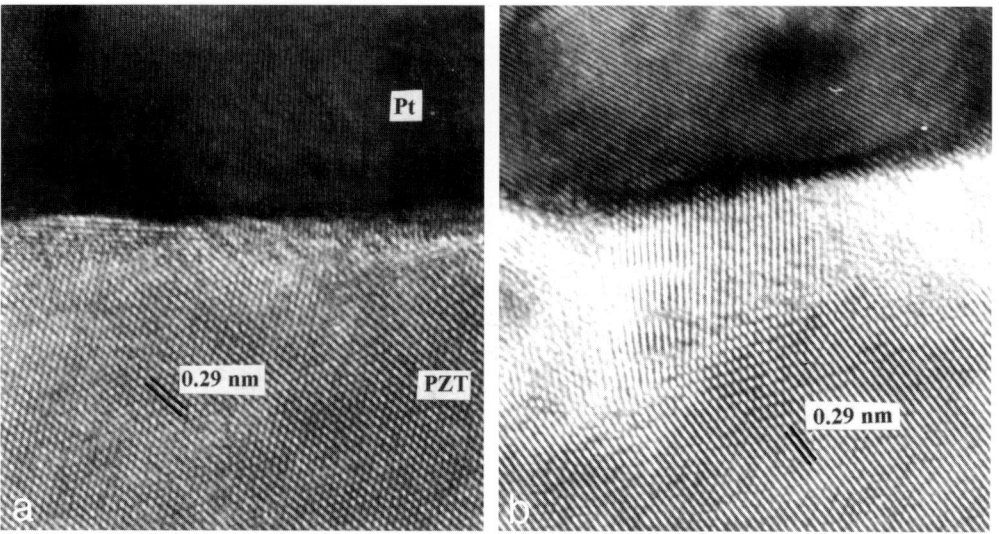

Fig. 3 High-resolution electron micrographs of the top electrode film interface showing (a) a region in which there is direct contact of the Pt to the PZT and (b) a similar area except the Pt/PZt interface contains second phase.

of a discontinuous second phase. Figure 3 shows high-resolution electron micrographs of the top electrode/film interface. Figure 3a shows a region in which there is direct contact of the Pt to the PZT. Figure 3b shows a similar area except the Pt/PZT interface contains some of the discontinuous second phase. The second phase is nanocrystalline and EDS analysis reveals the region to be Pb deficient and Zr rich compared with the bulk.[8] These data suggest that the interfacial layer is a pyrochlore/fluorite-structured phase.[8] The transformation mechanisms from amorphous to perovskite occurs via the formation of an intermediate pyrochlore/fluorite structure compound.[5] In general, the pyrochlore/fluorite is metastable and transforms readily to the perovskite phase. However, if, locally, insufficient Pb is present, the pyrochlore/fluorite can be stabilised in a Zr-rich, Pb deficient form.[5, 8] Commonly, this is found at the surface of a film usually near grain boundaries where Pb loss might be expected.[5] The top electrode is sputtered onto the film after crystallisation and consequently pyrochlore/fluorite crystals become trapped at the top electrode/film interface. The pyrochlore/fluorite has $\varepsilon_r < 100$ whereas the bulk has $\varepsilon_r \sim 1000$. The pyrochlore/fluorite may therefore act as a region where the field is concentrated and may initiate dielectric breakdown. Elimination of the pyrochlore phase is essential if reproducible properties are to be achieved. The use of a Pb, Ti rich final 'covercoat' has proved useful in improving the surface quality and preventing pyrochlore entrapment.[9]

SPUTTERED FILMS (40/60)

Figure 4 is a dark field (DF) image showing the general microstructure of a film deposited by *in situ* reactive sputtering. The films have a columnar grain structure perpendicular to the Pt electrode approximately 0.5 μm wide. The microstructure of films deposited by *in situ* reactive sputtering is discussed in more detail in Ref. 10. However, one of the major problems with sputtered films is controlling the orientation of the film so that the polar axis, eg (001) for tetragonal, is perpendicular to the substrate surface.[10] Pure $PbTiO_3$ generally adopts a (001) orientation when deposited, irrespective of the deposition technique and orientation of the substrate.[10] Figure 5 is a TEM image of a sample in which a thin layer of $PbTiO_3$ has been deposited prior to the deposition of the bulk of the film. The thin layer acts as a (001) oriented seed for the nucleation of (001) oriented 40/60 PZT.

MOCVD (50:50)

Figure 6 is a TEM image showing part of a cross-section through a PZT film deposited by MOCVD. The deposition parameters are discussed in more detail in Ref. 7. The film exhibits a high crystalline quality with no second phase visible either in the bulk

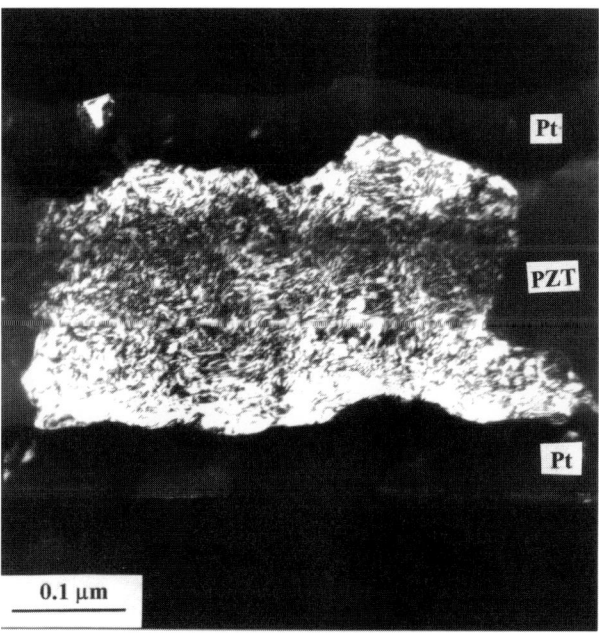

Fig. 4 Dark field (DF) image showing the general microstructure of a film deposited by *in situ* reactive sputtering.

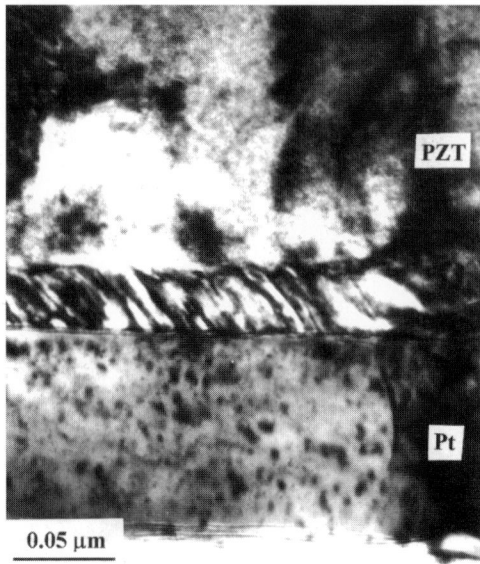

Fig. 5 BF TEM image of a sample in which a thin layer of PbTiO$_3$ has been deposited prior to the deposition of the bulk of the film.

Fig. 6 TEM image showing a part of a cross-section through a PZT (50:50) film deposited by MOCVD.

of the film or at the electrode/film interfaces. Figure 7 shows HRTEM images of the top electrode film interface before and after a top electrode post deposition anneal at 500°C in O_2 for 30 min.[7] In the non-annealed film, a thin amorphous layer can be observed at the film/top electrode interface whereas in the annealed sample the layer is absent. The small area of material typically examined by TEM casts some doubt as to whether this improvement in contact is consistent throughout the sample. However, the annealed films show an improvement in the squareness of the hysteresis loops but this may be related to changes in the stress state within the sample as the material is thermally cycled.[7]

Multilayer Sol–gel Thin Films (53/47)

Although sol–gel routes can be used to produce films of about 300 nm in thickness relatively easily, thicker films are difficult to achieve since cracks develop in the as deposited sol–gel on drying. One method of using sol–gel to deposit thicker films (1–10 μm) is to multilayer. In this deposition route, a film between 100–200 nm is deposited and crystallised into the perovskite state. The crystallised film then acts as a substrate for the deposition a further sol–gel layer which, in turn, is crystallised. In this manner the film thickness can be built up to the required level. Figure 8 is a TEM cross-section showing a multilayered film deposited using an acetic acid based route by the group of Dr S. J. Milne at the University of Leeds.[11] A dense film is observed but interface regions between the multilayers are still present. The interfacial regions

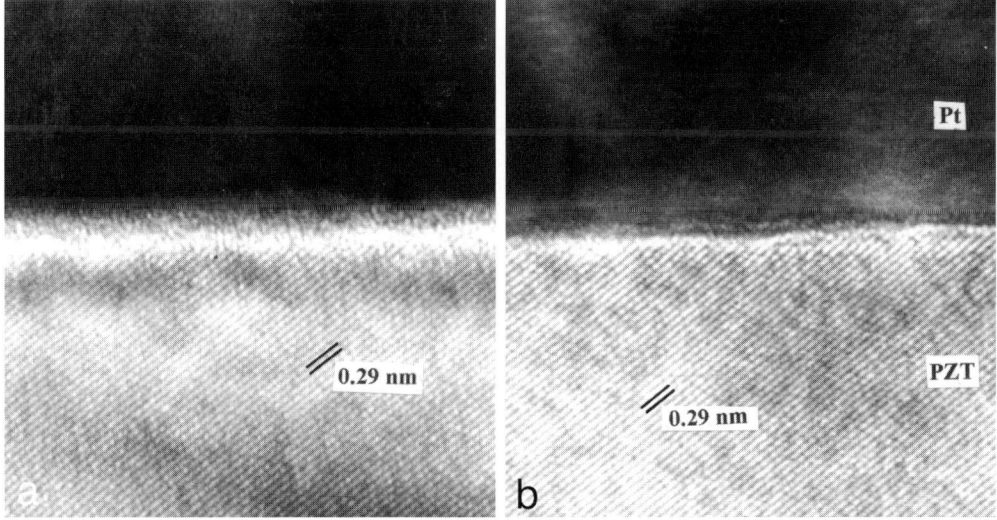

Fig. 7 HRTEM images of the top electrode/film interface (a) before and (b) after a top electrode post deposition anneal at 500°C in O_2 for 30 mins.[7]

Fig. 8 BF TEM cross-section showing a multilayer film deposited by an acetic acid based route by the group led by Dr S. J. Milne, Department of Materials, University of Leeds.

sometimes exhibit second phase and are often Ti deficient compared with the bulk. This observation is not surprising since in effect they are crystallised surfaces (thus Zr rich as discussed previously) upon which further deposition has occurred.

CONCLUSIONS

In order to interpret and obtain reproducible properties in PZT based thin films. A full understanding is required of the relationship between deposition parameters and the microstructure. In particular, interfaces in the FEstack need to be examined in great detail to determine whether detrimental layers and phases are present.

ACKNOWLEDGEMENTS

The authors would like to thank Professor N. Setter, Dr K. Brooks and Dr T. Maeder at the Ecole Polytechnique Federale de Lausanne (EPFL), Switzerland for the manufacture of sol–gel and sputtered films. The authors are indebted to Dr G. J. M. Dormans and M. De Keiser at Philips, Eindhoven, Holland for providing films deposited by MOCVD. Multilayer films were provided by Dr S. J. Milne and Dr D. Hind at the Department of Materials Science, University of Leeds, UK. Part of this work was performed at the Labo. Ceramique, EPFL and part at the Department of Engineering Materials, University of Sheffield. Dr Reaney would like to acknowledge the support of the Fond Nationale de Recherche in Switzerland and EPSRC grant GR/M33501.

REFERENCES

1. A. J. Moulson and J. M. Herbert: *Electroceramics*, Chapman and Hall, London, 1992.
2. B. Jaffe, W. R. Cooke and H. Jaffe: *Piezoelectric Ceramics*, Academic Press, New York, NY, 1971.
3. N. M. Shorrocks, A. Patel, M. J. Walker and A. D. Parsons: *Microelectron. Eng.*, 1994, **29**, 59–66.
4. K. D. Budd, S. K. Dey and D. A. Payne: *Br. Ceram. Proc.*, 1985, **36**, 107.
5. K. G. Brooks, I. M. Reaney, R. Klissurska, Y. Huang, L. Bursill and N. Setter: *J. Mater. Res.*, 1994, **9** (10), 2540–2553.
6. T. Maeder, P. Muralt, M. Kohli, A. Kholkin and N. Setter: *Ceramic Films and Coatings*, Institute of Materials, London, 1995.
7. G. J. M. Dormans, M. De Keiser, P. J. V. Veldhoven, D. M. Frigo, J. E. Holewijn, G. P. M. V. Mie and C. J. Smit: *Chem. Mater*, 1993, **5**, 448.
8. I. M. Reaney, K. Brooks, R. Klissurska, C. Pawlaczyk and N. Setter: *J. Am. Ceram. Soc.*, 1994, **77** (5), 1209–1216.
9. I. M. Reaney, D. V. Taylor and K. G. Brooks: *J. Sol–Gel Sci. Technol.*, 1998, **13**, 813–820.
10. I. M. Reaney, T. Maeder and P. Mural: *Ceramic Films and Coatings*, Institute of Materials, London, 1995, 219–229.
11. S. J. Milne: Unpublished results.

Grain Boundary Problems in Textured Superconducting Oxides

J. S. ABELL

School of Metallurgy and Materials, University of Birmingham, Edgbaston, Birmingham B15 2TT, UK

ABSTRACT

The significance of grain boundaries in determining the overall properties of the new superconducting oxides is discussed. Grain boundaries restrict current flow between grains and therefore require characterisation and subsequent control. The crystallographic nature of the boundaries can be studied by electron back scattered diffraction, the resistance to magnetic flux penetration can be imaged by magneto-optic techniques, and the variation in electrical properties can be revealed by voltage contrast. The combination of these mapping methods provides a powerful tool for determining the influence of grain boundaries on the macroscopic superconducting behaviour.

INTRODUCTION

The intrinsic superconducting properties of the high temperature superconducting oxides can only be effectively realised in single crystals and epitaxial thin films. Both these specimen geometries are necessarily limited to relatively small scale and two-dimensional configurations: single crystals adopt a platelet growth morphology but are difficult to grow greater than a few mm in extent, and, in order to grow epitaxially, thin films require suitable single crystal substrates, which in turn are limited in size. The properties of large scale 1–d, 2–d and 3–d artefacts, therefore, are mainly determined by extrinsic factors resulting from the particular processing route employed and the associated prevailing microstructure. Most resultant microstructures, therefore, contain grain boundaries which are, perhaps, the most dominant feature controlling the overall superconducting behaviour.

The obvious advantage of an increased critical temperature in the superconducting oxides is partially offset by the corresponding decrease in the coherence length, an intrinsic parameter which determines the spatial extent of superconducting order. The reduction in this order parameter makes the nature and width of grain boundaries and other defects of major significance in determining macroscopic superconducting behaviour. Supercurrents have great difficulty crossing wide, high angle boundaries, less so with narrow, low angle boundaries. These 'weak links' are also particularly susceptible to magnetic field penetration, which can also limit superconducting performance.

The marked anisotropy of these compounds means that appropriately textured microstructures exhibit better properties than random polycrystalline material. Most of these phases adopt a layered orthorhombic or tetragonal crystal structure with a large c/a,b ratio. Conductivity is significantly greater in the a–b plane than along the c-axis and so the first required crystallographic element of the texture is to ensure that the c-axis is perpendicular to the required current path. The second desired element is then some degree of in-plane alignment achieved by the reduction of rotation about the c-axis between grains to a minimum. This so-called bi-axial texture provides a microstructure with low misorientations across grain boundaries, thus providing minimal impedance to current flow.

The spatial distribution of textured grains and the misorientation between grains can be mapped by electron back scattered diffraction (EBSD), and the corresponding distribution of magnetic flux by magneto-optic (MO) observations. The spatial distribution of critical currents in thin film structures can be observed by electron beam induced voltage (EBIV) contrast in the scanning electron microscope. Using a combination of these mapping procedures, characterisation and consequent control of grain boundary formation in these complex materials can be achieved.

EXPERIMENTAL METHODS

ELECTRON BACK SCATTER DIFFRACTION

Localised texture measurement within individual grains of a macroscopically textured material has become possible through developments in electron back scatter diffraction in the SEM. Variations in texture of such microstructures – or microtextures – and misorientations across grain boundaries can be mapped by automated scanning during diffraction data collection using a process now known as crystal orientation mapping (COM). This technique, which offers grain size and distribution together with the local crystallography, is of particular significance for these materials since no chemical etchants are available to delineate grain boundaries.

The experimental arrangement showing the juxtaposition of specimen stage, beam and camera is given in Fig. 1. The electron beam impinges on a tilted sample in the SEM and a diffraction pattern is formed on a phosphor screen which is viewed by a low light camera. The pattern is stored and indexed by comparison with standard patterns as shown in the figure. The beam can then be moved or scanned over the rest of the sample. The beam can be scanned within a given field of view, or large area scans or composite mosaics across large specimens can be achieved via a motorised stage. The data can then be displayed in several different ways. Maps of the position and characteristics of the grain boundaries in the sample can be generated from the crystal orientation data. Colours can be interactively assigned to the boundaries according to the misorientation angle and superimposed on the secondary electron image. The grain boundary misorientation can be quantified and the orientation measurements made on the sample can be plotted on pole figures. The system used in the present work is an Oxford Instruments OPAL fitted to a JEOL 840 with LaB_6 filament.

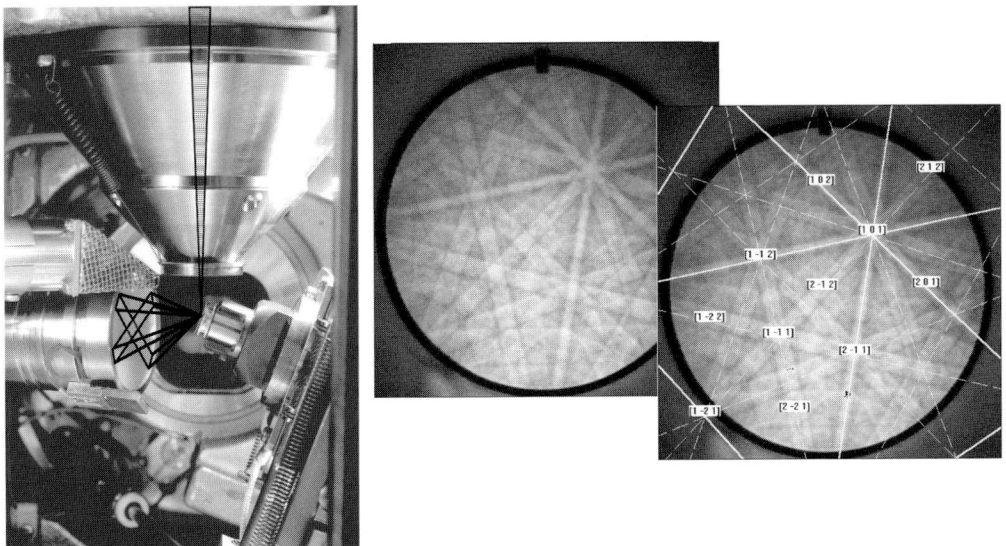

Fig. 1 Experimental arrangement for EBSD with example of diffraction pattern and computer fit.

MAGNETO-OPTIC IMAGING

One of the most important parameters in determining the quality of superconducting materials is their behaviour in the presence of a magnetic field. The magnetic properties of high-T_c materials are well documented, but techniques available to visualise flux distributions in superconducting samples are limited and are mostly indirect. Such methods include magnetic powder decoration, neutron scattering and Hall probe measurements. Magneto-optic (MO) imaging allows flux patterns to be directly visualised on a microscopic scale.

The rotation of the plane of polarisation of a transmitted beam of light in a Faraday active material in the presence of a magnetic field allows the mapping of flux density and distribution in a superconductor under current bias or applied magnetic field conditions. Thus, flux penetration characteristics and transport current flow can be investigated in a variety of sample configurations. The angle of rotation α is given by

$$\alpha = VlH \tag{1}$$

where V is a constant for a particular MO film, l its thickness and H the applied field.

A schematic of the basic arrangement is shown in Fig. 2. White light (A), plane polarised by a polarising filter (B), enters the magneto-optically active layer (MOL). This consists of an epitaxial iron garnet thin film (F), grown on a transparent substrate of $Gd_3Ga_5O_{12}$ (C) by liquid phase epitaxy. A thin reflective layer (D) of Ti/Al deposited on the film doubles the rotation angle by increasing the distance l in Eqn. 1 to $2l$. The whole layer is mounted on the sample (E) to be studied. The finite dis-

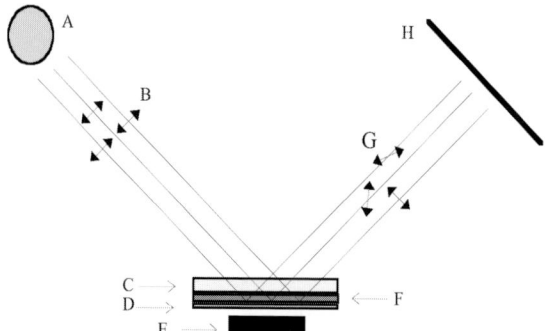

Fig. 2 Schematic of the magneto-optic system. The labelled components are described in the text.

tance between the sample and MOL due to surface roughness will limit the spatial resolution observed. Any magnetic field present in the sample and MOL will cause the plane polarised light to be rotated (G), which, when passed through the analyser filter set crossed with respect to the polariser, will be detected by an imaging device (H). With no magnetic field present, light will not undergo rotation and will, therefore, be blocked by the analyser. In this case any regions in the MOL where no field is present will be imaged as dark areas, whereas bright areas will be caused by regions with field present. The higher the flux, the greater the rotation and the brighter the contrast.

The iron garnet MO layer was secured on the polished surface of the sample within an Oxford Instruments continuous flow liquid helium optical cryostat, mounted on a Nikon polarising microscope. An external magnetic field normal to the plane of the MO layer and, therefore, parallel to the c-axis of the sample was supplied by a copper-wound, water-cooled electromagnet. A maximum field of 180 mT was generated at the sample surface.

ELECTRON BEAM INDUCED VOLTAGE CONTRAST

Electron Beam Induced Voltage (EBIV) contrast allows the spatial distribution of critical current density in thin film superconducting structures to be imaged with micron resolution. The structure is held close to its superconducting/normal transition on a low temperature stage in an SEM. With a dc bias current applied, the electron beam acts as a local source of heat, which can be sufficient to drive that region into the normal regime with a finite resistance, thus generating a local voltage which is measured. As the beam is scanned across the structure contrast is developed between superconducting and resistive regions so that variations in critical properties can be mapped spatially. If the structure is completely homogeneous voltage should be detected in the whole sample as either the temperature or bias current is increased slightly, but if local areas are 'weaker' than others, voltage will be generated preferentially in those regions. With appropriate circuitry and a frame store a voltage contrast map can be obtained synchronously with the conventional secondary electron image. A schematic of the experimental arrangement is shown in Fig. 3.

EXAMPLES OF GRAIN BOUNDARY IMAGING

The marked anisotropy in the electrical and magnetic properties of the cuprate superconductors provides the determining factor in the choice of processing route for superconducting devices and components for particular applications. The aim in most cases is to induce as much crystallographic alignment into the material compatible with the shape and size of the final product. Sample configurations range from single crystals and epitaxial layers, through wires and tapes, to bulk monolithic geometries. In all cases the degree of alignment or texture is critical in determining the critical current and magnetic behaviour. In textured material, a knowledge of the orientation distribution of the grains and the associated misorientation angle at grain boundaries is essential to the processing – microstructure – property relationships in these complex materials.

Examples of grain boundary observations on a variety of different sample types of the two most widely studied cuprates, $YBa_2Cu_3O_7$ (YBCO) and $Bi_2Sr_2Ca_2Cu_3O_8$ (BSCCO) are given below.

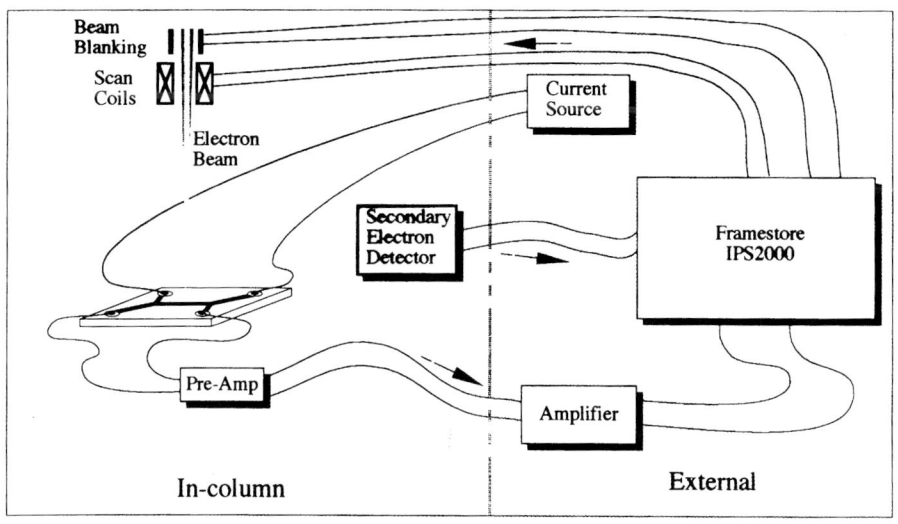

Fig. 3 Schematic of the EBIV system.

TEXTURED YBCO THICK FILMS

Textured thick films of YBCO can be prepared by melt processing (a programmed thermal cycle which includes exceeding the peritectic temperature of 1025°C) a screen printed coating on polycrystalline YSZ.[1] In this case, the predominantly (001) texture derives from the preferred spherulitic solidification morphology of the YBCO phase. Local deviations from the desired microtexture can be identified by COM, as shown in Fig. 4a. Components of a particular spherulite have misorientations <8° between them, but larger deviations occur between individual spherulites, as indicated by the three groups of orientations in Fig. 4b.[2]

EPITAXIAL YBCO THIN FILMS

The growth of epitaxial single layers of YBCO on a variety of oxide substrates is now well established by PVD techniques such as evaporation and pulsed laser deposition (PLD). However, the fabrication of multi-layer structures for particular device configurations can present problems of maintaining epitaxy through the layers, particularly if the initial layers are patterned and growth of subsequent layers has to take place over steps in the layer underneath. The presence of low angle grain boundaries or other defects in the film at such steps can degrade the superconducting properties and seriously effect the efficiency of performance of the final device.

The spatial distribution of critical currents in thin film structures and devices can be

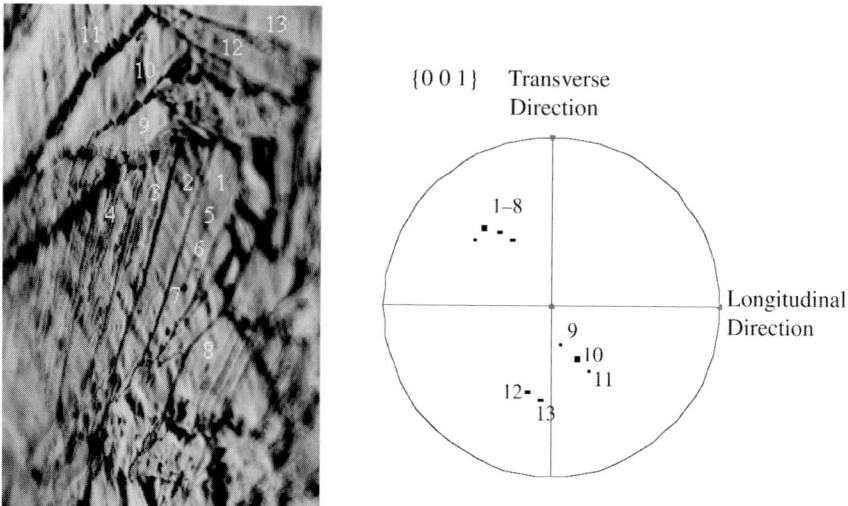

Fig. 4 Micrograph of locations on a YBCO thick film spherulite (a) and the corresponding orientations plotted on a pole figure (b).

investigated by means of electron-beam induced voltage contrast in the SEM, as described above.[3] A comparison of secondary electron and voltage contrast images of a superconducting thin film microbridge is shown in Fig. 5. In this case, the thin film from which the microbridge has been patterned has been grown over steps produced artificially in the underlying substrate to study the effect of growth on surfaces inclined to the lattice-matched crystallographic single crystal surface (100 in this case).[4] An SEM image of the bridge is shown in Fig. 5a and the corresponding EBIV image in 5b. Light contrast in the EBIV image corresponds to positive voltage showing that the microbridge is becoming preferentially normal at each of the steps across which the track passes. The critical current densities of that part of the film which has grown over the steps are inferior to the rest of the track. Preferential magnetic flux invasion at the steps is also observed by MO imaging as seen in Fig. 5c, with total flux penetration at higher field in Fig. 5d.

TEM of the cross-section of such a sample (Fig. 6a) reveals the presence of low angle boundaries at the angles of the step in the substrate. The change of orientation at the step cannot be accommodated by the growing film without generating a grain boundary. The grain boundary can in itself be current limiting if it is a tilt boundary between two c-axis oriented films as depicted in the schematic of Fig. 6b, or the reduced Jc is due directly to the region of unfavourably oriented film between the boundaries so that the current has to flow along the c-axis as shown in Fig. 6c. It has

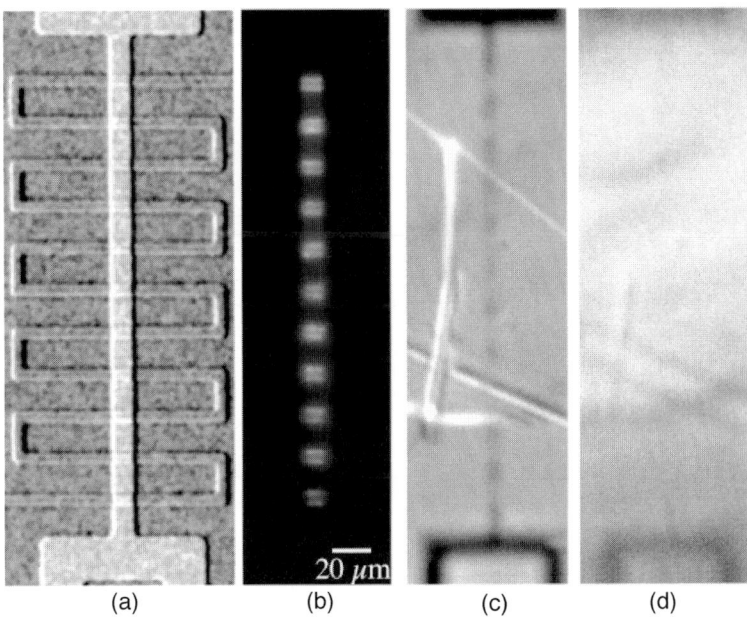

Fig. 5 SEM image (a), EBIV image (b) and MO images at 65 mT (c) and 250 mT (d) of a patterned YBCO microbridge over steps in the substrate. All images at the same magnification.

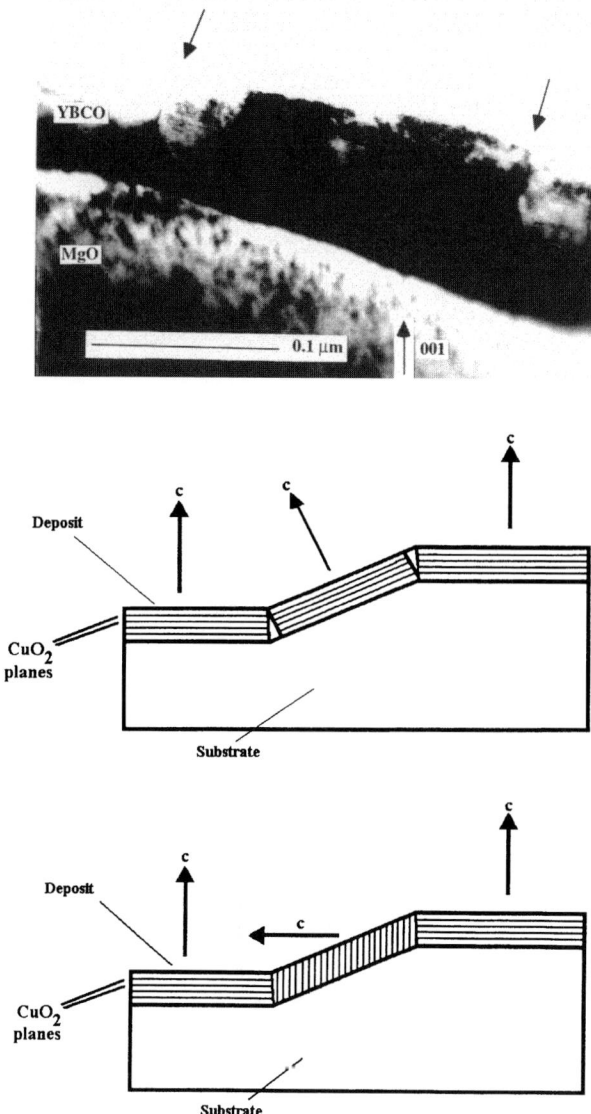

Fig. 6 TEM micrograph of YBCO layer over a step in the substrate indicating (a) a low angle tilt grain boundary. Schematic of low (b) and high (c) angle grain boundary formed in a film grown over a substrate step.

been found that such boundaries can be eliminated by first coating the substrate with a sacrificial layer of PrBCO, a non-superconducting analogue of YBCO with better lattice-matching than the underlying substrate.

Bi-axially Textured YBCO Coated Tapes

The fabrication of high current carrying practical conductors for power engineering applications is one of the major technological challenges facing the superconductivity research and development community. The manufacture of kilometre length wires and tapes with reproducible, homogeneous properties will open up many potential technical opportunities. Thermo-mechanically textured, silver-sheathed BSCCO tape production is a maturing technology with useable electrical and magnetic properties (see below), but a major step towards the so-called second generation of wires and tapes has been demonstrated by the development of bi-axially textured, open YBCO tapes on flexible metallic substrates.[5]

The technology involves the deposition of films on pre-textured substrates designed to provide an appropriate bi-axial template for the superconducting layer to copy. Ni and its alloys at present offer the best combination of cube texture, lattice matching and thermal expansion coefficient for basal plane alignment of YBCO, but chemical incompatibility necessitates the deposition of intermediate buffer layers. The choice here is wide but, of compatible oxide layers, CeO2 and YSZ offer the best combination. A typical architecture employed in fabricating YBCO bi-axial textured composite tapes is shown in Fig. 7 A complex growth procedure involving PVD techniques has been evolved in order to maintain the microtexture through the intermediate layers into the final superconducting layer. On a macroscopic level the desired growth is pseudo-epitaxial but is genuinely epitaxial within each grain. Misorientations between grains should be as low as possible.

COM plays an essential part in assessing the efficiency of transfer of bi-axial microtexture from the base substrate through the buffer layers to the final YBCO layer. An example of the microtexture observed on a base Ni substrate is shown in Fig. 7. The

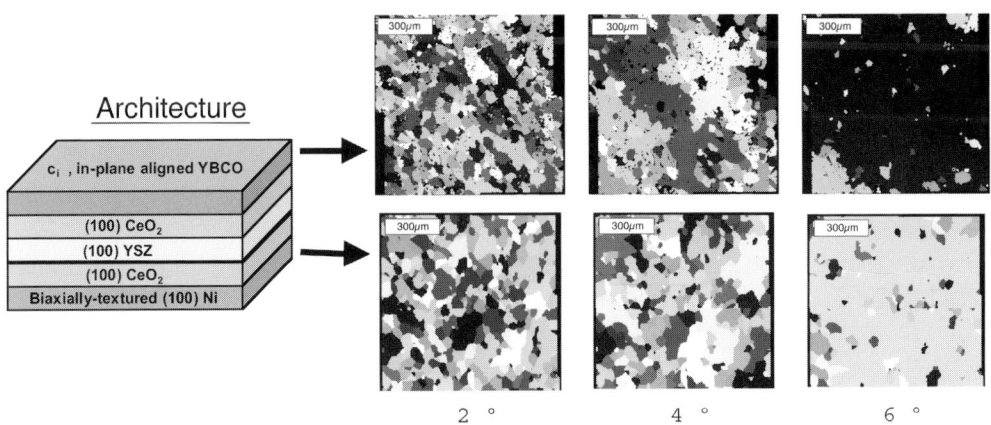

Fig. 7 Typical bi-axial tape architecture with misorientation maps of the base substrate and final layer.

software colours the grains the same if they are the same orientation within a certain grain boundary misorientation. Thus Fig. 7 shows in grey scale the grains which are misoriented to within 2° and 4°, but relaxing this to 6° shows the majority of the grains to be of similar orientation. Similar images can be obtained for the intermediate buffer layers to monitor the efficiency of copying the texture from the base substrate through to the final superconducting layer, also shown in Fig. 7. There is evidence in the 4° map of an improvement in texture in the final layer, which is confirmed in the 6° map. Nearly all the grains in the field of view (1 × 1 mm) are aligned to within 6°. Certainly, as far as supercurrents are concerned, this map of the microstructure offers a percolative current path through the sample. There is no significance in the contrast in grey scales between the upper and lower pattern; the uniformity within a map is the significant characteristic.

At present it has not proved possible to correlate these observations with MO imaging of flux distribution in these composite tapes due to the magnetic signal from the Ni substrates, but it is hoped that such comparisons will prove possible in the future with the use of non-magnetic flexible textured base substrates.

THERMO-MECHANICALLY TEXTURED BSCCO TAPES

Superconducting Ag-sheathed $(Bi,Pb)_2Sr_2Ca_2Cu_3O_x$ [Bi(Pb)-2223] tapes, produced by the oxide-powder-in-tube (OPIT) technique, have demonstrated the capacity to carry large currents at liquid nitrogen temperature for large scale engineering applications.[6] However, problems of reproducibility and maintaining high critical current values while scaling up to practical lengths still remain. Current flow in practical conductors such as Bi2223/Ag clad tapes and YBCO based thick films is markedly inhomogeneous, as established by hall probe and magneto-optic measurements. The basic reason seems to be that the processing of such complex materials at both nanometer and macroscopic scales is inherently complex. Magneto-optical imaging has been used[7] to directly demonstrate that the transport J_c of the BSCCO-2223 tapes is more determined by their connectivity than by flux pinning and that an independent and much larger limiting influence on the J_c comes from unhealed cracks produced by deformation during the processing of the tapes.

The texture in these tapes derives from the thermo-mechanical processing which consists of repeated cycles of rolling and pressing followed by annealing. Microstructural characterisation of the Bi(Pb)-2223/Ag tapes is important for understanding the complex chemical phase relations in the Bi(Pb)-2223 system, and in particular the nature of the Bi-2223 phase and its texture development. The grains of the oxide are essentially two-dimensional in morphology and this platelet nature with the large face parallel to the a–b plane, responds to the mechanical deformation by adopting a highly aligned microstructure consisting of parallel platelets. Connectivity between aligned grains is then achieved during the heat treatment. The degree of alignment can be measured by transmission x-ray diffraction,[8] but COM is clearly inappropriate due to the presence of the sheath. MO imaging, however, can provide an indication of the effectiveness of the processing by imaging the magnetic response of the material. Fig. 8 shows the plan-view MO image in applied field of three tapes prepared under different conditions.[9] The images are dominated by opposite contrast regions running along the length of the tape. Dark regions shield magnetic flux very well and consist of well-aligned and well- connected grains, whereas light contrast reveals the presence of flux penetrating at cracks and weaker grain boundaries. MO images can also be recorded while the tape is carrying a transport current due the field generated. This allows the current distribution to be mapped and also correlated with microstructural features such as grain boundaries and the core/sheath interface.

ARTIFICIALLY ENGINEERED GRAIN BOUNDARIES IN MELT PROCESSED YBCO

Bulk material geometries are attracting considerable attention as magnetic bearings, inductive components in medium size power engineering applications, and as permanent magnet materials (trapping up to 5 T at 77 K). The flux shielding and trapping behaviour are strongly dependent on the crystallographic alignment, grain size and distribution, a combination of which properties can only be provided by COM.

The fabrication of large scale artefacts from melt textured YBCO for magnetic components in power engineering applications is hindered by the difficulty of growing large single domain samples with uniform properties. The strength of the trapped field depends on the magnitude and the length scale over which the supercurrent flows, requiring large grained pseudo-crystal material with large dimensions and a high critical current density. Bulk melt textured material grown by a seeding route usually exhibits significant degradation of properties for samples greater than a few cm in diameter.[10] For larger artefacts, the possibility exists of joining sub-components,

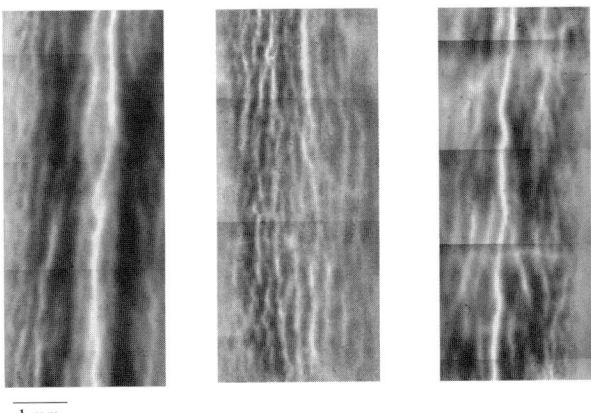

1 mm

Fig. 8 MO images of BSCCO tapes with bright contrast corresponding to the presence of flux.

315

if the quality of the joints does not limit the current flow significantly. The current flow across such boundaries can be assessed by suitable contact placement[11] but the magnetic flux distribution in the vicinity of these joints also requires characterisation. Magneto-optic imaging has been used to compare the magnetic flux penetration of such boundaries with that of the surrounding bulk material.

Artificially engineered grain boundaries with misorientation less than 3° were produced between single grains of melt textured YBCO as described elsewhere.[12] An optical micrograph of the sample studied by MO imaging is shown in Fig. 9a. The engineered join (A) runs along the entire width of the sample and is intercepted by cracks (C). As a comparison, another grain was joined using insulating GE varnish (B), representing the worst case for an engineered join. The high field behaviour of the artificial joins was compared with the bulk material. Any defects present in the join will allow magnetic flux to penetrate into the join before the bulk regions. Conversely, a good join should show no real difference in behaviour to the surrounding bulk regions. Note that both the a and c axes are in the plane of the MO layer, allowing the field behaviour in these directions to be compared.

The sample was field-cooled to 80 K in an applied field of 17 mT. Figure 9b is a magneto-optic image of the sample in its remnant state. The dark regions in the bulk of the sample represent pinned flux in the direction of the applied field (out of the page). The insulating join (B), not being superconducting, does not contain any trapped flux, hence allows pinned flux from the bulk to return to the underside of the sample, completing the magnetic circuit. Thus, the dark contrast within the join corresponds to return flux going into the page with the intervening bright contrast representing flux free areas.

The majority of the artificial grain boundary (line A) behaves very similarly to the varnished join, hence acts as an insulating join at this temperature and field. However, the region of the grain boundary labelled D shows no appreciable difference in trapped flux compared to that of the bulk regions. This indicates a strongly coupled

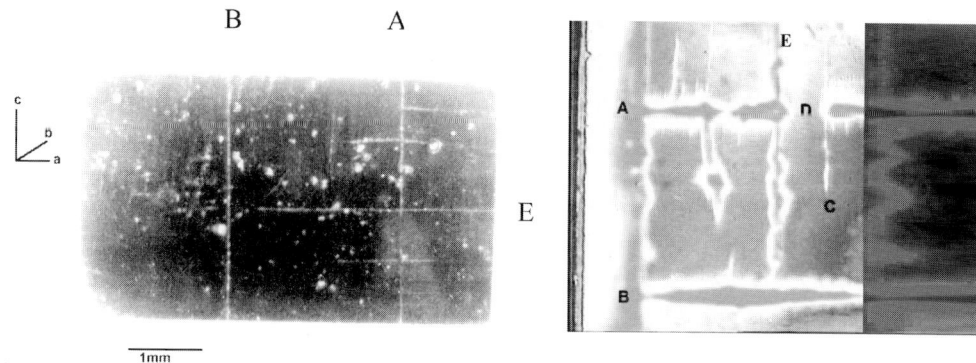

Fig. 9 Optical image (a) and MO image (b) of bulk melt processed YBCO with artificial grain boundary.

grain boundary that can support the current flowing in the c–a plane that is generating the detected field. Running vertically, parallel to the a-direction, through-cracks can be seen allowing some field to return. These can also be seen optically (Fig. 9a). The successful joining of melt processed YBCO bulk material offers the opportunity of fabricating large scale magnetic components.

CONCLUSIONS

Mapping techniques involving the microstructure and crystallography, electrical and magnetic properties of superconducting oxides are proving to be a powerful combination in the assessment and characterisation of a wide variety of specimen configurations and are already having a significant impact on the eventual technological exploitation of these complex materials.

ACKNOWLEDGEMENTS

I would like to thank members of the superconducting materials group in the School who performed much of the work presented here. The bulk YBCO sample was kindly supplied by the IRC, Cambridge and the MO images form part of a collaboration with the University of Wisconsin. Financial support from EPSRC and Oxford Instruments is gratefully acknowledged.

REFERENCES

1. J. B. Langhorn and J. S. Abell, *Supercon. Sci. Techn.,* 1998, **11**, 751.
2. T. C. Shields, J. S. Abell and A. Drake, *IOP Conf. Ser.,* 1995, **147**, 425.
3. C. A. Hollin, J. S. Abell, S.W. Goodyear, N. G. Chew and R. G. Humphreys, *Appl. Phys. Lett.,* 1994, **64**, 918.
4. M. Zamboni, S. A. L. Foulds, S. Koishikawa, K. Matsumoto, M. Murakami and J. S. Abell, *IOP Conf. Ser.,* 1997, **158**, 85.
5. D.P. Norton *et al., Science,* 1996, **274**, 755.
6. S. X. Dou and H. K. Liu, *Supercon. Sci. Techn.,* 1993, **6**, 297.
7. X. Y. Cai, A. Polyanskii, Q. Li, G. N. Riley and D. C. Larbalestier, *Nature,* 1998, **392**, 906.
8. J. Jiang, T.C. Shields, J. S. Abell and G. Bushnell-Wye, *Physica,* 1998, **C 306**, 91.
9. J. Jiang, T. C. Shields, J. S. Abell, A. Polyanskii, D. M. Feldman and D. C. Larbalestier, *IEEE Trans Appl. Sup.,* 1999, **9**, 1812.
10. D. A. Cardwell, *Mater. Sci. Eng.,* 1998, **B53**, 1.
11. R. A. Doyle, A. Bradley, W. Lo and D. Cardwell, *Appl. Phys. Lett.,* 1998, **73**, 117.
12. W. Lo, D. A. Cardwell, A. Bradley, R. A. Doyle, Y. H. Shi and S. Lloyd, *IEEE Trans. Appl. Supercon.,* 1999, **9**, 2042

High Spatial Resolution Chemical Microanalysis in the TEM: Some Pitfalls

I. P. JONES

School of Metallurgy and Materials, The University of Birmingham, Edgbaston, Birmingham B15 2TT, UK

INTRODUCTION

This two day conference in celebration of Ray Smallman's 70th birthday coincides with the opening of the new joint Birmingham–Loughborough–Warwick transmission electron microscope. The theme of the conference, which Ray himself suggested, is 'grain boundaries'. There is a particularly intimate connection here with our new microscope, which is an FEI F20 200 kV TEM equipped with a field emission electron source. The field emission aspect means that an intense and very narrow beam of electrons can be created and directed at the specimen. In conjunction with ancillary equipment – the parallel Electron Energy Loss Spectrometer and the Energy Dispersive X-ray analyser – this means that chemical analysis of a region on the specimen half a nanometre across (~2 atoms) is possible. This is two orders of magnitude better than was possible with the previous generation of TEMs. It is a fact that by far the most common topic in the several dozen projects which formed our joint (Birmingham–Loughborough–Warwick) case for support concerns the analysis of grain boundaries. Hence the connection.

The other papers in this volume give a good idea of the range of materials and phenomena where grain boundaries are controlling. It is clearly important to be able to characterise their geometry and chemistry. Previously to carry out high spatial resolution chemical analysis of grain boundaries, either the specimen had to be fractured across the grain boundary (not always possible) and then analysed by some surface technique (eg Auger Spectrometry, not always the most easily quantifiable of methods) or, alternatively, atom probe microscopy could be used – again a time-consuming and difficult technique which is not always feasible. The advent of field emission TEM is of extreme significance to grain boundary chemical analysis.

The purpose of this paper is to point out some of the problems which may befall, not from any negative point of view, but simply so that an awareness of such problems may enable their better avoidance. This will be even more important with the next generation of STEMs[1,2] with their sub-Angstrom probes.

THE TECHNIQUE

The technique and its quantification are apparently simple and straightforward (Fig. 1). A narrow beam of electrons is fired through a specimen along a grain boundary. In the X-ray method a proportion of the characteristic X-rays, which are given off isotropically, are admitted to a solid state EDX detector. Their intensities are measured and then converted to a composition either by using a standard, or (somewhat less accurately) by calculation. In the EELS method the transmitted electron energies are analysed using a prism and a parallel array of CCD detectors. The main differences from the EDX method are:

- better detectability
- more difficult quantification
- more exacting requirement for thin foils
- only the forward scattered electrons are accepted
- more difficulties with one element obscuring another.

So there are pluses and minuses: sometimes one technique is better, sometimes the other.

These are the principles; in practice a series of problems can arise:

I   Beam spreading and microscope artefacts
IIA   Surface etching
IIB   Surface plating   }Specimen preparation
IIIA Bulk radiation damage   }Radiation
IIIB Differential surface spattering   }Damage

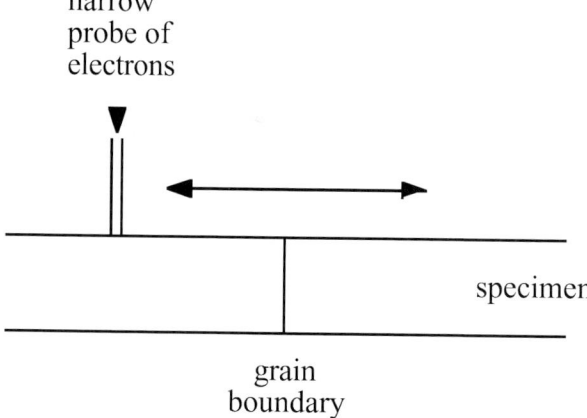

narrow
probe of
electrons

specimen

grain
boundary

Fig. 1 Electron beams can now be made so narrow (5×10^{-10} m) that the chemical composition of a grain boundary may be measured.

I BEAM SPREADING AND MICROSCOPE ARTEFACTS

The spreading of the electron beam within the specimen (Fig. 2) is fairly well understood (eg Hutchings *et al.*[3]) and may be calculated and unfolded from the measured profile to give the true chemical profile (eg by Fourier filtering: Jones[4]).

What is not so well appreciated is that the Rutherford ($\equiv$ Schrödinger) scattering cross-sections which are generally employed are totally inappropriate for heavier elements. (This is equally true for SEM beam spreading, in fact.) For elements with $Z > 30$, Mott ($\equiv$ Dirac) cross-sections must be used (see Fig. 3).[6]

Beam spreading does not affect EELS : it is an EDX problem only. Extraneous signals from the interaction between the electron beam and the microscope, most commonly X-rays irradiating the whole specimen rather than that part defined by the electron beam, are not really a problem for modern TEMs and can in any case be removed by measuring and subtracting a hole count spectrum.[5]

Again, what is not commonly recognised is the more insidious effect of electrons scattered by the specimens hitting the top of the lower pole piece and sending back a shower of electrons and X-rays.[8,9] I say 'insidious' because this is a non-trivial problem and because it cannot be removed via a hole spectrum.

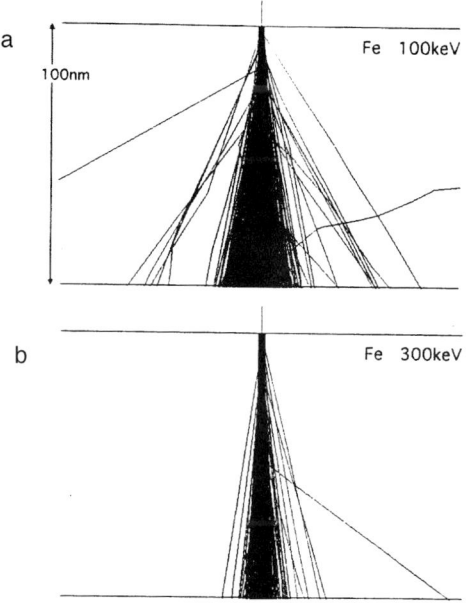

Fig. 2 Electron beam spreading in a thin foil.[5]

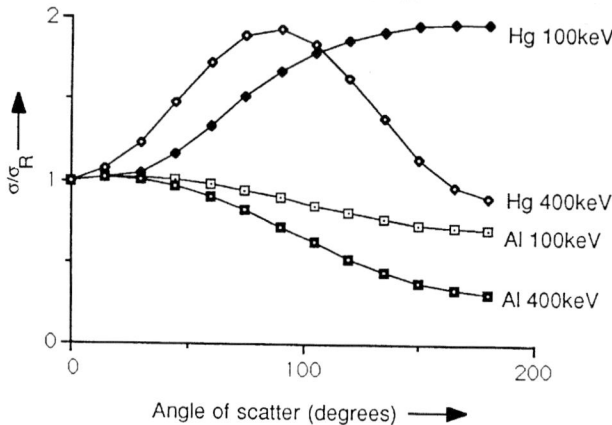

Fig. 3 The ratio of Mott to Rutherford cross-sections (σ/σ_R) for nuclear scattering of electrons.[7]

IIA SURFACE ETCHING AND IIB SURFACE PLATING

Electrochemical and ion beam specimen preparation techniques result in the problems illustrated in Fig. 4.

The presence of a 'polishing layer' (the same term is applied to both electrochemical polishing and ion beam etching) may be revealed by a plot of X-ray intensities versus specimen thickness: a very useful plot indeed. The layer may be largely removed by a low voltage (eg 2 kV) ion beam wipe.[10]

Etching at a grain boundary can be detected via a change in the Bremsstrahlung signal (careful: this can come from a change in composition as well).

IIIA BULK RADIATION DAMAGE

Because the grain boundary is a point defect sink, if the electron beam causes radiation damage, then radiation induced segregation will occur via inverse Kirkendall or size effects (Fig. 5). Displacement damage in metals is fairly simply predictable.[6] Ionisation and some of the damage mechanisms in semiconductors are more insidious.

Alternatively, particularly when one component is displaced and the other not, preexisting composition profiles may be changed. A classic example is that of phosphorus at grain boundaries in steels.[11]

Figure 6 shows that the results of the conventional way of taking a profile and the much more accurate result obtained by starting with the beam on the boundary and then taking alternate points either side. In the former case, the beam evaporates the

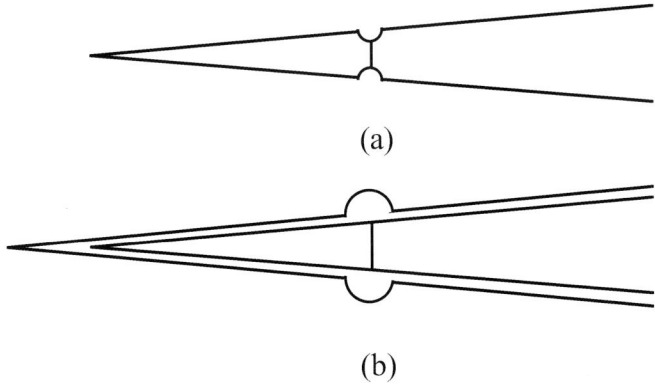

(a)

(b)

Fig. 4 (a) Preferential etching at a grain boundary; (b) polishing film on TEM specimen.

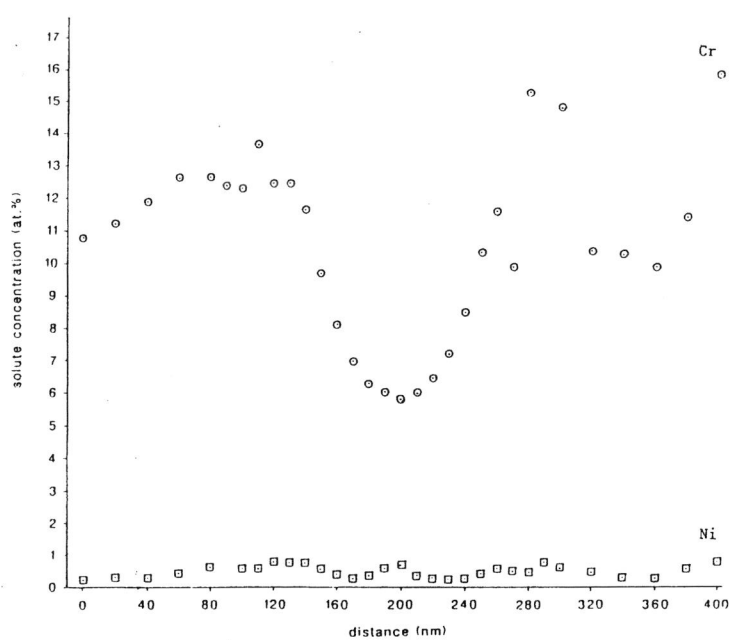

Fig. 5 Chemical gradients about a grain boundary built up by electron irradiation. The grain boundary is at 200 nm on the abscissa.[12]

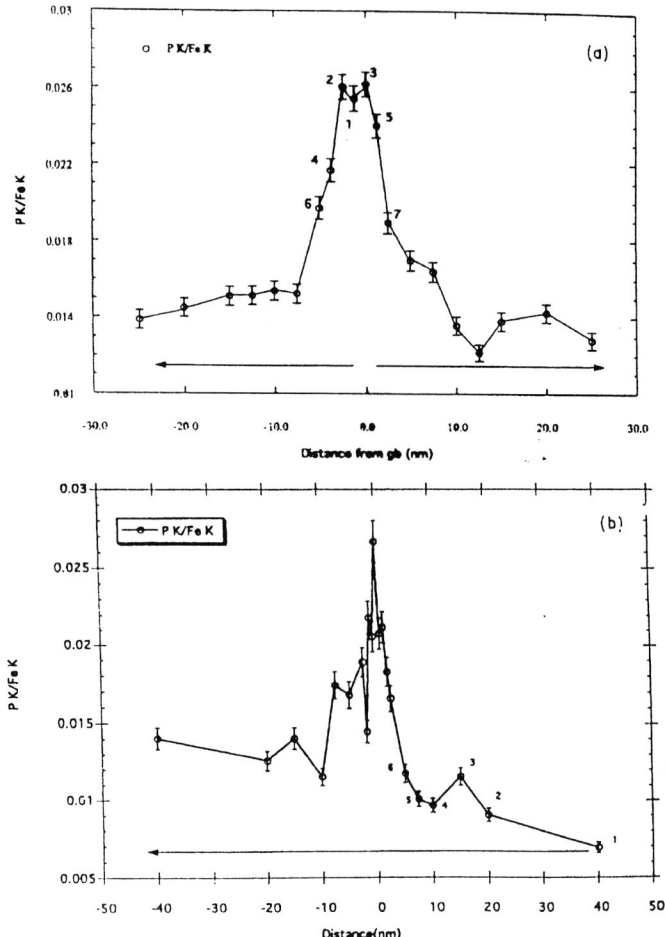

Fig. 6 Phosphorus segregation to a grain boundary in a steel. The numbers indicate the order in which the measurements were taken. Note how much better the profile in (a) is than that in (b).

phosphorus off the boundary and effectively pushes it in front of it. (This problem does not arise with PEELS with its shorter counting times.)

IIB DIFFERENTIAL SURFACE SPUTTERING

Surface sputtering takes place at much lower beam voltages than bulk radiation damage (roughly at E_s, the sublimation energy[13] (See Table 1)). Prolonged exposure

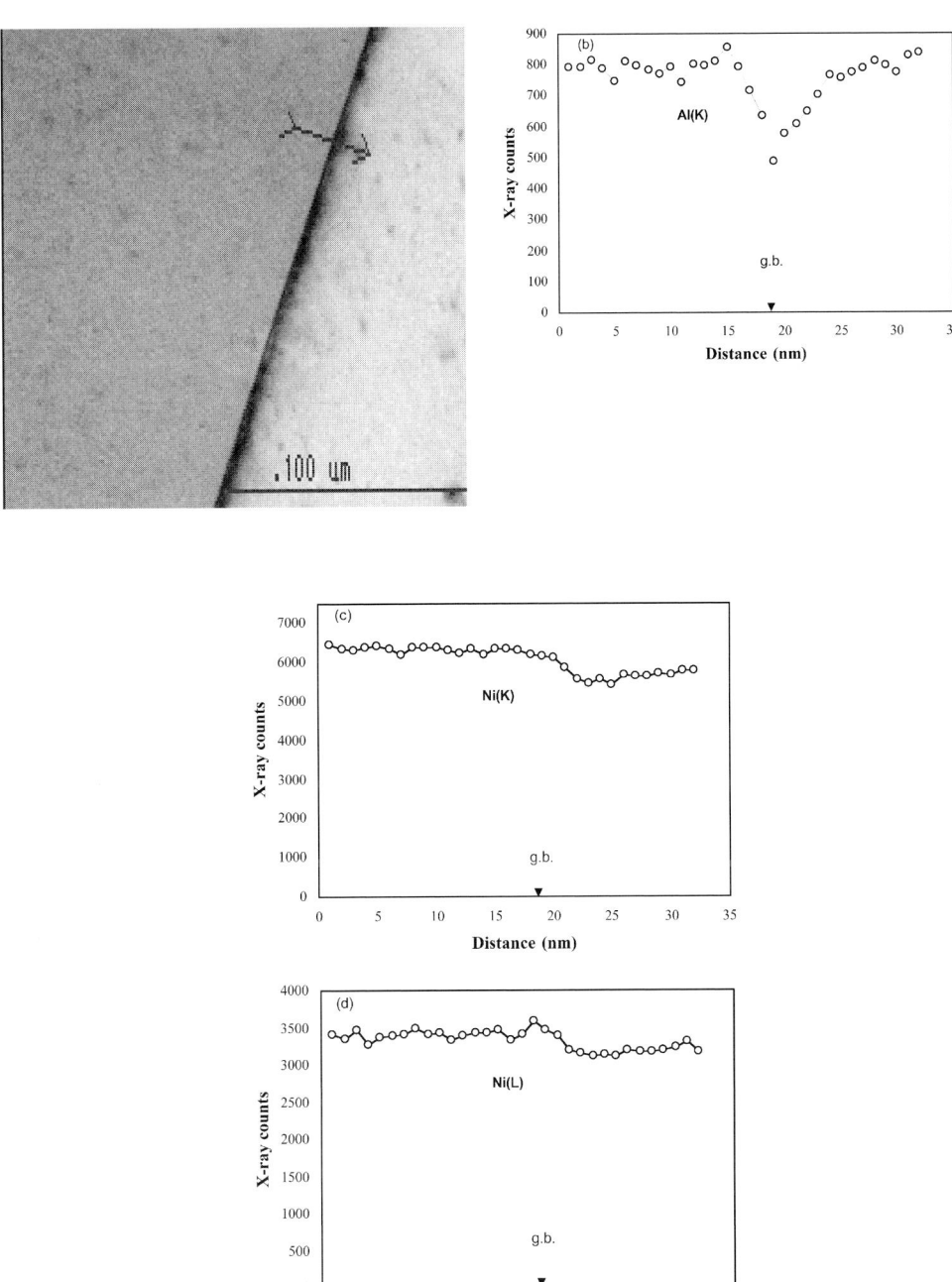

Fig. 7 Segregation profiles around a grain boundary in Ni₃Al.[15]

to an intense electron beam causes the lighter element to be removed more quickly than the heavier.[14] Thus the composition changes and the sputtering is said to be 'differential'. Recent work suggests it may be doubly differential, taking place preferentially at eg a grain boundary. Figure 7 shows concentration profiles around a grain boundary in Ni_3Al which we believe were created by the electron beam.[15] This problem can be prevented by coating with a thin layer of eg carbon.[14]

None of these problems is insurmountable, but it is important to be aware of their possible occurrence.

REFERENCES

1. O. L. Krivanek, N. Dellby, A. J. Spence, R. A. Camps and L. M. Brown, Institute of Physics Conference Series **153**, *EMAG 97,* 1997, 35.
2. M. Haider, S. Uhlemann, F. Schwan, H. Rose, B. Kabius and K. Urban, Nature 1998, **392**, 769.
3. R. Hutchings, I. P. Jones, M. H. Loretto and R. E. Smallman, *Electron Microscopy Toronto,* 1978, **1**, 544.
4. I. P. Jones, G. J. Mahon, A. W. Nicholls, M. J. Yates and Y. G. Zhang, *Analytical Electron Microscopy,* 1984, 34–36.
5. D. B. Williams and C. B. Carter, *Transmission Electron Microscopy,* Plenum, 1996.
6. C. R. Bradley, ANL Report 88–48, 1988.
7. J. A. Doggett and L. V. Spencer, *Phys. Rev.,* 1956, **103**, 1597.
8. J. N. Chapman, C. L. Gray, B. W. Roberston and W. A. P. Nicholson, *X-ray Spectrom.,* 1983, **12**, 153.
9. Y. H. Li and M. H. Loretto, *Journal of Microscopy,* 1993, **170**, 259–264.
10. J. M. Pountney and M. H. Loretto, *Electron Microscopy,* 1980, **3**, 180.
11. D. Ozkaya, J. Yuan, L. M. Brown and P. E. J. Flewitt, *J. Microscopy,* 1995, **180**, 300.
12. G. J. Mahon, A W. Nicholls, I. P. Jones, C. A. English and T. M. Williams, *Proceedings of a Symposium held at Berkeley Nuclear Laboratories on 23rd Sept 1986,* edited by D. I. R Norris (CEGB Technology, Planning Research Division) 1987, 99.
13. F. Seitz, *Discussion Faraday Soc.,* 1949, **5**, 571.
14. D. A. Muller, S. Subramanian, P. E. Batson, J. Silcox and S. L. Sass, *Acta Met.,* 1996, **44**, 1637.
15. P. Shang, R. Keyse, I. P. Jones and R. E. Smallman, *Phil. Mag. A.,* 1999, **79**, 2539.

Table 1 Threshold electron accelerating voltages for bulk radiation damage (between 4 and 5 E_s) and for surface spattering ($\sim E_s$).[6]

Atomic Number	Element	Atomic Mass	Sublimation Energy (E_s) cV	Threshold Accel. Voltage at E_s	Threshold Accel. Voltage at $2E_s$	Threshold Accel. Voltage at $4E_s$	Threshold Accel. Voltage at $5E_s$
3	Li. Lithium	6.939	1.670	5.3	10.5	20.7	25.7
4	Be. Beryllium	9.012	3.360	13.6	26.9	52.5	64.9
5	B, Boron	10.811	6.030	28.9	56.3	107.5	131.6
6	C, Carbon	12.010	7.430	39.2	75.7	142.7	173.7
11	Na, Sodium	22.990	1.120	11.6	22.9	44.9	55.6
12	Mg, Magnesium	24.312	1.520	16.6	32.6	63.4	78.2
13	Al, Aluminum	26.980	3.420	40.4	78.1	147.0	178.9
14	Si, Silicon	28.090	4.680	56.7	108.3	200.3	242.1
15	P, Phosphorus	30.974	3.440	46.4	89.3	166.9	202.6
16	S, Sulfur	32.060	2.870	40.3	77.9	146.6	178.4
19	K, Potassium	39.100	0.929	16.3	32.1	62.4	76.9
20	Ca, Calcium	40.080	1.850	32.7	63.6	120.8	147.6
21	Sc, Scandium	44.956	3.920	74.8	141.1	256.7	308.4
22	Ti, Titanium	47.900	4.860	96.9	180.3	322.5	385.1
23	V, Vanadium	50.942	5.330	111.5	205.9	364.7	434.1
24	Cr, Chromium	51.996	4.110	89.5	167.3	300.9	360.0
25	Mn, Manganese	54.938	2.940	68.9	130.5	238.6	287.2
26	Fe, Iron	55.847	4.310	99.9	185.6	331.3	395.3
27	Co, Cobalt	58.933	4.440	107.8	199.5	354.1	421.9
28	Ni, Nickel	58.710	4.460	107.9	199.6	354.3	422.1
29	Cu, Copper	63.546	3.500	92.9	173.3	310.8	371.5
30	Zn, Zinc	65.380	1.350	38.7	74.9	141.3	172.1
31	Ga, Gallium	69.72	2.840	83.4	156.5	282.7	338.8
32	Ge, Germanium	72.590	3.880	115.3	212.5	375.4	446.5
33	As, Arsenic	74.922	3.140	97.8	182.0	325.2	388.3
34	Sc, Selenium	78.960	2.350	78.5	147.7	267.9	321.5
37	Rb, Rubidium	85.470	0.850	32.1	62.4	118.6	144.9
38	Sr, Strontium	87.620	1.700	63.9	121.3	222.8	268.7
39	Y, Yttrium	88.906	4.400	154.8	279.8	483.8	571.5
40	Zr, Zirconium	91.220	6.310	216.4	381.8	645.6	754.3

Author Index

Subject Index